The Future of Post-Human Acoustics

The Future of Post-Human Acoustics

A Preface to a New Theory of Sound and Silence

Peter Baofu

Viva Books

New Delhi | Mumbai | Chennai | Kolkata | Bengaluru | Hyderabad | Kochi | Guwahati

First Indian Edition 2012

VIVA BOOKS PRIVATE LIMITED

- 4737/23 Ansari Road, Daryaganj, New Delhi 110 002
  E-mail: vivadelhi@vivagroupindia.net, Tel. 42242200
- 76, Service Industries, Shirvane, Sector 1, Nerul, Navi Mumbai 400 706
  E-mail: vivamumbai@vivagroupindia.net, Tel. 27721273, 27721274
- Jamals Fazal Chambers, 26 Greams Road, Chennai 600 006
  E-mail: vivachennai@vivagroupindia.net, Tel. 28290304, 28294241
- B-103, Jindal Towers, 21/1A/3 Darga Road, Kolkata 700 017
  E-mail: vivakolkata@vivagroupindia.net, Tel. 22836381, 22816713
- 7, Sovereign Park Apartments, 56-58, K. R. Road, Basavanagudi, Bangalore 560 004
  E-mail: vivabangalore@vivagroupindia.net, Tel. 26607409, 26607410
- 101-102 Mughal Marc Apartments, 3-4-637 to 641, Narayanguda, Hyderabad 500 029
  E-mail: vivahyderabad@vivagroupindia.net, Tel. 27564481, 27564482
- Beevi Towers, First Floor, SRM Road, Kaloor, Kochi 682 018, Kerala
  E-mail: vivakochi@vivagroupindia.net, Tel: 0484-2403055, 2403056
- 232, GNB Road, Beside UCO Bank, Silpukhuri, Guwahati 781 003
  E-mail: vivaguwahati@vivagroupindia.net, Tel: 0361-2666386

www.vivagroupindia.com

Published by arrangement with
Cambridge International Science Publishing
7, Meadow Walk, Great Abington
Cambridge CB1 6AZ, UK

ISBN : 978-81-309-1995-9

Published by Vinod Vasishtha for Viva Books Private Limited, 4737/23, Ansari Road, Daryaganj, New Delhi-110 002. Printed and bound by Syndicate Binders, Noida.

To Those in the Future World Beyond Sound and Silence

BOOKS ALSO BY PETER BAOFU

Category I: Social Sciences and Related Fields

- 38. *The Future of Post-Human Education* (2011)
- 37. *The Future of Post-Human War and Peace* (2010)
- 36. *The Future of Post-Human Law* (2010)
- 35. *The Future of Post-Human Organization* (2009)
- 34. *The Future of Post-Human Mass Media* (2009)
- 33. *The Rise of Authoritarian Liberal Democracy* (2007)
- 32. *Beyond the World of Titans, and the Remaking of World Order* (2007)
- 31. *Beyond Capitalism to Post-Capitalism* (2005)
- 30. Volume 1: *Beyond Democracy to Post-Democracy* (2004)
- 29. Volume 2: *Beyond Democracy to Post-Democracy* (2004)
- 28. *The Future of Capitalism and Democracy* (2002)

Category II: Formal Sciences and Related Fields

- 27. *The Future of Post-Human Formal Science* (2010)
- 26. *The Future of Post-Human Language* (2009)
- 25. *The Future of Post-Human Geometry* (2009)
- 24. *The Future of Post-Human Mathematical Logic* (2008)
- 23. *The Future of Information Architecture* (2008)

Category III: Natural Sciences and Related Fields

- 22. *The Future of Post-Human Geology* (2010)
- 21. *Beyond Cosmology to Post-Cosmology* (2010)
- 20. *The Future of Post-Human Personality* (2010)
- 19. *The Future of Post-Human Sexuality* (2010)
- 18. *The Future of Post-Human Urban Planning* (2009)
- 17. *The Future of Post-Human Engineering* (2009)
- 16. *The Future of Post-Human Unconsciousness* (2008)
- 15. *The Future of Complexity* (2007)
- 14. *Beyond Nature and Nurture* (2006)
- 13. *The Future of Post-Human Space-Time* (2006)
- 12. *The Future of Post-Human Consciousness* (2004)

Category IV: The Humanities and Related Fields

- 11. *The Future of Post-Human Humor* (2011)
- 10. *Beyond Ethics to Post-Ethics* (2011)
- 9. *The Future of Post-Human Religion* (2010)
- 8. *The Future of Post-Human Chess* (2010)
- 7. *The Future of Post-Human Martial Arts* (2009)
- 6. *The Future of Post-Human Creative Thinking* (2009)
- 5. *The Future of Post-Human Knowledge* (2008)
- 4. *The Future of Aesthetic Experience* (2007)
- 3. *Beyond Civilization to Post-Civilization* (2006)
- 2. Volume 1: *The Future of Human Civilization* (2000)
- 1. Volume 2: *The Future of Human Civilization* (2000)

CONTENTS

Part Three: Silence

Part Four: Conclusion

TABLES

Category III: Visions on the Mind

Category IV: Visions on History

Category IX: Visions on Culture

Acknowledgments

For those who had read any previous books of mine, it is not surprising that this book too is written to challenge conventional wisdom and to replace it with an alternative original way of thinking—or more specifically in the current context, a new theory to understand the future fate of acoustics (in relation to sound and silence).

In this light, this book receives no external funding nor help from any formal organization or institution, because of its political incorrectness—as this is something that I often emphasized in all my previous books.

My only reward is that joyful wonder of ideas in a way not conceived before in history.

In any event, and as always, I bear the sole responsibility for the ideas presented in this book.

ABBREVIATIONS

ALD = Peter Baofu. 2007. *The Rise of Authoritarian Liberal Democracy: A Preface to a New Theory of Comparative Political Systems*. Cambridge, England: Cambridge Scholars Publishing, Ltd.

BCIV = Peter Baofu. 2006. *Beyond Civilization to Post-Civilization: Conceiving a Better Model of Life Settlement to Supersede Civilization*. NY: Peter Lang Publishing, Inc.

BCPC = Peter Baofu. 2005. *Beyond Capitalism to Post-Capitalism: Conceiving a Better Model of Wealth Acquisition to Supersede Capitalism*. NY: The Edwin Mellen Press.

BCOS = Peter Baofu. 2010. *Beyond Cosmology to Post-Cosmology: A Preface to a New Theory of Different Worlds*. Cambridge, England: Cambridge International Science Publishing, Ltd.

BDPD1 = Peter Baofu. 2004. Volume 1. *Beyond Democracy to Post-Democracy: Conceiving a Better Model of Governance to Supersede Democracy*. NY: The Edwin Mellen Press.

BDPD2 = Peter Baofu. 2004. Volume 2. *Beyond Democracy to Post-Democracy: Conceiving a Better Model of Governance to Supersede Democracy*. NY: The Edwin Mellen Press.

BEPE = Peter Baofu. 2011. *Beyond Ethics to Post-Ethics: A Preface to a New Theory of Morality and Immorality*. Charlotte, NC: Infomration Age Publishing.

BNN = Peter Baofu. 2006. *Beyond Nature and Nurture: Conceivng a Better Way to Understand Genes and Memes*. Cambridge, England: Cambridge Scholars Publishing, Ltd.

BWT = Peter Baofu. 2007. *Beyond the World of Titans, and the Renaking of World Order: A Preface to a New Logic of Empire-Building*. Cambridge, England: Cambridge Scholars Publishing, Ltd.

FAE = Peter Baofu. 2007. *The Future of Aesthetic Experience: Conceiving a Better Way to Understand Beauty, Ugliness and the Rest*. Cambridge, England: Cambridge Scholars Publishing, Ltd.

FC = Peter Baofu. 2007. *The Future of Complexity: Conceiving a Better Way to Understand Order and Chaos.* London, United Kingdom: World Scientific Publishing Co.

FCD = Peter Baofu. 2002. *The Future of Capitalism and Democracy.* MD: The University Press of America.

FHC1 = Peter Baofu. 2000. Volume 1. *The Future of Human Civilization.* NY: The Edwin Mellen Press.

FHC2 = Peter Baofu. 2000. Volume 2. *The Future of Human Civilization.* NY: The Edwin Mellen Press.

FIA = Peter Baofu. 2008. *The Future of Information Architecture: Conceiving a Better Way to Understand Taxonomy, Network, and Intelligence.* Oxford, England: Chandos Publishing (Oxford) Limited.

FPHA = Peter Baofu. 2011. *The Future of Post-Human Acoustics: A Preface to a New Theory of Sound and Silence.* Cambridge, England: Cambridge International Science Publishing, Ltd.

FPHC = Peter Baofu. 2004. *The Future of Post-Human Consciousness.* NY: The Edwin Mellen Press.

FPHCHESS = Peter Baofu. 2010. *The Future of Post-Human Chess: A Preface to a New Theory of Tactics and Strategy.* Cambridge, England: Cambridge International Science Publishing, Ltd.

FPHCT = Peter Baofu. 2009. *The Future of Post-Human Creative Thinking: A Preface to a New Theory of Invention and Innovation.* Cambridge, England: Cambridge Scholars Publishing, Ltd.

FPHE = Peter Baofu. 2009. *The Future of Post-Human Engineering: A Preface to a New Theory of Technology.* Cambridge, England: Cambridge Scholars Publishing, Ltd.

FPHEDU = Peter Baofu. 2011. *The Future of Post-Human Education: A Preface to a New Theory of Teaching and Learning.* Cambridge, England: Cambridge International Science Publishing, Ltd.

FPHFS = Peter Baofu. 2010. *The Future of Post-Human Formal Science: A Preface to a New Theory of Abstraction and Application.* Cambridge, England: Cambridge Scholars Publishing, Ltd.

FPHG = Peter Baofu. 2009. *The Future of Post-Human Geometry: A Preface to a New Theory of Infinity, Symmetry, and Dimensionality.* Cambridge, England: Cambridge Scholars Publishing, Ltd.

FPHGEOL = Peter Baofu. 2010. *The Future of Post-Human Geology: A Preface to a New Theory of Statics and Dynamics.* Cambridge, England: Cambridge International Science Publishing, Ltd.

FPHH = Peter Baofu. 2011. *The Future of Post-Human Humor: A Preface to a New Theory of Joking and Laughing.* Cambridge, England: Cambridge International Science Publishing, Ltd.

FPHK = Peter Baofu. 2008. *The Future of Post-Human Knowledge: A Preface to a New Theory of Methodology and Ontology.* Oxford, England: Chandos Publishing (Oxford) Limited.

FPHL = Peter Baofu. 2009. *The Future of Post-Human Language: A Preface to a New Theory of Structure, Context, and Learning.* Cambridge, England: Cambridge Scholars Publishing, Ltd.

FPHLAW = Peter Baofu. 2010. *The Future of Post-Human Law: A Preface to a New Theory of Necessity, Contingency, and Justice.* Cambridge, England: Cambridge Scholars Publishing, Ltd.

FPHMA = Peter Baofu. 2009. *The Future of Post-Human Martial Arts: A Preface to a New Theory of the Body and Spirit of Warriors.* Cambridge, England: Cambridge Scholars Publishing, Ltd.

FPHML = Peter Baofu. 2008. *The Future of Post-Human Mathematical Logic: A Preface to a New Theory of Rationality.* Cambridge, England: Cambridge Scholars Publishing, Ltd.

FPHMM = Peter Baofu. 2009. *The Future of Post-Human Mass Media: A Preface to a New Theory of Technology.* Cambridge, England: Cambridge Scholars Publishing, Ltd.

FPHO = Peter Baofu. 2009. *The Future of Post-Human Organization: A Preface to a New Theory of Communication, Decision-Making, and Leadership..*Cambridge, England: Cambridge Scholars Publishing, Ltd.

FPHP = Peter Baofu. 2010. *The Future of Post-Human Personality: A Preface to a New Theory of Normality and Abnormality.* Cambridge, England: Cambridge International Science Publishing, Ltd.

FPHR = Peter Baofu. 2010. *The Future of Post-Human Religion: A Preface to a New Theory of Spirituality.* Cambridge, England: Cambridge International Science Publishing, Ltd.

FPHS = Peter Baofu. 2010. *The Future of Post-Human Sexuality: A Preface to a New Theory of the Body and Spirit of Love-Makers.* Cambridge, England: Cambridge Scholars Publishing, Ltd.

FPHST = Peter Baofu. 2006. *The Future of Post-Human Space-Time: Conceivng a Better Way to Understand Space and Time.* New York: Peter Lang Publishing, Inc.

FPHU = Peter Baofu. 2008. *The Future of Post-Human Unconsciousness: A Preface to a New Theory of Anomalous Experience.* Cambridge, England: Cambridge Scholars Publishing, Ltd.

FPHUP = Peter Baofu. 2009. *The Future of Post-Human Urban Planning: A Preface to a New Theory of Density, Void, and Sustainability.* Cambridge, England: Cambridge Scholars Publishing, Ltd.

FPHWP = Peter Baofu. 2010. *The Future of Post-Human War and Peace: A Preface to a New Theory of Aggression and Pacificity.* Cambridge, England: Cambridge Scholars Publishing, Ltd.

• PART ONE •

Introduction

CHAPTER 1
INTRODUCTION—THE IMPORTANCE OF ACOUSTICS

> Xenophon says that there is no sound more pleasing than one's own praises.
> —Plutarch (BT 2010)

The Questionability of a Conventional Wisdom

Is sound really so pleasant in its nature that it can flatter oneself to the point that, as Exnophon once said, "there is no sound more pleasing than one's own praises"? (BT 2010)

This positive idea of sound can be compared with a similarly flattering view of its opposite (that is, silence), in that, as Thomas Carlyle once wrote, "Speech is of Time, Silence is of Eternity." (BT 2010a)

Contrary to these equally wonderful views about sound and silence (and other claims as will be discussed in the book), acoustics (in relation to sound and silence) is neither possible nor desirable to the extent that the respective ideologues (on the opposing sides of the fence) would like us to believe.

Needless to say, one should not misleadingly treat this questionability of the conventional wisdom about the nature of sound and silence as a suggestion that acoustics is therefore a worthless field of study, or that some fields of study (related to acoustics) like music, mathematical acoustics, psychoacoustics, architectural acoustics, and so on are to be ignored. Obviously, neither of these extreme views is reasonable either.

Instead, this book provides an alternative (better) way to understand the future of acoustics, especially in the dialectic context of sound and silence—while learning from different approaches in the literature but without favoring any one of them (nor integrating them, since they are not necessarily compatible with each other).

In other words, this book offers a new theory (that is, *the multilateral theory of acoustics*) to go beyond the existing approaches in the literature on acoustics in an original way.

If successful, this seminal project is to fundamentally change the way that we think about acoustics, from the combined perspectives of the mind, nature, society, and culture, with enormous implications for the human future and what I originally called its "post-human" fate.

Acoustics in Relation to Sound and Silence

As a starting point, it is important to define the word "acoustic," which is "from the Greek word *ἀκουστικός* (*akoustikos*), meaning 'of or for hearing, ready to hear' and that from *ἀκουστός* (*akoustos*), 'heard, audible,' which in turn derives from the verb *ἀκούω* (*akouo*), 'I hear.'" (WK 2010)

With this definition in mind, acoustics therefore refers to the study of sound—and, for that matter, silence (which is "the relative or total lack of audible sound," dependent on the sound perception of an organic species or an electrical equipment). (WK 2010c)

So, "[a] scientist who works in the field of acoustics is an acoustician while someone working in the field of acoustics technology may be called an acoustical or audio engineer." (WK 2010)

For clarification, there are three major aspects of acoustics as a field of study to be remembered at the outset.

Firstly, acoustics is an interdisciplinary science.

The reason is that "[t]he application of acoustics can be seen in almost all aspects of modern society with the most obvious being the audio and noise control industries." (WK 2010)

Secondly, acoustics deals with both audible and non-audible kinds of sound.

The reason is that "acousticians had extended their studies to frequencies above and below the audible range," and "it became conventional to identify these frequency ranges as 'ultrasonic' and 'infrasonic' respectively, while letting the word 'acoustic' refer to the entire frequency range without limit." (WK 2010)

And thirdly, acoustics deals with both sound and silence.

The reason is that sound can be unwanted; for instance, when sound is "unwanted," it is known as "noise." (WK 2010a) This kind of sound (like noise) is unwanted, because "[n]oise can block, distort, change or interfere with the meaning of a message in human, animal and electronic communication." (WK 2010b)

Thus, "noise control" (or "noise regulation") is a major concern in acoustics (especially, though not exclusively, for "noise control industries") and is thus related to the study of silence (although "noise control" does not necessarily mean that sound has to be absolutely eliminated to the point of utter silence). Consequently, the distinction between sound and silence can be one of degree.

With these three clarifications in mind—let us look more closely at the nature of (a) sound and (b) silence, as explained below, in that order, and summarized in *Table 1.1*.

Sound

The study of sound in physical acoustics concerns five major steps about "the generation, propagation and reception of mechanical waves and vibrations," as shown below: (WK 2010)

Cause → Generation → Propagation → Reception → Effect
(Transduction)   (Transduction)

These five steps "can be found in any acoustical event or process." (WK 2010)

In the first step about cause—the source of sound varies, relative to both "natural and volitional" circumstances, be they related to humans, animals, natural events, or technological products. (WK 2010)

In the second and fourth steps about generation and reception—the concern here is about "transduction," which is "the process that convert[s] energy from some other form into acoustic energy, producing the acoustic wave." (WK 2010)

In the third step about propagation—"[t]here is one fundamental equation that describes acoustic wave propagation, but the phenomena that emerge from it are varied and often complex. The wave carries energy throughout the propagating medium. Eventually this energy is transduced again into other forms, in ways that again may be natural and/or volitionally contrived." (WK 2010)

And in the final step about effect—the result "may be purely physical or it may reach far into the biological or volitional domains. The five basic steps are found equally well whether we are talking about an earthquake, a submarine using sonar to locate its foe, or a band playing in a rock concert." (WK 2010)

With these five steps in mind, the phenomena of progagation and transduction can be elaborated a bit more, in what follows, in that order.

The Propagation of Sound Waves

The propagation of sound waves is the central stage in an acoustical process and can be described in terms of two kinds, namely, (a) "pressure levels" and (b) "frequency," as shown below.

(a) Pressure Levels

Sound can be understood in terms of pressure levels (especially in relation to the degrees of loudness).

The term "sound pressure" refers to "the difference, in a given medium, between average local pressure and the pressure in the sound wave." (WK 2010a) And "sound pressure level" then refers to "the loudness of the…disturbances [due to the difference between the two pressures] and "is measured on a logarithmic scale in decibels." (WK 2010)

In this way, sound is "a sequence of waves of pressure which propagates through compressible media such as air or water….Sound can propagate through solids as well, but there are additional modes of propagation…." (WK 2010a)

In other words, through air and water (as in gases, plasma, and liquids), "[s]ound is transmitted…as longitudinal waves, also called compression waves. Through solids, however, it can be transmitted as both longitudinal waves and transverse waves. Longitudinal sound waves are waves of alternating pressure deviations from the equilibrium pressure, causing local regions of compression and rarefaction, while transverse waves (in solids) are waves of alternating shear stress at right angle to the direction of propagation." (WK 2010a)

For instance, "[i]n fluids such as air and water, sound waves propagate as disturbances in the ambient pressure level. While this disturbance is usually small, it is still noticeable to the human ear. The smallest sound that a person can hear, known as the threshold of hearing, is nine orders of magnitude smaller than the ambient pressure." (WK 2010)

In any event, "[d]uring their propagation, waves can be reflected, refracted, or attentuated by the medium. The behaviour of sound propogation is generally affected by three things," as shown below: (WK 2010a; JHU 2010)

- "A relationship between density and pressure. This relationship, affected by temperature, determines the speed of sound within the medium."

- "The propogation is also affected by the motion of the medium itself. For example, sound moving through wind. Independent of the motion of sound through the medium, if the medium is moving, the sound is further transported."
- "The viscosity of the medium also affects the motion of sound waves. It determines the rate at which sound is attenuated. For many media, such as air or water, attenuation due to viscosity is negligible."

But there is a special case, in that, "when sound is moving through a medium which does not have constant physical properties, it is refracted (either dispersed or focused)." (WK 2010a; JHU 2010)

This dependence of sound waves on the medium is important enough. For instance, the speed of sound is not exactly constant, as it depends on the nature of the medium, because "the speed of sound is proportional to the square root of the ratio of the elastic modulus (stiffness) of the medium to its density. Those physical properties and the speed of sound change with ambient conditions." (WK 2010a)

Consider this illustration: "[T]he speed of sound in gases depends on temperature. In 20 °C (68 °F) air at the sea level, the speed of sound is approximately 343 m/s (1,230 km/h; 767 mph) using the formula...$v = (331 + 0.6T)$ m/s....In fresh water, also at 20 °C, the speed of sound is approximately 1,482 m/s (5,335 km/h; 3,315 mph). In steel, the speed of sound is about 5,960 m/s (21,460 km/h; 13,330 mph)." (WK 2010a; TQ 2010)

In addition, "[t]he speed of sound is also slightly sensitive (a second-order anharmonic effect) to the sound amplitude, which means that there are nonlinear propagation effects, such as the production of harmonics and mixed tones not present in the original sound." (WK 2010a)

(b) Frequency

Besides pressure levels (especially in relation to the degrees of loudness, as described above)—sound waves can also be understood in terms of frequency (especially in relation to the degrees of pitch).

For instance, "[p]hysicists and acoustic engineers tend to discuss sound pressure levels in terms of frequencies, partly because this is how our ears interpret sound. What we experience as 'higher pitched' or 'lower pitched' sounds are pressure vibrations having a higher or lower number of cycles per second. In a common technique of acoustic measurement, acoustic signals are sampled in time, and then presented in more meaningful forms such as octave bands or time frequency plots. Both these popular methods are used to analyze sound and better understand the acoustic phenomenon." (WK 2010)

In general, the spectrum of sound waves "can be divided into three sections: audio, ultrasonic, and infrasonic," as explicated below: (WK 2010)

- The Audio Range—"falls between 20 Hz and 20,000 Hz. This range is important because its frequencies can be detected by the human ear. This range has a number of applications, including speech communication and music." (WK 2010) But other animals can have different audio ranges; for instance, "dogs can perceive vibrations higher than 20,000 Hz." (WK 2010a)
- The Ultrasonic Range—"refers to the very high frequencies: 20,000 Hz and higher. This range has shorter wavelengths which allows better resolution in imaging technologies. Medical applications such as ultrasonography and elastography rely on the ultrasonic frequency range." (WK 2010)
- The Infrasonic Range—refers to "the lowest frequencies….These frequencies can be used to study geological phenomena such as earthquakes." (WK 2010)

Nowadays, there are sophisticated instruments which can be used to study the propagation of sound waves. For instance, "the Spectrum analyzer facilitate visualization and measurement of acoustic signals and their properties. The Spectrogram produced by such an instrument is a graphical display of the time varying pressure level and frequency profiles which give a specific acoustic signal its defining character." (WK 2010)

The Transduction of Sound Waves

But the propagation of sound waves is only a step in an acoustical process, since there is also "transduction," which often requires a device known as "transducer" to work.

A "transducer" is "a device for converting one form of energy into another. In an acoustical context, this usually means converting sound energy into electrical energy (or vice versa). For nearly all acoustic applications, some type of acoustic transducer is necessary." (WK 2010)

Good examples of a transducer are "loudspeakers, microphones, hydrophones and sonar projectors. These devices convert an electric signal to or from a sound pressure wave." (WK 2010)

Transducers make good use of different transduction principles like "electromagnetism, electrostatics and piezoelectricity. The transducers in most common loudspeakers (e.g., woofers and tweeters), are electromagnetic devices and generate waves using a suspended diaphragm driven by an electromagnetic voice coil, sending off pressure waves. Electret microphones and condenser microphones employ electrostatics. As the sound wave strikes the microphone's diaphragm, it moves and induces a voltage change. The ultrasonic systems used in medical ultrasonography employ piezoelectric transducers. These are made from special ceramics in which elastic vibrations and electrical fields are interlinked through a property of the material itself." (WK 2010)

Silence

Besides sound—acoustics also deals with "silence," which is "the relative or total lack of audible sound," dependent on the sound perception of an organic species or an electrical equipment. (WK 2010c)

Silence in Different Contexts

But silence can be understood in different ways, dependent on different contexts.

Firstly, "the word silence may also refer to any absence of communication, even in media other than speech." (WK 2010c; AHD 2000)

Secondly, silence, alternatively but in the opposite direction, can be understood as "total communication, in reference to non verbal communication and spiritual connection." (WK 2010c)

Thirdly, silence can also be understood in the context of linguistic anaysis. For instance, "[i]n discourse analysis, brief absences of speech mark the boundaries of prosodic units used by speakers. Silence in speech can be the result of hesitation, stutters, self-correction, or the deliberate slowing of speech for the purpose of clarification or processing of ideas. These are short silences. Longer pauses in language occur in interactive roles, reactive tokens, or turn-taking." (WK 2010c)

And finally, silence can also be understood in the context of cultural norms. For illustration, "[a]ccording to cultural norms, silence can be interpreted as positive or negative. For example, in a Christian Methodist faith organization silence and reflection during the sermons might be appreciated by the congregation, while in a Southern Baptist church, silence might mean disagreement with what is being said, or perhaps disconnectedness from the congregated community. Deaf people function in a completely silent culture." (WK 2010c)

The rest of the book will further address issues like this in the context of acoustics (while relating the analysis of silence to that of sound too).

The Diverse Sub-Fields of Acoustics

This brief introduction to the topics of sound and silence in acoustics is only a beginning to understand the nature of acoustics. The next business then is to be aware of the diverse sub-fields of acoustics as a field of study.

In accordance to the official PACS (Physics and Astronomy Classification Scheme) classification system, there are at least "seventeen major subfields of acoustics....These have been grouped into three domains: physical acoustics, biological acoustics and acoustical engineering," as shown below: (WK 2010)

- "Physical Acoustics"
    - –Ex: "aeroacoustics"
    - –Ex: "general linear acoustics"
    - –Ex: "nonlinear acoustics"
    - –Ex: "structural acoustics and vibration"
    - –Ex: "underwater sound"
- "Biological Acoustics"
    - –Ex: "bioacoustics"
    - –Ex: "musical acoustics"
    - –Ex: "physiological acoustics"
    - –Ex: "psychoacoustics"
    - –Ex: "speech communication"
- "Acoustical Engineering"
    - –Ex: "acoustic measurements and instrumentation"
    - –Ex: "acoustic signal processing"
    - –Ex: "architectural acoustics"
    - –Ex: "environmental acoustics"
    - –Ex: "transduction"
    - –Ex: "ultrasonics"
    - –Ex: "room acoustics"

Of course, there can be other sub-fields of acoustics which are not listed above.

For instance, there is the so-called "Lindsay's Wheel of Acoustics" which was proposed by R. Bruce Lindsey (1964); in this alternative listing, "the scope of acoustics start[s]...from the four broad fields of Earth Sciences, Engineering, Life Sciences, and the Arts. The outer circle lists the various disciplines one may study to prepare for a career in acoustics. The inner circle lists the fields within acoustics that the various disciplines naturally lead to." (K. Gee 2010)

Therefore, the above listing by the official PACS classification system is more illustrative than exhaustive.

However, as a way to advertise what is to come, this book will further address the different topics in these listings, not just at the physical and biological levels of analysis but also at the cultural and societal levels too.

The Theoretical Debate

This introduction to the diverse sub-fields of acoustics and my suggestion to broaden its inquiry to include cultural and societal levels of analysis too (not just focusing only at the physical and biological levels) then bring us to the larger issue concerning the theoretical debate on sound and silence in acoustics.

Although there can be many different theoretical approaches in the debate—there are five main ones to be summarized hereafter, and they can be called, in the absence of better words, namely, (a) the *mental* argument, (b) the *natural* argument, (c) the *societal* argument, (d) the *cultural* argument, and (e) the *multilateral* argument (which is my original contribution to the debate), to be elabroated hereafter, respectively.

Lest misunderstanding occurs, three clarifications should be made here, as this is something that I regularly emphasize in my previous books, whenever a theoretical debate is summarized for a topic, *almost verbatim*.

Firstly, there can be other arguments besides the first four here, that is, the "mental," the "natural," the "societal," and the "cultural" arguments (as indicated above). The advantage to select these four here has to do with their illustrative representation of the diverse theories in the literature (which are deemed sufficient for the purpose at hand in this book).

Secondly, the examples (theories) given to illustrate each argument are also supposed to be illustrative, not exhaustive. After all, the arguments will be addressed again in the rest of the book.

And thirdly, the four arguments are not necessarily mutually exclusive. For instance, those who advocate the "mental" argument can also consider the "cultural" argument, although they do not focus on the latter.

And the reverse is also true, in that those who make the "cultural" argument can also consider the "mental" argument, although they do not focus on the latter.

In other words, their disagreement is often one in degree, not in kind.

The Mental Argument

With these three clarifications in mind—the first major theoretical approach is called, in the absence of better words, the *mental* argument, which focuses, relatively speaking of course, more on the biological and psychological aspects of sound and silence in acoustics.

Therefore, the word "mental" in the title of the argument is suggestive, because it refers to the focus on the biological and psychological aspects of sound and silence so understood.

But the mental argument has different issues of concern, and different defenders of the argument offer their own distinctive stands on the issues, depending on their own interests.

For illustration, consider these major versions of the mental argument, namely, in the absence of better words, (a) the *psychological* model, and (b) the *genetic* model—to be summarized below, respectively.

The Psychological Model

The first version of the mental argument has much to do with the "psychological" model (or more precisely, the "psychoacoustic" model), which refers to "the scientific study of sound perception. More specifically, it is the branch of science studying the psychological and physiological responses associated with sound (including speech and music)." (WK 2010d)

Instead of treating sound as a physical or mechanical process of wave propagation (as in physical acoustics, which was already described in *Sec. 1.2*)—the psychoacoustic model focuses on the "sensory and perceptual" aspects of sound. (WK 2010d)

So, when studying sound, a psychoacoustician will analyze instead how, "within the ear," the sound "is transformed into neural action potentials. These nerve pulses then travel to the brain where they are perceived. Hence, in many problems in acoustics,...both the ear and the brain are involved in a person's listening experience." (WK 2010d)

For instance, in 1933, Harvey Fletcher and Wilden Munson at Bell Labs did some work on "equal-loudness contours" in psychoacoustics, and their aim was to identify "the sound pressure level (dB), over the range of audible frequencies, which are perceived as being of equal loudness." (WK 2010d) Fletcher and Munson succeeded in the identification by "using pure tones reproduced via headphones, and the data they collected are called Fletcher-Munson curves. Because subjective loudness was difficult to measure, the Fletcher-Munson curves were averaged over many subjects." (WK 2010d)

The Genetic Model

The second version of the mental argument has to do with the "genetic" model, in that it focuses on the genetic basis of hearing sound.

For instance, there had been studies which showed that "[h]earing loss can be inherited. Both dominant genes and recessive genes exist which can cause mild to profound impairment. If a family has a dominant gene for deafness it will persist across generations because it will manifest itself in the offspring even if it is inherited from only one parent. If a family had genetic hearing impairment caused by a recessive gene it will not always be apparent as it will have to be passed onto offspring from both parents. Dominant and recessive hearing impairment can be syndromic or nonsyndromic. Recent gene mapping has identified dozens of nonsyndromic dominant (DFNA#) and recessive (DFNB#) forms of deafness." (WK 2010e; E. Lynch 1997)

A good example is the study by M. Schraders (2010) and his colleagues to show that "mutations in PTPRQ are a cause of autosomal-recessive nonsyndromic hearing impairment." (WK 2010e)

However, for the critics, both the "psychoacoustic" and the "genetic" versions of the mental argument have their own major problems. For illustration, consider two major criticisms below.

Firstly, one major criticism of both the "psychoacoustic" and the "genetic" versions of the mental argument is that they are too exclusive, in ignoring other versions of the mental argument (that is, other than the version that each model focuses on).

And secondly, another major criticism of both the "psychoacoustic" and the "genetic" versions of the mental argument is that they are too narrow, in failing to take into account of other arguments (like the natural, societal, and cultural ones, as will be introduced later in this section).

The Natural Argument

The second theoretical approach is called, in the absence of better words, the *natural* argument, which focuses, relatively speaking of course, more on the natural aspects of sound and silence in the state of nature.

Therefore, the word "natural" in the title of the argument is suggestive, because it refers to the focus on the natural aspects of sound and silence in the state of nature so understood.

Like the mental argument—the natural argument has different issues of concern, and different defenders of the argument offer their own distinctive stands on the issues, depending on their own interests.

For illustration, consider two major versions of the natural argument, namely, in the absence of better words, (a) the *mechanical* model, and (b) the *mathematical* model—to be summarized below, respectively.

The Mechanical Model

The first version of the natural argument has much to do with the "mechanical" model," which, unlike the "psychoacoustic" model in the mental argument (as described in the previous sub-section), focuses instead on the "physical or mechanical" process of sound waves.

As already described in *Sec. 1.2* (and will not be repeated here)—the mechanical model looks into the five major steps of acoustics, like the physical properties of transduction and propagation in nature, for instance.

The Mathematical Model

The second version of the natural argument has much to do with the "mathematical" model.

Both the "mechanical" and the "mathematical" models have much in common, in that both make good use of mathematics to understand acoustics.

But there is a major difference of degree between them, in that the "mathematical" model is more obsessed than the "mechanical" model, relatively speaking of course, with the use of mathematics in depicting the reality of sound and silence. The distinction between the two models is officially known as the difference between "physical acoustics" and "mathematical acoustics" as two distinct sub-fields in the literature.

For instance, a good illustration of the mathematical model in action concerns the measurement of sound pressure, which is "the difference, in a given medium, between average local pressure and the pressure in the sound wave. A square of this difference (i.e., a square of the deviation from the equilibrium pressure) is usually averaged over time and/or space, and a square root of this average provides a root mean square (RMS) value." (WK 2010a)

Now, sound pressure can then be measured "as a level on a logarithmic decibel scale. The sound pressure level (SPL) or Lp is defined" in what follows: (WK 2010a)

$$Lp = 10 \log_{10} (P^2/Pref^2) = 20 \log_{10} (P/Pref) \text{ dB}$$

where "*P* is the root-mean-square sound pressure," "*Pref* is a reference sound pressure," and *dB* is the sound pressure level to be expressed in decibels. (WK 2010a)

It is important to note that "[c]ommonly used reference sound pressures, defined in the standard ANSI S1.1-1994, are 20 μPa in air and 1 μPa in water. Without a specified reference sound pressure, a value expressed in decibels cannot represent a sound pressure level." (WK 2010a)

Elegant this equation looks like for sure—but, for the critics, both the "mathematical" and the "mechanical" versions of the natural argument have their own major problems too. For illustration, consider three major criticisms below.

Firstly, one major criticism of the "mathematical" version of the natural argument is that mathematics may yield elegant equations, but they do not necessarily represent reality as it is, in a complete way. I already addressed this issue, among others, in *The Future of Post-Human Mathematical Logic* (2008).

Secondly, another major criticism of both the "mathematical" and the "mechanical" versions of the natural argument is that they are also exclusive, in not giving equal weight to other versions of the natural argument (that is, other than the one that each model focuses on).

And thirdly, one more major criticism of both the "mathematical" and the "mechanical" versions of the natural argument is that they too are narrow, in not giving sufficient consideration of other arguments (like the mental, societal, and cultural ones, as introduced above and below in this section).

The Societal Argument

The third theoretical approach is called, in the absence of better words, the *societal* argument, which focuses, relatively speaking of course, more on the societal aspects of sound and silence in acoustics.

In this sense, the word "societal" in the title of the argument is suggestive, because it refers to the societal aspects of sound and silence so understood.

Like the mental and natural arguments—the societal argument has different issues of concern, and different defenders of the argument offer their own distinctive stands on the issues, depending on their own interests.

For illustration, consider these two major versions of the societal argument, namely, in the absence of better words, (a) the *architectural* model and (b) the *legal* model—to be summarized in what follows, respectively.

The Architectural Model

The first version of the societal argument has to do with the "architectural" model (or more precisely, the "architectural-acoustic" model), which refers to "the science of noise control within buildings." (WK 2010f)

In the older days, "[t]he first application of architectural acoustics was in the design of opera houses and then concert halls"; but nowadays, "[a]rchitectural acoustics includes room acoustics, the design of recording and broadcast studios, home theaters, and listening rooms for media playback." (WK 2010f)

For instance, in interior space acoustics, acousticians try to engage in noise control by targeting three ways to solve sound problems, namely, the so-called "ABCs" as shown below: (WK 2010f)

- "A = Absorb {via drapes, carpets, ceiling tiles, etc.)"
- "B = Block (via panels, walls, floors, ceilings and layout)"
- "C = Cover-up (via sound masking)"

Of course, in many cases, "all three of these are recommended to achieve optimal results," but it is of common knowledge that "C = Cover-up by increasing background sound produces the most dramatic improvement in speech privacy—with the least disruption and typically the lowest cost." (WK 2010f)

The Legal Model

And the second version of the societal argument has to do with the "legal" model, which refers to the law for "the right to quiet enjoymnet" in the common law tradition. (WK 2010g)

In other words, "under the common law, persons in possession of real property (land owners, lease holders etc.) are entitled to the quiet enjoyment of their lands....If a neighbour interferes with that quiet enjoyment, either by creating smells, sounds, pollution or any other hazard that extends past the boundaries of the property, the affected party may make a claim in nuisance." (WK 2010g; W. Prosser 1966)

For illustration with a specific example, "[u]nder English law, unlike in the USA, it is no defence that the claimant 'came to the nuisance': the 1879 case of Sturges v Bridgman is still good law, and a new owner can bring a claim in nuisance for the existing activities of a neighbour." (WK 2010g)

However, for the critics, both the "architectural" and "legal" versions of the societal argument have some major problems. For illustration, consider three major criticisms, as shown below.

Firstly, one major problem with the "legal" model is that what is true in one legal tradition may not be so in another, as shown in the example above about "nuisance" under English law as opposed to American law.

Secondly, another major problem with both the "architectural" and "legal" versions of the societal argument is that they are too restrictive, in ignoring other versions of the societal argument (that is, other than the one that each theory focuses on).

And thirdly, one more major problem with both the "architectural" and "legal" versions of the societal argument is that they are also too narrow, in excluding other arguments (like the mental, natural, and cultural ones, as introduced above and below in this section).

The Cultural Argument

The fourth theoretical approach is called, in the absence of better words, the *cultural* argument, which focuses, relatively speaking of course, more on the cultural aspects of sound and silence in acoustics.

Therefore, the word "cultural" in the title of the argument is suggestive, because it refers to the focus on the cultural aspects of sound and silence so understood.

Like the mental, natural, and societal arguments—the cultural argument has different issues of concern, and different defenders of the argument offer their own distinctive stands on the issues, depending on their own interests.

For illustration, consider these two major versions of the cultural argument, namely, in the absence of better words, (a) the *musical* model and (b) the *spiritual* model—to be summarized in what follows, respectively.

The Musical Model

The first version of the cultural argument has to do with the "musical" model, which refers to the use of sound and silence in music.

For instance, in 1952, American experimental composer John Cage (1912–1992) composed a famous three-movement composition known as "4′33″ (pronounced Four, thirty-three),…for any instrument (or combination of instruments)." (WK 2010h)

In this musical piece with both sound and silence, the part of silence is intriguing, because "the score instructs the performer not to play the instrument during the entire duration of the piece throughout the three movements (the first being thirty seconds, the second being two minutes and twenty-three seconds, and the third being one minute and forty seconds)." (WK 2010h; J. Pritchett 2006; L. Solomon 1998; R. Kostelanetz 2003)

On the other hand, there is also the part of sound, in that, "[a]lthough commonly perceived as 'four minutes thirty-three seconds of silence,' the piece actually consists of the sounds of the environment that the listeners hear while it is performed." (WK 2010h; W. Fetterman 1996; J. Lienhard 2003)

The Spiritual Model

The second version of the cultural argument has to do with the "spiritual" model, which refers to a way of life which emphasizes the use of silence (or quietness) for meditation and contemplation.

For instance, "centering prayer is a popular method of contemplative prayer or Christian meditation, placing a strong emphasis on interior silence." (WK 2010i)

Centuring prayer has multiple origins, however, because "most authors trace its roots to the contemplative prayer of the Desert Fathers of early Christian monasticism, to the Lectio Divina tradition of Benedictine monasticism, and to works like *The Cloud of Unknowing* and the writings of St. Teresa of Avila and St. John of the Cross." (WK 2010i)

But the origins of "the 'Centering Prayer' movement in modern Catholicism and Christianity can be traced to several books published by three Trappist monks of St. Joseph's Abbey in Spencer, Massachusetts in the 1970s: Fr. William Meninger, Fr. M. Basil Pennington and Abbot Thomas Keating." (WK 2010i; CO 2006)

However, for the critics, both the "musical" and "spiritual" versions of the cultural argument have some major problems. For illustration, consider a few major criticisms, as shown below.

Firstly, one major problem with the "musical" model is that what is considered as a good piece of music in terms of its sound and silence may not be so for someone else. For instance, in the example about the piece by Cage, "4′33″ became Cage's...most controversial composition" for those who are critical to it. (WK 2010h; J. Pritchett 2006)

Secondly, another major problem with the "spiritual" model is that what is desirable for the followers of the movement to practice centuring prayer is not necessarily so for the Catholic Church, because "[s]ome critics have argued that centering prayer contains practices that were warned against by the document *Aspects of Christian meditation*, issued by the Vatican's Congregation for the Doctrine of the Faith, led by then Cardinal Joseph Ratzinger." (WK 2010i; ACL 2007)

Thirdly, still another major problem with both the "musical" and "spiritual" versions of the cultural argument is that they are also restrictive, in ignoring other versions of the cultural argument (that is, other than the one that each theory focuses on).

And fourthly, one more major problem with both the "musical" and "spiritual" versions of the cultural argument is that they are also too narrow, in excluding other arguments (like the mental, natural, and societal ones, as introduced above and below in this section).

The Multilateral Argument

And fifthly, unlike the previous four arguments (viz., the mental, the natural, the societal, and the cultural)—the fifth one to be introduced here can be called, in the absence of better words again, the *multilateral* argument, which learns from all other theoretical approaches (which include the previous four arguments and also other examples as will be introduced in the rest of this book), without, however, favoring any of them, nor trying to integrate them (as they are not necessarily compatible with each other, as the introduction of the theoretical debate in the previous sub-sections has shown).

The Multilateral Theory of Acoustics

My *multilateral* argument can be more precisely called *the multilateral theory of acoustics*, since it is to propose a new theory of sound and silence.

My multilateral theory of acoustics has four major distinct features.

Firstly, my theory makes good use of all theoretical approaches in the literature on sound and silence, be they about the mental argument, the natural argument, the societal argument, or the cultural argument—especially from the most comprehensive combined perspectives of the mind, nature, society, and culture (as will be clear in Chapter Two and Chapter Three).

Secondly, just like many other theories of mine in my previous books, my theory here does not heavily favor any specific theory over others in the literature, nor trying to integrate them (as they are not necessarily compatible with each other). For this reason, the word "multilateral" in the title of the theory is suggestive.

Thirdly, my theory treats the issue of acoustics in the distinctive dialectic context of sound and silence, such that one is not to be reduced as part of the analysis of the other.

And fourthly, it contains seventeen major theses, namely, (a) the first thesis: the absoluteness-relativeness principle, (b) the second thesis: the predictability-unpredictability principle, (c) the third thesis: the explicability-inexplicability principle, (d) the fourth thesis: the preciseness-vagueness principle, (e) the fifth thesis: the simpleness-complicatedness principle, (f) the sixth thesis: the openness-hiddenness principle, (g) the seventh thesis: the denseness-emptiness principle, (h) the eighth thesis: the slowness-quickness principle, (i) the nineth thesis: the expansion-contraction principle, (j) the tenth thesis: the theory-praxis principle, (k) the eleventh thesis: the convention-novelty principle, (l) the twelveth thesis: the evolution-transformation principle, (m) the thirteenth thesis: the symmetry-asymmetry principle, (n) the fourteenth thesis: the softness-hardness principle, (o) the fifteenth thesis: the regression-progression principle, (p) the sixteenth thesis: the sameness-difference principle, and (q) the seventeenth thesis: the post-human creativeness—to be elaborated in the rest of the book, with a summary in the concluding chapter.

Of course, as this is something that I often emphasized in my previous books, other principles (besides the 16 as cited above) are also relevant, but these 16 are the most relevant in the current case study (in terms of the number of citations of each principle in the book).

Even then, in some cases, the difference between any two given principles, for instance, in terms of the number of citations in a book, is rather small, so the reason in those cases is more aesthetic (than otherwise), because it looks nicer to list only 16 theses for 16 principles (than 22 theses for all of the 22 principles) in the Table of Contents.

This is true, even if different studies of the same kind can yield different views about the degree of relevance for each principle, depending on the specific nature of a research in question, needless to say. So, if a different author analyzes the same subject matter in a different way, the relevance of the principles will be different.

With this clarification in mind—the seminal project here, if successful, will fundamentally change the way that we think about sound and silence in acoustics, from the combined perspectives of the mind, nature, society, and culture, with enormous implications for the human future and what I originally called its "post-human" fate.

Theory and Meta-Theory

The *multilateral* argument, or *the multilateral theory of acoustics*, presupposes some kind of meta-theory, in special relation to methodology and ontology—just as all theories, either explicitly or implicitly, require their own versions of meta-theory.

The Classification of Theory and Meta-Theory

Theory and meta-theory, broadly speaking, can be further classified into different categories, as shown below:

- On Theory
  - —Ex: in relation to the mind
  - —Ex: in relation to nature
  - —Ex: in relation to society
  - —Ex: in relation to culture

- On Meta-Theory
  - —Ex: in relation to methodology
  - —Ex: in relation to ontology

Each category, in its turn, can then be further sub-divided into different levels of analysis—which will be elaborated in the next two sections (and summarized in *Table 4.43*).

With this reminder in mind, I already proposed (in my previous books) my distinctive contribution to meta-theory, namely, methodology (known as "sophisticated methodological holism") and ontology (known as "existential dialectics").

The next two sections are to introduce my meta-theory—that is, existential dialectics and sophisticated methodological holism, respectively—to be repeated, *often verbatim* (with only a few updated revisions), from my previous works, as this is something that I often do in each new book of mine as an introduction to some background information for the convenience of the readers.

In the process, I shall also introduce my numerous new theories (48 so far in 39 books, as of early 2011) in relation to the four categories of the mind, nature, society, and culture in the context of theory (as also summarized in *Table 4.43*).

A Unified Theory of Everything

In the end, my distinctive contribution to both meta-theory and theory serves as a foundation to unify all domains of knowledge for an overarching understanding of all forms of reality (by way of some ontological principles at the meta-theoretical level and the comprehensive perspectives of the mind, nature, society, and culture at the theoretical level—as will be explained in the next two sections).

This is so, without commiting the sins of reductionism and reverse-reductionism (as often seen in many holistic approaches, with the fad of systems approach as a most recent notorious example, as already critically explained in *The Future of Post-Human Formal Science*, or *FPHFS*), since my distinctive contribution makes good use of different schools of thought without favoring any of them nor trying to integrate them (as they are not necessarily compatible with each other), so as to include (or adjust for) heterogeneity, conflict, subjectivity, and complexity, for example.

In this sense, the word "unified" does not have to mean an integrative approach, in a narrow sense, nor a systems approach, in a broad one, both of which I have critically rebuked (in my previous books), while learning from them.

This overarching project (consisting of numerous theories of mine, as will be summarized in the next section), thus, constitutes my grand unifed theory of everything in the world and beyond.

The Logic of Existential Dialectics

To start, the ontology of existential dialectics can be summarized in this section—*often verbatim* (with a few updated revisions), from my previous works, as this is something that I regularly do in each new book of mine as an introduction to some background information for the convenience of the readers.

After all, this book, like all others of mine, is in conversation with all previous ones of mine, for the final aim to converge all of them into a larger grand project on the future of intelligent life, both here on earth and there in deep space unto multiverses.

The summary can be organized in relation to (a) the conception of existential dialectics (or its ontics), (b) the syntax of existential dialectics (or its ontomethodology and ontologic), (c) the semantics of existential dialectics (or its ontosemantics), and (d) the pragmatics of existential dialectics (or its ontopragmatics), respectively hereafter—again *often verbatim*.

This means that, for those readers who had read some of my previous books, the summary below is really not much new (*almost verbatim*, with some updated revisions).

But for those who have never read my previous books, it is a good review of them.

The Conception of Existential Dialectics (or Its Ontics)

This ontology was originally designated as "existential dialectics" in *Beyond Capitalism to Post-Capitalism* (herein abbreviated as *BCPC*), although it was already analyzed in other books of mine like the 2 volumes of *The Future of Human Civilization* (*FHC*), *The Future of Capitalism and Democracy* (*FCD*), *The Future of Post-Human Consciousness* (*FPHC*), and the 2 volumes of *Beyond Democracy to Post-Democracy* (*BDPD*).

Lest any confusion occurs, it is important to stress at the outset that the word 'existential' in "existential dialectics" has nothing to do with Existentialism, which I rebuked in *FHC*, *FCD*, and also *FPHC*. Instead, the word here merely refers to the existence of intelligent life (both primitive and advanced) in a broad sense.

Some Basic Concepts

The conception of existential dialectics (or its ontics) makes use of different concepts (like "sets," "elements," "relations," "operations," "functions," "truth values," "axioms," "postulates," and "principles"—as shown in *Table 4.3*), which are important for the understanding of any logic of ontology.

That clarified—existential dialectics, as a language of ontology, can be analyzed in different ways that a language, as an analogy, is often analyzed, namely, in relation to phonology ("the study of patterns of a language's sounds"), phonetics ("the study of the physical aspects of sounds of human language"), morphonology ("the study of the internal structure of words"), syntax ("the study of how words combine to form grammatical sentences"), semantics ("the study of the meaning of words [lexical semantics] and fixed word combinations [phraseology]"), and pragmatics ("the study of how utterances are used...literally, figuratively, or otherwise...in communicative acts"). (WK 2007)

But since my theory of existential dialectics makes use of the English language for communication here, it does not propose a new way to make the patterns of a language's sounds (as in phonology), to study the physical aspects of sounds (as in phonetics), or even to strive for a new organization of the internal structure of words for the English language (as in morphonology).

For this reason, the language of existential dialectics to be analyzed hereafter is not concerned with the phonological, phonetic, and morphonological aspects of the English language. Instead, the analysis will explore the syntax, semantics, and pragmatics of existential dialectics as a language of ontology to improve our understanding of reality.

The Ontic Dispute

But first, there is an ontic dispute to be addressed, which is related to what I called "the dilemma of ontology" as introduced earlier in *Sec. 1.4* of *The Future of Post-Human Knowledge* (*FPHK*).

A good point of departure concerns what I want to call *the ontic dispute*, on how to understand the very nature of ontology.

Traditionally, scholars in the field used to study the conception of ontology in terms of its entities (e.g., material objects, abstract numbers), properties (e.g., duration, plurality), and relationships (e.g., causation).

This is a good starting point, but it suffers from different problems. Consider three examples of bias below, which are quite common in the literature, and the recent version (of treating concepts and principles in ontology on the basis of "plurality," "dynamism," "duration," "interaction," "life," "consciousness," and "volition") by Reginald Firehammer (2005) only constitutes a latest addition to this old (impoverishing) tradition.

Firstly, the classification of entities, properties, and relationships is too material-centric, in often focusing more on the existence of material entities than on non-material ones.

Secondly, the classification is too anthropo-centric, in often favoring the factor of human agency in depicting reality.

And lastly, the classification is too confused among the three categories relating to each other, in often lumping them together without adequate consideration of their intricated interrelationships (both in relation to each and in relation to all).

A good alternative is precisely what I want to offer in this debate, that is, a new classification of ontological entities, properties, and their relationships in terms of *method*, *structure*, *process*, *agency*, and *outcome* instead. Let me explicate what these words mean below.

In relation to "method"—it concerns how ontology is to be studied. In this sense, the word 'method' here should not be confused with the different usage of the same term 'method' as a technique of investigation in the context of methodology (as already introduced in *Sec. 1.2* of *FPHK*).

In relation to "structure"—it concerns what the general patterns of entities, properties, and relationships are.

In relation to "process"—it concerns how these general patterns change over time.

In relation to "agency"—it concerns how agency affects any change of these general patterns over time (without putting, however, too much emphasis on the role of human agency).

And in relation to "outcome"—it concerns what the outputs of this change over time are.

The Syntax of Existential Dialectics (or Its Ontomethodology and Ontologic)

The syntax of existential dialectics, analogously speaking, refers here to the structure of an ontology which can be studied by way of combining ideas into complex relationships like ontological principles to depict reality in the world—and can be analyzed in seven sub-sections, namely, (a) the selection criteria, (b) further clarifications on classification, (c) ontomethodology, (d) ontologic, (e) the principles as short cuts, (f) the principles as family resemblances, and (g) the dialectic constraints imposed by the principles, respectively hereafter.

The Selection Criteria

To start, not any pair of relationship can be chosen as an ontological principle in existential dialectics; otherwise, there could be billions of them under the sky.

Five illustrations below suffice to clarify this point.

(a) Firstly, an ontological principle cannot *overlap* with other principles, in that it is too closely related with other ones.

Of course, one cannot totally rule out any relationship between two entities, especially in a complex world of everything being linked to everything else—but the comparison here is relative, not absolute.

In this sense, the flexibility-inflexibility pair cannot be an ontological principle, since it has something closely in common with other ones (like the simpleness-complicatedness principle, for instance).

As an illustration, something which is flexible in interpretation is more likely to allow multiple viewpoints and thus makes the overall picture more complicated than otherwise (simple).

(b) Secondly, an ontological principle cannot be *redundant* in relation to other principles, in that it somehow duplicates other principles.

In this sense, for instance, the directness-indirectness pair cannot be an ontological principle, because it somehow duplicates other ontological principles (like the preciseness-vagueness principle, for instance).

As an illustration, something which is direct means to be right to the point and does not need to go to an unnecessarily long loop—so it is less vague, in being clearer about the thing in question.

(c) Thirdly, an ontological principle cannot be *derived* from other ones—as if it is a child-parent relationship.

In this sense, the convenience-inconvenience pair cannot be an ontological principle, because it can be derived from other ones (like the slowness-quickness principle, for instance).

For instance, something which is convenient already implies that it can be obtained quickly (rather than slowly)—all things considered.

(d) Fourthly, an ontological principle cannot be *trivial* (or *parochial*)—as if it is only one component competing with other ones for inclusion in a set.

In this sense, the consistency-inconsistency pair cannot be an ontological principle, because its validity is limited (or parochial), as it can be easily challenged and replaced by opposing major (not minor) theories (e.g., the correspondence theory of truth, the pragmatic theory of truth, the semantic theory of truth, the disclosedness theory of truth, and whatnot), not just in accordance to the consistency (or coherence) theory of truth (for consistency).

Thus, the consistency argument is only one among different opposing arguments in relation to the logic of what constitutes "truth."

(e) And fifthly, an ontological principle cannot be *one-sided*, that is, part of one side (category) of a dialectic pair.

In this sense, the extroversion-introversion pair cannot be an ontological principle, since it is part of the side (category) of "normality" in the normality-abnormaltiy pair (because both extroversion and introversion are two different "normal" personality traits in psychology).

In addition, the normality-abnormality pair is already related to the symmetry-asymmetry principle, the density-emptiness principle, the convention-novelty principle, and the same-difference principle (as will be explained in the section on "the principles as family resemblances" later).

Although these examples are not exhaustive, they illustrative the selectiveness of any ontological principle to be existentially dialectic.

Further Clarifications on Classification

Lest any misunderstanding carelessly occurs, some further clarifications on classification should be made here (and summarized in *Table 4.8*).

(a) Firstly, the total number of ontological principles is unknown, to be discovered later, as our knowledge of the world becomes more advanced.

Consequently, the principles as introduced in my books are not exhaustive, with new ones being added, whenever more of them are discovered in later research.

At least, future generations can pick up where I leave off and continue the discovery.

In this sense, my work should be treated as a pioneering effort for the development of a systematic, comprehensive analysis of a new general ontology for the future of knowledge.

(b) Secondly, the principles are not rigidly classified, as they can be reclassified in a different way.

For instance, the preciseness-vagueness principle is classified under the category of "structure" but can be reclassified under the category of "method," although in so doing, it has a different meaning in the context of method.

The same logic applies to the same-difference principle under the category of "outcome," which can be reclassified under the category of "structure," although in so doing, once more, it has a different meaning in the context of structure.

(c) Thirdly, the principles are to be understood together, not that each prinicple is to be treated independently of others. After all, the principles are mutually constraining, in that they work together as a whole.

(d) Fourthly, the principles are relevant to all subject matters, but some principles are, relatively speaking of course, more useful to some subject matters than others under certain conditions—as implied in the symmetry-asymmetry principle.

(e) And fifthly, the principles constitute only two levels of analysis, this time, at the ontological and methodological levels—while other levels of analysis (from the perspectives of the mind, nature, society, and culture) are also needed, in order to understand reality in its totality (as explained in the section on "sophisticated methodological holism," so as to avoid the dual dangers of reductionism and reverse-reductionism).

Ontomethodology

With these selection criteria and clarifications in mind—the syntax of existential dialectics can be divided into two main smaller sections for analysis, since they are related to each other in the context of some fundamental principles, that is, (a) ontomethodology and (b) ontologic, to be addressed hereafter, respectively.

In the context of ontomethodology, some ontological principles are proposed here in relation to the nature of methodology, which however is tied up with the general issue of ontology in the context of knowledge.

In this sense, ontomethodology (in the context of these ontological principles in the category of method) are related with ontologic in general (in the context of ontological principles in general).

Some good instances of ontological principles involving ontomethodology include the formalness-informalness principle, the predictability-unpredictability principle, the explicability-inexplicability principle, the partiality-totality principle, and the absoluteness-relativeness principle, which will be introduced in conjunction with ontological principles in the ontologic of existential dialectics in general hereafter.

Ontologic

With this clarification of the relationship between ontomethodology and ontologic in mind (which is comparable to the relationship between methodology and ontology as described in *Sec. 1.3* of *FPHK*)—the first three principles in existential dialectics were introduced in *BCPC*. And more principles were later added to the logic in my subsequent books (like *FC*, *FAE*, and others), as explained below.

More specifically, in *BCPC*, I proposed three principles (based on previous works of mine, not just something out of the blue) for the ontological logic, namely, (a) the regression-progression principle on the direction of history, (b) the symmetry-asymmetry principle on the relationships among existents, and (c) the change-constancy principle on the alteration of things—or in a more elegant term, the dynamics of space-time (in *FC*).

In *The Future of Complexity* (*FC*), I added three more principles, on top of the three aforementioned, namely, (d) the partiality-totality principle on the relationships between parts and whole, (e) the order-chaos principle on the pattern of things, and (f) the predictability-unpredictability principle on the occurrence of things—as part of the ontological logic of existential dialectics.

In *The Future of Aesthetic Experience* (or simply *FAE*), another principle is added, namely, (g) the evolution-transformation principle (on the multiple kinds of agency).

Also in *FAE*, I then reclassified the 7 principles into four categories, namely, (i) in relation to method, for the partiality-totality principle and the predictability-unpredictability principle, (ii) in relation to process, for the change-constancy principle and the order-chaos principle, (iii) in relation to agency, for the symmetry-asymmetry principle and the evolution-transformation principle, and (iv) in relation to outcome, for the regression-progression principle.

In *The Rise of Authoritarian Liberal Democracy* (*ALD*), I added two more principles, namely, (h) the softness-hardness principle on the force of agency (which is to be classified under the category about agency) and (i) the same-difference principle on the metamorphosis of change (which is to be classified under the category about outcome).

In *The Future of Information Architecture* (*FIA*), I introduce three additional principles, that is, (j) the simpleness-complicadness principle on the interconnection among things, (k) the preciseness-vagueness principle on the refinement of things, and (l) the slowness-quickness principle on the speed of change—with the first two in relation to structure and the third in relation to process.

In *The Future of Post-Human Unconsciousness* (*FPHU*), I added three more principles, that is, (m) the openness-hidden principle on the detection of things, in relation to structure, (n) the explicability-inexplicability principle on the underlying mechanisms of things, in relation to method, and (o) the expansion-contraction principle on the growth of things, in relation to process.

In *The Future of Post-Human Knowledge* (*FPHK*), I added one more principle, that is, (p) the absoluteness-relativeness principle on the multiplicity of entities, in relation to method.

In *The Future of Post-Human Mathematical Logic* (*FPHML*), I added one more principle, that is, (q) the formalness-informalness principle on the formal requirements of systems, in relation to method.

In *The Future of Post-Human Engineering* (*FPHE*), I added one more principle, that is, (r) the theory-praxis principle on the duality of knowledge, in relation to agency.

In *The Future of Post-Human Creative Thinking* (*FPHCT*), I added one more principle, that is, (s) the convention-novelty principle on the nature of creative thinking, in relation to agency.

In *The Future of Post-Human Geometry* (*FPHG*) here, I added one more principle, that is, (t) the finiteness-transfiniteness principle on the nature of numbers, in relation to structure.

In *The Future of Post-Human Urban Planning* (*FPHUP*), I added one more principle, that is, (u) the denseness-emptiness principle on the distribution of entities in space, in relation to structure.

And in *The Future of Post-Human Humor* (*FPHH*) here, I add one more principle, that is, (v) the seriousness-playfulness principle on the extent of seriousness, in relation to agency.

With this update, there are twenty-two principles so far in existential dialectics, that is, five principles for the category in relation to method (viz., the formalness-informalness principle, the partiality-totality principle, the predictability-unpredictability principle, the explicability-inexplicability principle, and the absoluteness-relativeness principle), five principles for the category in relation to structure (viz., the finiteness-transfiniteness principle, the simpleness-complicatedness principle, the preciseness-vagueness principle, the openness-hiddenness principle, and the denseness-emptiness principle), four principles for the category in relation to process (viz., the change-constancy principle, the slowness-quickness principle, the order-chaos principle, and the expansion-contraction principle), six principles for the category in relation to agency (viz., the theory-praxis principle, the convention-novelty principle, the symmetry-asymmetry principle, the evolution-transformation principle, the softness-hardness principle, and the seriousness-playfulness principle), and two principles for the category in relation to outcome (viz., the regression-progression principle and the same-difference principle).

There are thus twenty-two principles in existential dialectics—so far (as summarized in *Table 4.4*).

The Principles as Short Cuts

Yet, these principles should be treated with caution, lest misunderstanding occurs, since they do not constitute rigid dualities (or dichotomies).

The reason is that each pair in an ontological principle consists of two opposites, which are, however, merely short cuts both for multiple variations and degrees, as well as for different interactions with multiple other entities unlike them.

In this light, each pair can end up having hundred (or even thousand, if not more) different versions, which interact with hundred (or even thousand, if not more) other entities—as summarized in *Table 4.5*.

There are two clarifications here.

Firstly, .this conception of shortcuts is not mutually exclusive nor absolute, in that the opposites can come in all shapes and sizes, with different degrees. For instance, by analogy, just as there are different degrees of the two colors "white" and "black"—there are likewise different degrees of the opposites in each ontological pair, to the extent that there can be multiple entities (not only two) interacting with each other in each pair.

And secondly, this conception of shortcuts do not ignore other possible entities in interacting with the two opposites (with their different versions) in each pair. For instance, by analogy, there are not only the two colors "white" and "black" as opposites, since there are other colors too besides them like "yellow," "green," "purple," or else. The same logic can be applied to each ontological pair, in that they also interact with other entities, not with only two of them (with their different versions).

Therefore, with these two clarifications in mind—each pair in an ontological principle serves only as an abbreviation for something more complicated and, therefore, although it contains two opposites, it should not be confused as a duality (dualism).

Existential dialectics rejects any dualism (or dichotomy) as too rigid and instead allows the multiplicity of entities, to the extent that between the two opposites in each pair exist many other alternatives to choose from. They are named in that short form for aesthetic elegance only, instead of listing all possible entities between the two opposites in the title (which would be too cumbersome and sound extremely awkward).

To be dialectic is to go beyond any rigid dichotomy and transcend into something different altogether in the long haul. One may be tempted to call the dialectic logic here with a different name like existential "multilectics" (instead of "dialectics"), but this naming is incorrect (or even misleading), for the two reasons aforestated.

After all, all the possible entities between any two opposites in an ontological pair are still about the two opposites, although each ontological pair come in all shapes and sizes (with different degrees) and do not exist by themselves but also interact with other entities unlike them. The virtue of revealing an ontological pair is to show how they relate within themselves (in multiple versions) and also interact with others unlike them (also in multiple versions).

This will be clear in a later sub-section titled "the dialectic constraints imposed by the principles."

The Principles as Family Resemblances

In addition, each ontological principle is generic, with some other comparable ontological pairs to be put in the same family (like a *family resemblance*)—as summarized in *Table 4.6*.

Ludwig Wittgenstein (1953) once suggested the idea of "family resemblance" in explaining different games classified under the same family called *games*.

Why should, for instance, playing football and chess as playing "games," when it is well understood that football is not the same as chess? The answer is that, although each game is different and has different rules, many of them (though not all) share, more or less, some commonalites (e.g., scoring as necessary for winning).

And this is so, even though some games share more than some others in any given selection of critieria, and no two games are exactly identical. So, Wittgenstein's point here is that there is no essential core which is common to all games, and the best that one can look for is some characteristics which are common to many (but not all) games. (A. Biletzki 2006)

(a) Appropriate Family Resemblances

In the context of existential dialectics, I want to distinguish two types of family resemblances, and, in the absence of better words, let me call them (a) *appropriate* family resemblances and (b) *inappropriate* family resemblances.

In regard to appropriate family resemblances—good examples include some of the pairs as already explained in an earlier section on selection criteria, like the flexibility-inflexibility pair, which can be put in the family resemblance of the simpleness-complicatedness principle, although the two pairs are not exactly identical. Likewise, the directness-indirectness pair can be put in the family resemblance of the preciseness-vagueness principle, although, again, the two pairs are not exactly identical.

In this sense, which specific pair in a family should be used to designate the name of the family can be at times a bit arbitrary, but with good reason.

For illustration, in the context of method, the partiality-totality principle can take the different form like individualisticness-holisticness, just as the explicability-inexplicability principle can take the different form like underlyingness-regularness—although each two pairs are not exactly identical and have slightly different meanings and usages.

A more comprehensive listing of this family resemblance for all other principles is shown in *Table 4.6* on the syntax of existential dialectics.

In fact, this section on family resemblances overlaps with the semantics of existential dialectics, which will be introduced later in the next section (and summarized in *Table 4.11*).

(b) Inappropriate Family Resemblances

However, there are some family resemblances which are not appropriate, and let me distinugish three of them by calling them, in the absence of better words, (b1) *inadequate* family resemblances, (b2) *compound* family resemblances, and (b3) *one-sided* family resemblances (as summarized in *Table 4.10*).

(b1) In regard to "inadequate" family resemblances—there are some pairs which are not adequate, because they fail the test of selection criteria (as already pointed out in an earlier section), especially in relation to the criterion of not being *trivial* (or *parochial*)—as if it is only one component competing with other ones for inclusion in a set.

For instance, the *consistency-inconsistency* pair is inadequate, because it is only one component competing with other ones for inclusion in the larger set of the formalness-informalenss principle (like the *completeness-incompleteness* pair and the *soundness-unsoundness* pair, as discussed in my book titled *The Future of Post-Human Mathematical Logic*, or *FPHML* in short).

The same is true for the *cognition-noncognition* pair, which overlaps with both the completeness-incompleteness pair and the soundness-unsoundness pair in the larger set of the formalness-informalenss principle.

(b2) In regard to "compound" (or non-atomic) family resemblances—there are some pairs which are not appropriate for a different reason, because they fail the test of selection criteria (as already pointed out in an earlier section), especially in relation to the criterion of not *overlapping* with other principles, in that it is too closely related with other ones.

For instance, the *normality-abnormality* pair is compound (and thus inappropriate), because it overlaps with several other principles, like the symmetry-asymmetry principle, the density-emptiness principle, the convention-novelty principle, and the same-difference principle, for instance (as already discussed in *FPHP*).

The normality-abnormality pair overlaps with the symmetry-asymmetry principle, because being "abnormal" (say, for those with mental illnesses) is often treated not equally by others who are "normal." So, the relationship reveals the symmetry-asymmetry principle.

The normality-abnormality pair also overlaps with the density-emptiness principle, because being "abnormal" (say, for those with mental illnesses) is often not as dense in the overall (not local) distribution of a population as those who are "normal." So, the relationship reveals the density-emptiness principle.

The normality-abnormality pair also overlaps with the convention-novelty principle, because being "abnormal" (say, for those with mental illnesses) often exhibits non-conventional (non-conformist) ideas and behaviors when contrasted with those who are "normal." So, the relationship reveals the convention-novelty principle.

The normality-abnormality pair also overlaps with the same-difference principle, because being "abnormal" (say, for those with mental illnesses) shares some similarities in outcome when compared with those who are "normal" (in that both contribute to the control and regulation of human thoughts and behaviors in society, albeit in different ways). Yet, being "abnormal" also has something different from being "normal," in that abnormal thoughts and behaviors follow a different pattern when contrasted with those which are normal. So, the relationship reveals the same-difference principle.

(b2.1) There is a special case of compound family resemblances (as summarized in *Table 4.10*), which can be called, in the absence of better words, *one-sided and compound* family resemblances.

For instance, the extroversion-introversion pair is a good example of *one-sided and compound* family resemblances, because, firstly, it is "one-sided," in being part of the side (category) of "normality" in the normality-abnormality pair—and, secondly, the normality-abnormality pair is in turn "compounded," because it is related to the symmetry-asymmetry principle, the density-emptiness principle, the convention-novelty principle, and the same-difference principle (as already explained earlier in the section on selection criteria).

(b3) And in regard to "one-sided" family resemblances—there are some pairs which are not appropriate for another different reason, because they fail the test of selection criteria (as already pointed out in an earlier section), especially in relation to the criterion of not being *one-sided*, that is, part of one side (category) of a dialectic pair.

For instance, the extroversion-introversion pair cannot be an ontological principle, since it is part of the side (category) of "normality" in the normality-abnormaltiy pair (because both extroversion and introversion are two different "normal" personality traits in psychology).

By the same logic, the vision-blindness pair and the sound-silence pair cannot be ontological principles, because the pairs are part of the side (category) of "asymmetry" in the symmetry-asymmetry principle (since being able to see and hear for some individuals and being unable to see or hear for others are two contrastive abilities in the world of animals and humans, for instance).

These illustrations of indequate, compound, and one-sided family resemblances show how difficult it is to come up with an ontological principle in existential dialectics out of the blue, because many of them violate the selection criteria in one way or the other. It is all too easy to join two opposing words and simply make them into a pair, but it is all too difficult to pass the test of the selection criteria to be qualified as an ontological principle in existential dialectics.

Other examples of inappropriate family resemblances are shown in *Table 4.10* on "The Syntax of Existential Dialectics VII: Types of Inappropriate Family Resemblances."

The Dialectic Constraints Imposed by the Principles

The principles, as they constitute the syntax of existential dialectics, are dialectic in character, such that, when they are applied, they impose dialectic constraints on how reality is to be understood.

Consider, say, the symmetry-asymmetry principle as an illustration here, in order to summarize two main characters of the dialectic constraints in question (as summarized in *Table 4.7*).

Firstly, to be dialectic here is to go beyond the narrow dichotomies (and, for that matter, any rigid multi-dimensional classificatory scheme), be they about "self" vs. "world," "freedom" vs. "unfreedom," "barbarity" vs. "civilization," "individuality" vs. "communality," and so on.

One way to do so (to go beyond) is to consider them all in terms of co-existence (without favoring one over the rest). For instance, my theory of "post-civilization" (to be summarized later in the section on the pragmatics of existential dialectics) is to go beyond barbarity and civilization in terms of understanding barbarity and civilization as being co-existent.

And the same logic can be said in relation to my theories of "post-democracy," "post-capitalsm," and others (also to be introduced later in the section on the pragmatics of existential dialectics), in regard to freedom vs. unfreedom, equality vs. inequality, communality vs. individuality, spirituality vs. materiality, and so on.

But to consider them all (in the dichotomies—and, for that matter, in any rigid multi-dimensional classifactory scheme) as co-existent is not the same as to imply that the opposing categories in any classificatory scheme are all equal, since, in accordance to the symmetry-asymmetry principle (as an illustration here), if they are equal in terms of being considered as co-existent, they are asymmetric in terms of being unequal in dialectic interaction (e.g., X can be more dominant than Y in case A, or Y is more dominant than X in case B), although in some special cases, they can be relatively equal (e.g., X and Y are relatively equal in case C).

For this reason, there are different versions of "post-democracy" and "post-capitalism" in my theories. As an illustration, in version I of the theory of post-democracy, freedom is more dominant than equality, whereas in version II of the theory of post-democracy, equality is more so than freedom.

But this "X more than Y" has to be understood in the context of dialectic logic (not in symbolic logic, as conventionally understood), in that both "X" and "Y" are important in post-democracy (in the context of dialectic logic), but in an asymmetry way.

By contrast, in symbolic logic, it often favors one over the other—be it in regard to privileging freedom over equality (in Fascism), favoring freedom relatively more than equality (in Liberal Democracy), or favoring equality relatively more than freedom (in Socialist Democracy). In regard to the latter two cases (about Liberal Democracy and Socialist Democracy), the difference between dialectic logic and symbolic logic can be one in degree, not in kind—in this sense, albeit not in other senses.

The same reasoning can be said about the relatonships between individuality and communality, between spirituality and materiality, and between formal legalism and informal legalism in the different versions of my theory of "post-capitalism."

Secondly, to be dialectic is to go beyond the narrow dichotomies (and, for that matter, any rigid multi-dimensional classificatory scheme) in another way, this time, in a transcendent way, that is, in exploring other possibilities or even other issues not considered within the narrow confines of narrow dichotomies (and, for that matter, any rigid multi-dimensional classificatory scheme).

As an analogy, in this second meaning, to go beyond the narrow dichotomy of "black" vs. "white" is not just to choose both "black" and "white" (as in the first meaning) but also to explore other options (e.g., "green," "purple," "blue," etc.—and, alternatively, "shade," "line," "curve," etc.).

By the same reasoning, to go beyond "democracy" is to transcend democracy (as in version III of my theory of "post-democracy") and to explore other possibilties of lifeforms (e.g., "floating consciousness," "hyper-spatial consciousness," etc., to live beyond the narrow obsession with freedom and equality).

This dialectic character of the principles in existential dialectics has important implications for the pragmatics of existential dialectics (as will be clear shortly, in the section on the pragmatics of existential dialectics).

The Semantics of Existential Dialectics (or Its Ontosemantics)

The syntax of existential dialectics so understood in terms of ontological principles only gives us the structure of ontology in the world, in an abstract (general) sense. These principles by themselves do not tell us the specific meanings in a given context.

In order to grasp the specific meanings of the principles in a given context, it is necessary to study the semantics of existential dialectics (as summarized in *Table 4.11*).

And the analysis of the ontological principles as family resemblances in the previous section on syntax is only a starting point (and thus overlaps a bit with the semantics of existential dialectics here).

For this reason, I have often gone to great lengths in my previous books on different subjects to explain the specific meanings of the principles when applied in different contexts.

Consider two examples below for illustration, which are not exhaustive.

In *FPHST*, for instance, I used the first three principles (i.e., the change-constancy principle, the regression-progression principle, and the symmetry-asymmetry principle) to propose "the perspectival theory of space-time," for a better way to understand space and time—especially, though not exclusively, in relation to future post-human history (as summarized in *Table 1.3*, *Table 1.4*, *Table 1.5*, and *Table 1.6*).

In so doing, I had to introduce concepts and theories specific to the field of physics and other related fields (e.g., "absolute space" and "absolute time" in "classical mechanics" and "relative space-time" in "the theory of relativity").

And in *BNN*, I also exploited the three principles to propose the "transcendent" approach to the study of genes and memes as a new way to understand the interaction between nature and nurture.

In so doing, I had to explore concepts and theories in the world of evolutionary biology (e.g., "mutation," "variation," "adaptation," "selection," and "inheritance" in classical evolutionary theory) and neural biology (e.g., "chromosome," "gene," "DNA," "RNA," "protein," "neuron," "neural network," and "behavior").

Other examples, besdies these two here, are given in other books of mine on different topics.

Therefore, this semantic dimension of existential dialectics is important, since, in each of my books when the ontological principles are used, each principle becomes relevant in a different meaning when contextualized within a given case study in question.

In this book, for instance, the principles are relevant in relation to a different subject matter (as summarized in the concluding chapter), and they can be shown in all shapes and sizes (since they are also to be treated as "short cuts," as indicated before), without, however, being reduced to "specific ontology" (as opposed to "general ontology"), as explained in *Table 4.9.*

The Pragmatics of Existential Dialectics (or Its Ontopragmatics)

Even the semantics of existential dialectics is not enough to understand reality, since semantics by itself does not tell us the nature of intentionality and interests in relation to the matrix of power formations and conflicting interests when existential dialectics as an ontology is applied.

Surely, one cannot completely separate the pragmatics of existential dialectics from its semantics (and for that matter, its syntax), but the point here is to give each a more elaborated analysis.

For this reason, the pragmatics of existential dialectics needs to be studied too, in special relation to the inclusion of power formation and conflicting interests in the application of the principles of existential dialectics (as summarized in *Table 4.12*).

It is thus no wonder that all my previous books have shown in different ways how and why these ontological principles and their theoretical applications can reveal the future world to come in a way that humans have never known, in a totally different battleground for competing human (and later post-human) interests by myriad groups fighting for their own versions of hegemony.

Perhaps a summary of my previous works in relation to existential dialectics and their contributions to my new theories as proposed over the years is deemed revealing here, for the further understanding of the pragmatics of existential dialectics (in close relation to the semantics of existential dialectics too).

That stressed—the following summary in four sub-sections is something that I often do in my previous books too, as a way to introduce the intellectual background of my ideas.

So, for those readers who had read my previous works, the following sub-sections serve as a reminder, as they are repeated hereafter *often verbatim* (with some updated changes, with the rest kept intact).

These sub-sections are, namely, (a) the two-way street connecting theory with meta-theory, (b) the theoretical application of existential dialectics, (c) direct and indirect applications of existential dialectics, and (d) multiple applications of existential dialectics—to be summaried herefafter, respectively.

The Two-Way Street Connecting Theory and Meta-Theory

However, lest any misunderstanding occurs, it is important to remember that the pragmatics of existential dialectics is not a one-way street (that is, using the ontological principles for theoretical insights in praxis) but a two-way one, namely, (a) from meta-theory to theory, and (b) from theory to meta-theory.

(a) On one side of the street, the ontological principles can inspire some theoretical insights in praxis, that is, in relation to some specific fields.

(b) On the other (opposing) side of the street, however, the study of a subject matter in the specific fields in turn reveals some more hitherto unknown ontological principles to be discovered and identified.

For this reason, three new principles were added in *FC*, one in *FAE*, two in *ALD*, and three in *FIA*, on top of the original three in *BCPC*, for example (as there are more)—after some research on the specific subject matters.

Both theory and meta-theory enrich each other in all my works over the years.

The Theoretical Application of Existential Dialectics

This two-way street connecting theory and meta-theory has contributed to my original construction of many theories in numerous books of mine on numerous fields in all domains of knowledge, which can be classified in terms of specific "meta-theories" (about epistemology, methodology, logic, ontology, etc.) and specific "theories" (about the mind, nature, culture, society, and the rest—as will be further explained later in the sub-section on "the holistic organization of an inquiry" in the section about "sophisticated methodological holism").

The final aim of my numerous new specific theories (on both theory and meta-theory) is to converge all of them (as proposed in my numerous books) into a grand project about the future of intelligent life, both here on earth and in deep space.

In this sense, all these theories of mine serve as a foundation to unify all domains of knowledge for an overarching understanding of all forms of reality (by way of some ontological principles and the comprehensive perspectives of the mind, nature, society, and culture), while allowing for heterogeneity, conflict, subjectivity, and complexity, for example.

With this grand project in mind—the summary of my specific theories on meta-theory are given in this section (on "existential dialectics") and the next (on "sophisticated methodological holism"), whereas the specific theories on theory will be summarized in the following categories of (a) society, (b) culture, (c) the mind, (d) nature, and (e) the rest, as shown below, in that order.

(a) Society

The category of "society" here can be divided into four sub-categories, i.e., (a1) social organizations, (a2) social institutions, (a3) social structure, and (a4) social systems (or trends).

(a1) Social Organizations

In relation to social organizations, I proposed a new thesis on the ambivalent nature of virtual organizations in the future, known as "the theory of e-civic alienation" in *FCD* (and summarized in *Table 4.19*).

And in *FPHO*, another theory called "the combinational theory of organization" is worked out, although it was first proposed in Ch.6 of *ALD*.

(a2) Social Institutions

There are different social institutions, be they about (a2.1) politics, (a2.2) economics, (a2.3) military affairs, (a2.4) mass media, (a2.5) law, and (a2.6) education, for instance.

(a2.1) In relation to politics, I offered different visions on (a2.1.1) political theory, (a2.1.2) comparative politics, and (a2.1.3) international relations, to cite three main ones.

(a2.1.1) In the context of political theory, my vision of a path-breaking political system in future times concerns the different forms of "post-democracy" to supersede democracy unto the post-human age.

My vision of post-democracy was called "the theory of post-democracy" (as summarized in *Table 4.22*, *Table 4.23*, and *Table 4.24*), whereas *Table 4.25* explains the distinctions among democracy, non-democracy, and post-democracy.

And *Table 4.26* stresses the multiple causes of the emergence of post-democracy, while *Table 4.27* clarifies some possible misunderstandings in regard to post-capitalism and post-democracy.

(a2.1.2) In the context of comparative politics, I also showed, in *ALD*, how and why authoritarianism can be an advanced form of liberal democracy (just as it has its own counterpart in non-liberal democracy), or what I originally called "authoritarian liberal democracy" (as summarized in *Table 4.21*), before it is to be superceded by "post-democracy."

(a2.1.3) In the context of international relations (both at the institutional and structural levels), I provided, in *BWT*, "the theory of cyclical progression of empire-building," for a better way to understand the logic of empire-building on earth and beyond—with the rise of what I originally envisioned as "hyper-empires" (like "the Chinese Union" and "the Indian Union"), "meso-empires" (like "the European Union" and "the North American Union") and "micro-empires" (like "the Middle Eastern Union" and "the Latin American Union") in the coming "post-post-Cold War era" as a visionary illustration, as summarized in *Table 4.20*.

In addition, in *FPHWP*, I proposed "the multifaceted theory of war and peace" to understand war and peace, in a small sense—or aggression and pacificity, in a broad sense.

(a2.2) In relation to economics, I also offered another vision, this time to understand the future forms of economic systems, that is, the different forms of "post-capitalism" to supersede capitalism unto the post-human age.

My vision of post-capitalism was referred to (in *FCD*, *BDPD* and *BCPC*) as "the theory of post-capitalism" (as summarized in *Table 4.28*, *Table 4.29*, *Table 4.30*, *Table 4.31*, and *Table 4.32*). And *Table 4.33* clarifies the differences among capitalism, non-capitalism, and post-capitalism, whereas *Table 4.34* shows multiple causes of the emergence of post-capitalism.

This vision on "post-capitalism" constitutes the third theoretical application of existential dialectics, this time, in that there is no wealth without poverty (or the wealth/poverty dialectics).

(a2.3) In relation to military affairs, I also proposed "the multifaceted theory of war and peace" (in *FCD*, although it was first so labeled in *FPHK*) for a new way to understand the causes of war—and, for that matter, the rise and fall of different forms of warfare over time.

This theory was further expanded in *FPHWP*, where I suggested a new way to understand war and peace, in a small sense—or aggression and pacificity, in a broad sense.

(a2.4) In relation to mass media, I also proposed "the totalistic theory of communication" in *The Future of Post-Human Mass Media* (or in short, *FPHMM*).

(a2.5) In relation to law, I also proposed "the reconstruction theory of law" in *The Future of Post-Human Law* (or in short, *FPHLAW*).

(a2.6) In relation to education, I propose "the heterodox theory of education" in *The Future of Post-Human Education* (or in short, *FPHEDU*).

(a3) Social Structure

In relation to social structure, I first proposed two additional theories, that is, "the theory of the cyclical progression of hegemony" in *FCD* (and also in *FPHC* and *BDPD*), and "the theory of the cyclical progression of empire-building" (in *BWT*, as already indicated above)—with important implications for the symmetry-asymmetry princple in existential dialectics, just to cite one example.

Then, in *BNN*, I also proposed "the theory of contrastive advantages" to understand the persistence of social structure by gender, race, ethnicity, or any other groupings.

(a4) Social Systems (or Trends)

In relation to social systems, there are several sub-categories for illustration, which can be about (a4.1) urbanization, (a4.2) technology, (a4.3) demography, (a4.4) the environment, and (a4.5) social trends as a whole (which also include the current phenomenon of globalization).

(a4.1) In relation to urbanization, I proposed "the contingent theory of urban planning" in *FPHUP* for a new way to understand density and void.

(a4.2) In relation to technology, I worked out "the ambivalent theory of technology" in *FPHE* for a new way to understand the nature of technology.

Some other books (like *FPHMM, FAE, FCD*, and *FHC*) also deal with technology, and a good example concerns "the synthetic theory of information architecture" in *FIA* on taxonomy, network, and intelligence, in conjunction with the rise of technology.

(a4.3) In relation to demography, I proposed, first in *FHC* (and later in other books like *FCD, FPHC, FPHU*, etc.), "the theory of post-humanity," in relation to the "post-human" age at some distant point of "after-postmodernity," long after human extinction, to be eventually superseded by "post-humans" of various forms.

Some good candidates of the post-human lifeforms include, for instance, "thinking robots," "thinking machines," "cyborgs," "genetically altered superior beings," "floating consciousness," "hyper-spatial consciousness," "unfolding unconsciousness," "the hyper-sexual body," and "the hyper-martial body." This post-human vision of mine was first originally worked out in *FHC* and further elaborated in both *FCD*, *FPHC*, *FPHU*, *FPHS*, and *FPHMA*, for instance.

Lest any confusion occurs, I need to emphasize two clarifications here about the term "post-human" as a neologism in my works.

Firstly, the word "post-human" here should *not* be confused with another term which looks similar but has a totally different meaning in the literature of postmodernism, namely, "post-humanism"—which constitutes a critique of "humanism" as tradionally understood (especially, though not exclusively, in relation to the idea of progress in science and reason in the Enlightenment project). (WK 2008)

And secondly, the word "post-human" here should also *not* be confused with a similar term which is used to champion the ideology of technology for the future co-existence between humans and cyborgs in "trans-humanism." (WK 2008a)

Instead, my term "post-human" in relation to "posthuman-ism" also rejects "transhumanism" (especially, though not exclusively, in relation to the promises of technology) and refers to something else altogether, that is, the future extinction of humans and its post-human successors in deep space and beyond unto multiverses.

Later, in *BEPE*, I went further and provided a more comprehensive critique of "transhumanism," together with other issues.

With these two clarifications in mind (as summarized in *Table 1.31*)—in the end, a most fundamental question about intelligent life now has an answer, in that, if asked, "What is the future of human civilization?"—my answer in *FCD* (89) is thus: "As addressed in Ch.7 of *FHC*, a later epoch of the age of after-postmodernity (that is, at some point further away from after-postmodernity) will begin, as what I called the 'post-human' history (with the term 'post-human' originally used in my doctoral dissertation at M.I.T., which was finished in November 1995, under the title *After Postmodernity*, still available at M.I.T. library, and was later revised and published as *FHC*). The post-human history will be such that humans are nothing in the end, other than what culture, society, and nature (with some luck) have shaped them into, to be eventually superseded by post-humans (e.g., cyborgs, thinking machines, genetically altered superior beings, and others), if humans are not destroyed long before then."

(a4.4) In relation to the environment, I also addressed the issue of sustainability, especially in *FHC*, *FCD*, and *FPHUP*. For instance, in *FPHUP*, I specifically proposed a new theory, "the contingent theory of urban planning," on the need to go beyond the contemporary obsession with sustainability, in the context of not only the environment but also other issues.

(a4.5) In relation to social trends as a whole, I proposed in Ch.9 of *FCD* (367-8), that "civilizational history will continue into the following cyclical progression of expansion, before it is to be superseded (solely as a high probability, since humans might be destroyed sooner either by themselves or in a gigantic natural calamity) by posthumans at some distant point in after-postmodernity (which I already discussed in *FHC*)" unto multiverses (different constellations of universes):

Local → Regional → Global →Solar →
Galactic → Clustery…→ Multiversal

The current phenomeon of "globalization" lies in the third category on the line above and was extensively analyzed in *FCD*, in close connection with "globalism" at the cultural level—in the context of "postmodernity" (which will be introduced later in the section on history).

In *BDPD*, this thesis of mine was referred to as "the theory of the cyclical progression of system integration and fragmentation," at the systemic level—in close relation to another thesis analyzed in Ch.9 of *FCD*, which was called, in the absence of better words, "the multifaceted theory of war and peace" in *FPHK* (as already indicated above).

In *FC*, I also proposed "the dialectic theory of complexity" to account for the factor of uncertainty, which can have a major role to play in the world.

(b) Culture

The category of "culture" here can also be divided into different sub-categories like (b1) aesthetics, (b2) religion, (b3) popular (mass) culture, (b4), logic, (b5) civilization as a whole, (b6) ethics, (b7) epistemology (methodology), (b8) metaphysics (ontology), and (b9) ideology.

However, (b7) epistemology (methodology), (b8) metaphysics (ontology), and (b9) ideology are already discussed elsewhere, be they about "meta-theory" (e.g., methodology and ontology under the categories of "existential dialectics" here and "sophisticated methodological holism" in the next section) or "theory" (e.g., ideology under the categories of "social institutions," as described earlier, and also of "civilization as a whole," as will be addressed shortly).

(b1) Aesthetics

In relation to aesthetics, I proposed "the transformative theory of aesthetic experience" in *FAE* about some great transformations of aesthetic experience in future history.

(b2) Religion

In relation to religion, I proposed "the comparative theory of religion" in *FHC* (and also in *FCD*, although it was first so labeled in *FPHK*).

And in *FPHR*, I provided a solid foundation of this theory with a better name, that is, "the comparative-substitutive theory of religion" for a new way to understand spirituality.

(b3) Popular (Mass) Culture

In relation to popular (mass) culture, there are some sub-categories for illustration, which can be, for example, (b3.1) mass culture in general and (b3.2) sports (like chess).

(b3.1) In relation to mass culture in general, I proposed, in *FHC*, a new way to understand mass culture in terms of some fundamental dilemmas, which is called, in the absence of better words, "the dualistic theory of mass culture" in *FPHO*.

(b3.2) In relation to sports (like chess), I proposed, in *FPHCHESS*, "the mediative-variative theory of chess" for a new way to understand tactics and strategy (in different traditions).

(b4) Logic

In relation to logic, I also worked out, in *FPHML*, "the contrastive theory of rationality" for a new way to understand rationality and knowledge (in the context of mathematical logic).

And in *FPHH*, I then proposed "the metamorphic theory of humor," with profound implications for the understanding of rationality.

(b5) Civilization as a Whole

In relation to civilization as a whole, I also argued, in *FHC*, no freedom without unfreedom (as shown in *Table 4.13*) and no equality without inequality (as shown in *Table 4.14*), especially in relation to the seven dimensions of life existence first worked out in *FHC* (i.e., the technological, the everyday, the true, the holy, the sublime/beautiful, the good, and the just).

Even more interestingly, in *BDPD* and later in *BCIV*, I further revealed a theoretical refinement of existential dialectics, this time, in arguing that there is no civilization without barbarity, with human civilization to be eventually superseded by what I originally analyzed as "post-human post-civilization" (which should not be confused with "post-human civilization"), in the context of the freedom/unfreedom and equality/inequality dialectics.

In the final analysis, civilization cannot live without barbarity and has to learn to co-exist with it in ever new ways. Preposteorus as this may seem to many contemporaries—it is no more imperative to preserve civilization than necessary to destroy barbarity, and the ideal of civilization is essentially bankrupt, to be eventually superseded by "post-civilization" (as summarized in *Tables 4.35-4.43*).

This freedom/unfreedom and equality/inequality dialectics holds in pre-moderny, modernity, postmodernity, and what I orignally called "after-postmodernity" in *FHC* and *FCD*—especially, though not exclusively, in the context of the emergence of "post-humans."

There are two specific theories about "after-postmodernity" here. The first theory about "after-postmodernity" is about the trinity of pre-modernity, modernity, postmodernity, and after-postmodernity at the cultural level—and was called, in *BDPD*, "the theory of the trinity of modernity to its after-postmodern counterpart," although it was first analyzed in *FHC* and later in *FCD* and also *BCIV* (as summarized in *Table 1.34*, *Table 1.35*, *Table 1.36*, and *Table 1.37* and will be elaborated further in the later sub-section on "the rest" about future history).

The second theory about "after-postmodernity" is about pre-modernity, modernity, postmodernity, and after-postmodernity at the historical level. For instance, a conclusion in *FHC*, which is shocking indeed for many contemporaries, is none other than that "[t]he post-human history will therefore mark the end of human history as we know it and, for that matter, the end of human dominance and, practically speaking, the end of humans as well. The entire history of human civilization, from its beginning to the end, can be summarized by four words, linked by three arrows (as already discussed in *FHC*)":

Pre-Modernity → Modernity →
Post-Modernity → After-Postmodernity

In *BDPD*, this thesis of mine was specifically called "the theory of the evolution from pre-modernity to after-postmodernity," at the historical level—as shown in *Table 1.34*, *Table 1.35*, *Table 1.36*, and *Table 1.37* (and will be elaborated further in the later sub-section on "the rest" about future history).

Therefore, "[t]he end of humanity in the coming human extinction is the beginning of post-humanity. To say an untimely farewell to humanity is to foretell the future welcome of post-humanity." (P. Baofu 2002: 89) This thesis of mine was known in *BDPD* as "the theory of post-humanity," at the systemic level—as already indicated in (a4.3) above.

(b6) Ethics

In relation to ethics, I proposed, in *BEPE*, "the theory of post-ethics" for a new way to understand morality and immorality.

(c) The Mind

The important vision about "post-humans" (as summarized earlier) brings us to the category of "the mind" here, which, as is often understood in everyday's usage, refers to the "brain" and the "body" (together with their interactions) and has different levels of consciousness which can be both "spiritual" and non-spiritual.

Therefore, the "mind" already presupposes the "brain," "body," and "spirit" (or "soul"), so that one does not fall into the intellectual trap of the "mind-body dualism" as debunked in modern (and especially, postmodern) philosophy (which I already went to great lengths to explain in *The Future of Human Civilization*).

That said—the category of "the mind" can be divided into two sub-categories, namely, in relation to (c1) biology and (c2) psychology.

(c1) Biology

In relation to biology, there are two concerns here, namely, (c1.1) genes and intelligence and (c1.2) the human body.

(c1.1) In relation to genes and intelligence, I worked out "the theory of contrastive advantages" (which was originally proposed in *FCD* and further worked out in *BNN*, as summarized in *Table 1.24*), to show the nature-nurture interactions of multiple levels in action (e.g., the biological, the psychological, the structural, the systemic, the cultural, and others) for humans and post-humans on earth and beyond, to the extent that different groups, be they on the basis of race, gender, ethnicity, class, age, or else, are not equal, on average, and have not been, nor will they be, even if there can be exceptions.

(c1.2) In relation to the human body, for instance, in *FPHMA*, I suggested "the expansive-contractive theory of martial arts," in which two great future transformations of the martial body were envisioned, that is, both "with the hyper-martial body" and "without the body" (as summarized in *Table 1.32*)—although this issue can also be classified under the category of "culture."

And in *FPHS*, I proposed "the theory of virtual sexuality," in which three great future transformations of the sexual body were envisioned, that is, "virtual pleasure," "the hyper-sexual body," and "the post-sexual floating mind" (as summarized in *Table 1.33*).

(c2) Psychology

In relation to psypchology (which also overlaps with biology too, however), I also proposed some greater transformations of the mind to come in the future, especially though not exclusively in the post-human age.

These greater transformations of the mind in psychology can be further sub-divided into different categories, namely, (c2.1) consciousness, (c2.2) unconsciousness (together with subconsciousness), (c2.3) creativity, (c2.4) learning, (c2.5) aggression (and pacificity), (c2.6) personality, and (c2.7) humor.

(c2.1) In relation to consciousness, I analyzed, in *FPHC*, the nature of consciousness in relation to the conceptual dimensions of consciousness (as shown in *Table 1.12*), the theoretical levels of consciousness (as shown in *Table 1.13*), the thematic issues of consciousness (as shown in *Table 1.14*), the different dimensions of human existence (viz., having-ness, belonging-ness, and being-ness) in relation to consciousness (as shown in *Table 1.15*, *Table 1.16*, *Table 1.17*, and *Table 1.18*).

In *BCPC*, the nature of consciousness was further explored in relation to cognition, emotion, and behavior, which can be in consciousness and other mental states (e.g., subconsciousness and unconsciousness), in the context of existential dialectics, as shown in *Table 1.19*, *Table 1.20*, *Table 1.21*, *Table 1.22,* and *Table 1.23*.

In fact, in *FCD*, I already proposed three different theories in relation to the limits of cognition, emotion, and behavior, namely, "the theory of cognitive partiality," "the theory of emotional non-neutrality," and "the theory of behavioral alteration."

Then, I argued that the existence of human consciousness will be superseded one day with "floating consciousness" (as first proposed in *FCD*) and "hyper-spatial consciousness" (as first proposed in *FPHC*) as a climax of evolution in consciousness, after the future extinction of human consciousness (as shown below):

Primordial consciousness → Human consciousness →
Post-human consciousness (with floating consciousness and hyper-spatial consciousness as a climax in the evolution of consciousness)

In *BDPD*, these latest theses of mine were called "the theory of floating consciousness" and "the theory of hyper-spatial consciousness," both at the cosmological and psychological levels (as summarized in *Table 1.26* and *Table 1.25*).

(c2.2) In relation to unconsciousness, for instance, in *FPHU*, the unconsious mind will undergo tremendous change, to the rise of what I originally called "unfolding unconsciouess" in relation to anomalous experience, as a climax of evolution in unconsciousness, in conjunction with the conscious mind (both human now and post-human later), especially at the psychological level, as shown below:

Primordial unconsciousness → Human unconsciousness →
Post-human unconsciousness (with unfolding unconsciousness
as a climax in the evolution of unconsciousness)

In *FPHU*, this thesis of mine is called "the theory of unfolding unconsciousness" in a small sense, or "the unfolding theory of anomalous experience," in a larger one, as shown in *Table 1.27* and *Table 1.28.*

(c2.3) In relation to creativity, for instance, in *FPHCT*, I also proposed "the comprehensive theory of creative thinking" to explore the role of creative thinking, in relation to technology and other factors, together with my original proposal of important creative techniques and traits (as shown in *Table 1.29*) and its double nature of desirability and undesirability (as shown in *Table 1.30*).

(c2.4) In relation to learning, for instance, in *FPHL*, I worked out "the multilogical theory of learning" for a new way to understand the nature of structure and context in language, in relation to learning and its future development in the context of the mind.

(c2.5) In relation to aggression (and pacificity), I proposed "the multifaceted theory of war and peace" in *FPHWP* (as mentioned above).

(c2.6) In relation to personality, I worked out "the contrarian theory of personality" in *FPHP* for a new way to understand normality and abnormality.

(c2.7) And in relation to humor, I worked out "the metamorphic theory of humor" in *FPHH* for a new way to understand joking and laughing.

(d) Nature

The category of "nature" here refers to the state of nature, which can be divided into some sub-categories like (d1) physics, (d2) cosmology, (d3) biology, (d4) geography, (d5) geology, and (d6) acoustics.

But since biology in (d3) overlaps with biology in (c1)—only phyiscs, cosmology, geography, and geology are addressed hereafter.

(d1) Physics

In relation to physics, I also proposed "the perspectival theory of space-time" in *FPHST* (based on *FPHC*) for a new way to understand space and time, as shown in *Table 1.3*, *Table 1.4*, *Table 1.5*, and *Table 1.7.*

In *FPHG* (based on *FPHC* and *FPHST*), I proposed "the selective theory of geometry" for a new way to understand geometry, with implications for the understanding of space-time in the context of infinity, symmetry, and dimensionality for future lifeforms that our world has never known (as shown in *Table 1.6*, *Table 1.7*, *Table 1.8; Table 1.9*, *Table 1.10,* and *Table 1.11*).

And in *FC*, I proposed "the dialectic theory of complexity" for a new way to understand order and chaos in the state of nature and beyond.

(d2) Cosmology

In relation to cosmology, my theories on physics (as indicated above) already have important implications for understanding the nature and the future of the universe (which was also addressed in *FHC*, among other works aforcited).

I also proposed "the theory of hyper-spatial consciousness" in *FPHC* (and also in *FPHG*) for a new way to understand the future evolution of the mind to enter into different dimensions of space-time that our current world has never known.

In addition, in *BCOS*, I proposed "the theory of post-cosmology" for a new way to understand the beginnings and ends of the cosmos and beyond.

(d4) Geography

In relation to geography, I proposed "the theory of the geopower of nature" in *ALD* for a different way to understand the relationships between geography and other entities (like the mind, culture, and society).

(d5) Geology

In relation to geology, I proposed "the resettlement theory of geology" in *FPHGEOL* for a different way to understand the relationships between statics and dynamics.

(d6) Acoustics

And in relation to acoustics, I propose here "the multilateral theory of acoustics" in *FPHA* for a different way to understand the relationships between sound and silence.

(e) The Rest

The category of "the rest" here refers to what cannot be strictly classified in any of the previous categories (i.e., society, culture, the mind, and nature). A good illustration of "the rest" refers to that which is, namely, (e1) historical.

(e1) Historical

Precisely here, I proposed "the theory of the evolution from pre-modernity to after-postmodernity" (as first worked out in *FHC* and later in *FCD* and *FPHC*), which overlaps—as a reminder from (b5)—with the issue concerning "civilization as a whole" in (b5).

For instance, in both *FCD* and *FPHC*, I worked out the structure of "post-human civilization" in terms of the trinity of after-postmodernity (i.e., "free-spirited after-postmodernity," "post-capitalist after-postmodernity," and "hegemonic after-postmodernity").

Both conceptually and theoretically, the trinity of after-postmodernity is a sequential extension to the trinity of modernity (i.e., "free-spirited modernity," "capitalist modernity,"and "hegemonic modernity") and the trinity of postmodernity (i.e., "free-spirited postmodernity," "capitalist postmodernity," and "hegemonic postmodernity") as first proposed in *FHC*.

And the trinity of pre-modernity (i.e., "pre-free-spirited pre-modernity," "pre-capitalist pre-modernity" and "hegemonic pre-modernity") was later conceived in *BCIV* to complete the historical set from pre-modernity to after-postmodernity.

In *BDPD*, this thesis about the trinity of pre-modernity, modernity, postmodernity, and after-postmodernity was collectively known as "the theory of the trinity of modernity to its after-postmodern counterpart," at the cultural level—as already so mentioned in (b5) earlier, and summarized in *Table 1.34*, *Table 1.35*, *Table 1.36*, and *Table 1.37*.

At the structural level, all these trinities are subject to the existential constraints (e.g., the freedom/unfreedom and equality/inequality dialectics in the context of "the cyclical progression of hegemony"), be the historical epoch in pre-modernity, modernity, postmodernity, or after-postmodernity in future times. Each of the historical epochs has its ever new ways of coming to terms with the ever new (different) mixtures of freedom/unfreedom and equality/inequality.

This is importantly so, not because, as is falsely assumed in conventional wisdom, one certain way is superior (or better) than another in terms of achieving more freedom and less unfreedom, or more equality with less inequality.

On the contrary, indeed, in each of the historical epochs, each increase of unfreedom greets each freedom achieved, just as each increase of inequality welcomes each equality achieved, albeit in ever new (different) ways. In *BDPD*, this thesis of mine was labeled as "the theory of the cyclical progression of hegemony," at the structural level, though it was first analyzed in *FCD*.

In *BDPD*, more theoretical applications of existential dialectics were further examined, in relation to five main features, in the context of the duality of oppression, namely, (a) that each freedom/equality achieved is also each unfreedom/inequality created, (b) that the subsequent oppressiveness is dualistic, both by the Same against the Others and itself and by the Others against the Same and themselves, (c) that both oppression and self-oppression can be achieved by way of downgrading differences (between the Same and the Others) and of accentuating them, (d) that the relationships are relatively asymmetric among them but relatively symmetric within them, even when the Same can be relatively asymmetric towards itself in self-oppression, and the Others can be likewise towards themselves, and (e) that symmetry and asymmetry change over time, with ever new players, new causes, and new forms, be the locality here on Earth or in deep space unto multiverses—as summarized in *Table.4.13*, *Table 4.14*, *Table 4.15*, and *Table 4.16*.

The same logic also holds both in relation to wealth and poverty (as addressed in *BCPC* and summarized in *Table 4.17* on the wealth/poverty dialectics) and in relation to civilization and barbarity (as addressed in *BCIV* and summarized in *Table 4.18* on the civilization/barbarity dialectics).

In *BDPD*, this thesis on existential dialectics was labeled as "the theory of existential dialectics."

Direct and Indirect Applications of Existential Dialectics

A different way to appreciate the usefulness of existential dialectics is by way of the analysis of its (a) direct and (b) indirect applications.

(a) In direct applications, on the one hand, the logic of existential dialectics can shed some theoretical insights on diverse phenomena in the world, and good instances are the usage of the principles of existential dialectics for the theoretical insights on the freedom/unfreedom dialectics, the equality/inequality dialectics, and the wealth/poverty dialectics—as introduced earlier in (b5) and (e) in the previous sub-section.

For instance, my books like *FPHST*, *BNN*, *FAE*, *FC*, *FIA*, *FPHU*, *FPHK*, *FPHML*, *FPHE*, *FPHMM*, *FPHCT*, *FPHG*, *FPHUP*, *FPHL*, *FPHO*, *FPHMA*, *FPHS*, *FPHLAW*, *FPHFS*, *FPHWP*, *FPHP*, *BEPE*, *BCOS*, *FPHGEOL*, *FPHCHESS*, *FPHR*, *FPHEDU*, *FPHH*, and *FPHA* use the principles to reveal some theoretical insights on the perspectives of space and time (as in *FPHST*), of nature and nurture (as in *BNN*), of beauty and ugliness (as in *FAE*), of order and chaos (as in *FC*), of taxonomy and network (as in *FIA*), of the unconscious mind (as in *FPHU*), of the nature of knowledge (as in *FPHK*), of mathematical logic (as in *FPHML*), of engineering (as in *FPHE*), of mass media (as in *FPHMM*), of creative thinking (as in *FPHCT*), of geometry (as in *FPHG*), of urban planning (as in *FPHUP*), of language (as in *FPHL*), of organization (as in *FPHO*), of martial arts (as in *FPHMA*), of sexuality (as in *FPHS*), of law (as in *FPHLAW*), of formal science (as in *FPHFS*), of aggression and pacificity (as in *FPHWP*), of normality and abnormality (as in *FPHP*). of morality and immorality (as in *BEPE*), of the beginnings and ends of the universe (as in *BOCS*), of statics and dynamics (as in *FPHGEOL*), of tactics and strategy (as in *FPHCHESS*), of secularness and sacredness (as in *FPHR*), of teaching and learning (as in *FPHEDU*), of joking and laughing (as in *FPHH*), and of sound and silence (as in *FPHA*).

(b) In indirect applications, on the other hand, the theoretical insights can further be used to reveal other phenomena directly from them (viz., the theoretical insights) and therefore indirectly from the principles themselves.

A good illustration is of course the use of the theoretical insights on the freedom/unfreedom and equality/inequality dialectics for the understanding of the civilization/barbarity dialectics.

This distinction between direct and indirect applications may sound a bit academic, since even in indirect applications, the phenomena under study can still be directly related back to the principles themselves.

In the previous example, as an illustration, the civilization/barbarity dialectics can be directly related to the principles of existential dialectics without the intermediate role of the freedom/unfreedom and equality/inequality dialectics.

Multiple Levels of Application

There is another issue to be clarified, however. In other words, the theoretical insights can be applied to multiple levels of analysis—even though, in a given example, it may refer to one level or a few only.

For instance, in the example concerning the freedom/unfreedom dialectics, it can be used at the structural level (e.g., in relation to the theory of the cyclical progression of hegemony), but it can also be exploited for other levels (e.g., the theory of post-capitalism at the institutional level).

All these levels of application should not be misleadingly construed, as I stress this before, as a one-way street (that is, to use the ontological principles for theoretical application at multiple levels of specific analysis) but a two-way one, in which theoretical insights in praxis, when studied in more specific contexts, can in turn refine the nature of existential dialectics (for example, with the addition of new principles).

Sophisticated Methodological Holism

The summary of my original meta-theory on methodology is provided below, again, to be repeated *often verbatim* from my previous works—as this is something I regularly do in each new book to introduce my past works, which are in conversation with this current one.

This means that, for those readers who had read some of my previous books, the summary below is really not much new (*almost verbatim*, with some updated revisions).

That clarified—this original contribution of mine to the study of methodology takes the form of an approach known as "the theory of methodological holism" or "methodological holism" in short, as already worked out in *FPHC* (2004).

I have made good use of this methodological approach of mine for all of my previous works.

The Confusion of Holism with Systems Approach

Yet, it is imperative to stress at the outset, as I often emphasized this in all my previous books, that my approach of "methodological holism" does not oppose or exclude "methodological individualism" (as some readers may be tempted to assume, as is conventionally understood) but actually includes it.

The confusion here has to do with equating holism with systems approach in general (be it about systems theory, chaos theory, complexity theory, or else), which I debunked, while learning something from it, as already elaborated in *The Future of Complexity* (2007)—and *The Future of Post-Human Formal Science* (2010) is another addition to this critique of the intellectual fad of systems approach.

In other words, my meta-theory of holism, or sophisticated methodological holism in short, makes a distinction between doing holistic work and doing systems work, for two reasons.

The first reason is that holism, in my theory, does not imply systems approach, since it rejects systems approach because of the myriad problems that the literature on systems approach has been unable to resolve, while learning something from it.

And the second reason is that my theory is related to the other theory of mine, that is, existential dialectics, which does not exclude methodological individualism (but includes it, because it simply goes beyond both reductionism and reverse-reductionism in a dialectic way).

For these reasons (and others too, as summarized in *Table 4.1*), my version of methodological holism is *sophisticated*—not *vulgar* as sometimes used by inapt scholars employing the same term.

The Ontological Constraints

Sophisticated methodological holism is subject to some ontological constraints, and five good examples include "the partiality-totality principle," "the predictability-unpredictability principle," "the explicability-inexplicability principle," "the absoluteness-relativeness principle," and "the formalness-informalness principle" in existential dialectics.

Of course, there can be more than five ontological principles in relation to method, but the five examples here are sufficient to illustrate the point in question.

With this caveat in mind—the first two were already analyzed in previous books of mine like *FC* and *FAE*, whereas the third one was addressed in *FPHU*, and the last two were accounted for in *FPHK* and *FPHML*, respectively.

For now, it suffices to show that sophisticted methodological holism, because of these ontological constraints on methodology, targets two major sins of methodology, namely, what I call, in the absence of better words, (a) *reductionism* and (b) *reverse-reductionism*—both of which come in all shapes and sizes, to be summarized below (and also shown in *Table 4.2*).

Against the Varieties of Reductionism

There are four versions of reductionism to be summarized here, which sophisticated methodological holism rejects, namely, (a1) conceptual, (a2) theoretical, (a3) methodological, and (a4) ontological forms of reductionism.

(a1) In conceptual reductionism, a good case in point concerns myriad dualities like mind vs. body, self vs. world, democracy vs non-democracy, and the like (as already addressed in *FHC*, *FPHC*, and *BDPD*, for instance).

(a2) In theoretical reductionism, an illuminating case study is best exemplified by what I originally called "the foundation fallacy" in *FPHST*, in any attempt to naively understand space-time from the physical perspective as the foundation and, consequently, to dangerously dismiss (or belittle) other perspectives.

In *FAE*, I elaborated further these versions of reductionism in the literature on aesthetics (e.g., form vs. content, representation vs. expression, critics vs. artists, and externalism vs. internalism).

In *FIA*, I revealed other forms of reductionism in the literature on information architecture (e.g., the constructivist argument and the representational argument).

In *FPHU*, I showed the persistent legacy of reductionism, this time, in the literature on anomalous experience (e.g., the obsession with physics, chemistry, and biology for explaining anomalous experience).

In *FPHE*, I examined another case of reductionism in action, in the context of engineering (e.g., technical constraints vs. normative constraints).

In *FPHMM*, I elaborated one more version of reductionism, in relation to the three domains of communication (e.g., the competing views on sending, connecting, and receiving).

In *FPHCT*, I explored another version of reductionism, in relation to invention and innovation (e.g., the bio-psychological argument vs. the socio-cultural arguments).

In *FPHG*, I identified another version of reductionism, in relation to infinity, symmetry, and dimensionality (e.g., the Euclidean argument vs. the non-Euclidean arguments).

In *FPHUP*, I analyzed another version of reductionism, in relation to density and void (e.g., the engineering argument and the ecology argument).

In *PFHL*, I examined another version of reductionism, in relation to structure and context (e.g., the structuralist argument and the contextualist argument).

In *PFHO*, I explored another version of reductionism, in relation to communcation, decision-making, and leadership (e.g., the rational-system argument vs. the natural-system argument vs. the open-system argument).

In *PFHMA*, I revealed another version of reductionism, in relation to the martial body and spirit (e.g., the spiritual argument vs. the materialist argument vs. the defensive argument).

In *PFHS*, I analyzed another version of reductionism, in relation to the sexual body and spirit (e.g., the naturalist argument vs. the constructivist argument).

In *PFHLAW*, I examined another version of reductionism, in relation to law (e.g., the necessity argument vs. the contengency argument).

In *FPHWP*, I show another version of reductionism, in relation to war and peace (e.g., the aggressivist argument vs. the pacifist argument).

In *BEPE*, I analyzed another version of reductionism, in relation to morality and immorality (e.g., the objectivist argument vs. the non-objectivist argument vs. the skeptical argument).

In *BCOS*, I examined another version of reductionism, in relation to the contested beginnings and speculative ends of the universe (e.g., the scientific argument vs. the religious argument vs. the esoteric argument vs. the metaphysical argument).

In *FPHP*, I showed another version of reductionism, in relation to normality and abnormality (e.g., the natural argument vs. the social argument vs. the cultural argument vs. the mental argument).

In *FPHGEOL*, I scrutinized another version of reductionism, in relation to statics and dynamics (e.g., the catastrophe argument vs. the uniformity argument vs. the revision argument).

In *FPHCHESS*, I showed another version of reductionism, in relation to tactics and strategy (e.g., the natural argument vs. the social argument vs. the cultural argument vs. the mental argument).

In *FPHR*, I revealed another version of reductionism, in relation to secularness and sacredness (e.g., the critical argument vs. the skeptical argument vs. the theist argument).

In *FPHEDU*, I examined another version of reductionism, in relation to teaching and learning (e.g., the teacher-centered argument vs. the student-centered argument vs. the balanced argument).

In *FPHH*, I explored another version of reductionism, in relation to joking and laughing (e.g., the natural argument vs. the social argument vs. the cultural argument vs. the mental argument).

And in *FPHA* here, I reveal another version of reductionism, in relation to sound and silence (e.g., the natural argument vs. the social argument vs. the cultural argument vs. the mental argument).

(a3) In methodological reductionism, a good illustration can be the debate between different versions of qualitative and quantitative methods (as already analyzed in *FC* and also *FHC*).

In *FPHML*, I examined similar reductionism, this time, in the literature on mathematical logic (e.g., the obsession with consistency, soundness, and completeness).

And in *FPHFS*, I explored the problems of reductionism in the context of formal science (e.g., the analytical argument, as opposed to the synthetic argument).

(a4) In ontological reductionism, an excellent instance involves another debate, this time between emergentism and reductionism in complexity theory (as addressed in *FC*) and also in psychology (as elaborated in *FPHC*, in the context of Being and Belonging).

But to target against the varieties of reductionism constitutes only a side of the same coin, and the other side concerns the varieties of reverse-reductionism.

Against the Varieties of Reverse-Reductionism

The other side of the same coin is the reverse version of reductionism, which is what I want to call, in the absence of better words, *reverse-reductionism*.

My sophisticated version of methodological holism targets against the varieties of reverse-reductionism (just as it also rejects the ones of reductionism).

Perhaps nothing expresses better the popularity of reverse-reductionim than the "anything-goes" mentality in postmodernism of our time, as shown in the following four versions.

(b1) In conceptual reverse-reductionism, any concept of "art" (e.g., fine art, applied art, outsider art, junk art) is welcome in postmodernism (as already analyzed in Ch.4 of *FHC*—and also in *FAE*).

(b2) In theoretical reverse-reductionism, a variety of art and literary theories co-exist. Take the case of literary studies, as there are now Literary Structuralism, Marxist Literary Criticism, New Criticism, Phenomenology, Hermeneutics, Language-Game Literary Criticism, Feminist Literary Criticism, Reception Theory, Reader Response Criticism, Poststructuralism, Semiotics, Pyschoanalytic Literary Criticism, just to cite some well-known ones, with no one being said to be better than any others (as detailedly analyzed in Ch.4 of *FHC*). (S. Raman 1997)

In *BNN*, I even introduced "the compromise fallacy" as another good illustration of theoretical reverse-reductionism, in misleadingly treating both genetic and environmental approaches as equally valid.

(b3) In methodological reverse-reductionism, multiple methodologies are deemed as acceptable in postmodernism (e.g., doing art without praxis, doing art with praxis, and doing art by sublation), as analyzed in Ch.4 of *FHC*.

And in *FPHFS*, I exposed the problems of reverse-reductionism in the context of formal science in relation to systems theory (e.g., the synthetic argument).

(b4) In ontological reverse-reductionism, no privileged ontology is allowed, and the door is open practically for anything in postmodernism (e.g., the equal status of the ontology of Being vs. that of Becoming, as already addressed in Ch.4 of *FHC*—and also in *FPHC*).

In *FAE*, I also introduced another version of reverse-reductionism, that is, "the pluralist fallacy," in the context of understanding aesthetic experience, for instance—although this fallacy has been committed not exclusively in relation to the ontological level (but also at the conceptual, theoretical, and methodological ones).

These dual dangers against reverse-reductionism (in this sub-section) and reductionism (in the previous sub-section) are something that sophisticated methodological holism rejects. And in *FPHFS*, I addressed these two problems further.

This point about the dual dangers is important enough, since many scholars often favor one against the other, but it is vital to target both.

The Holistic Organization of an Inquiry

With these dual dangers against reductionism and reverse-reductionism in mind—sophisticated methodological holism suggests that an inquiry of any given phenomenon is more complete, if treated in the context of a comprehensive analysis at all relevant levels, which challengingly encompass all the fields of human knowledge, ranging from the natural sciences and formal sciences through the social sciences to the humanities.

This section is something that I had already stressed in all my previous books and repeat (*often verbatim*) hereafter. But, for those readers who had read my previous books, this serves as a reminder.

With this reminder in mind—there are multiple ways to engage in a holistic inquiry with all relevent levels of analysis. Over the years, I have proposed different ways to fulfill this holistic methodological requirement, as repeated in my previous books.

Hereafter is a summary of four major ways, namely, (a) by discipline, (b) by domain, (c) by subject, and (d) by meta-analysis—all of which fulfill the holistic requirement of sophisticated methodological holism on a given issue.

(a) In a holistic organization by discipline, a good illustration is the multiple levels of analysis, namely, (a1) the micro-physical, (a2) the chemical, (a3) the biological, (a4) the psychological, (a5) the organizational, (a6) the institutional, (a7) the structural, (a8) the systemic, (a9) the cultural, (a10) the macro-physical (cosmological), and (a11) other relevant levels, which are either a combination of all these levels or the practical applications with a combination of them.

This holistic organization by discipline is used in *FCD* and *BCPC*, for instance.

(b) In a holistic organization by domain, some good candidates include the classification of different perspectives of analysis in relation to (b1) nature, (b2) the mind, (b3) society, and (b4) culture, as already worked out in *BCIV*, *FPHST*, *FC*, *FAE*, *FPHU*, *FPHK*, *FPHE*, and *FPHFS*.

Culture in (b4), in this re-classification, is the same as culture in (a9) and can be further divided into different sub-categories, with good examples like (a9i) epistemology (methodology), (a9ii) aesthetics, (a9iii) ethics, (a9iv) metaphysics (ontology), and (a9v) religion. For clarification, epistemology (methodology) in (a9i) and metaphysics (ontology) in (a9iv) overlap with the category of "meta-theory."

Society in (b3) includes the organizational in (a5), the institutional in (a6), the structural in (a7), and the systemic in (a8).

The mind (b2) has more to do with the chemical in (a2), the biological in (a3), and the psychological in (a4), although it can overlap with (a1), (b3) and (b4), for instance.

It should be stressed that the "mind" here also includes the categories like the "brain," "body," and the "spirit" (or the "soul")—whenever needed in a given context, so that one does not fall into the intellectual trap of the "mind-body dualism" in the history of modern (and especially, postmodern) philosophy, which I already went to great lengths to explain in *The Future of Human Civilization*. After all, in everyday's usage, the "mind" already implies the "brain" and the "body" (together with their interactions) and has different levels of consciousness which can be both "spiritual" and non-spiritual.

And nature in (b1) refers to the micro-physical in (a1) and the macro-physical (cosmological) in (a10), although it can also overlap with (a2), (a3) and (a4).

(c) In a holistic organization by subject, some excellent examples concern the analysis of consciousness in *FPHC* (i.e., on Having, Belonging, and Being); of civilization in *FHC* (i.e., the True, the Holy, the Everyday, the Technological, the Beautiful/Sublime, the Good, and the Just); of international politics in *BWT* (i.e., Hyper-Empires, Meso-Empires, and Micro-Empires); of nature and nurture in *BNN* (i.e., genes and memes); of information architecture in *FIA* (i.e., taxonomy and network); of mathematical logic in *FPHML* (i.e., consistency, soundness, and completeness); of communication in *FPHMM* (i.e., sending, connecting, and receiving); of creative thinking in *FPHCT* (i.e., invention and innovation); of geometry in *FPHG* (i.e., infinity, symmetry, and dimensionality); of urban planning in *FPHUP* (i.e., density and void), of language in *FPHL* (i.e., structure and context), and of organization in *FPHO* (i.e., communication, decision-making, and leadership), in *FPHMA* (i.e., the martial body and the martial spirit), in *FPHS* (i.e., the sexual body and the sexual spirit), in *FPHLAW* (i.e., necessity and contingency), in *FPHWP* (i.e., aggression and pacificity), in *FPHP* (i.e., normality and abnormality), in *BEPE* (i.e., morality and immorality), in *BCOS* (i.e., the beginnings and ends of the universe), in *FPHGEOL* (i.e., statics and dynamics), in *FPHCHESS* (i.e., tactics and strategy), in *FPHR* (i.e., secularness and sacredness), in *FPHEDU* (i.e., teaching and learning), in *FPHH* (i.e., laughing and joking), and in *FPHA* (i.e., sound and silence)—by incorporating all the relevent levels of analysis as cited above in each.

(d) And in a holistic organization by meta-analysis, two good cases in point involve the works on post-capitalism in *BCPC* and on authoritarian liberal democracy in *ALD*, with the classification of analysis in theory and meta-theory.

The essential point to remember here, as I thus stress in all my previous books, is that the multiple levels of analysis can be reorganized in many different ways, insofar as none of the levels (if relevant to an inquiry in question) is ignored or dismissed, to avoid the danger of reductionism (and for that matter, the one of reverse-reductionism, in the opposite direction).

Three Clarifications

Three clarifications are needed here, however, to avoid any hasty misunderstanding.

Firstly, the entities in each classification are not mutually exclusive. For instance, in the holisitc organization by domain, nature in (b1) can also be linked to the chemical in (a2), the biological in (a3), and the systemic in (a8). And the mind (b3) can alternatively be related to the micro-physical in (a1) and the chemical in (a2), for example.

Secondly, there is also the important factor of luck (or uncertainty in conventional usage), but it is already implied or allowed in each of the organizations of an inquiry—especially in relation to the predictability-unpredictability principle and the order-chaos principle in existential dialectics. This topic about uncertainty was already addressed in *FHC*—and more extensively, in *FC*, in the context of order and chaos.

And lastly, the comparison in a classification is not absolute, but relative, as there are often some shades of gray, not exactly black or white (figuratively speaking).

Some further clarifications and qualifications of sophisticated methodological holism are shown in *Table 4.2*.

Some Distinctive Usefulness

In the end, sophisticated methodological holism—when applied, especially though not exclusively, as illustrated in my numerous works—

can enrich the understanding of reality in some distinctive ways. Here are four examples (as revealed in each of my books).

Firstly, it provides a comprehensive analysis of a subject matter, from which much can be learned about reality, both in relation to the perspectives of the mind, nature, society, and culture.

Secondly, it offers a new classification of the subject matter.

Thirdly, it suggests some visions of the future in relation to the subject matter in question.

And fourthly, it proposes some insights on meta-theory (e.g., methodology and ontology) in general—with the clear understanding, however, of the dilemma of specific vs. general ontology (as shown in *Table 4.9* on the syntax of existential dialectics in the context of the dilemma of ontology).

For this very reason, all of these ways are important, without reducing one into the analysis of another. (*FPHCT*)

Chapter Outline

With this summary of my two original meta-theories (on methodology and ontology, *almost verbatim* from my previous works) in mind—this book is organized in four main parts, namely, (a) Part I on the introduction to this book, (b) Part II on sound, (c) Part III on silence, and (d) Part IV on the conclusion of the book.

The book is thus organized in four chapters.

The introductory chapter, that is, Chapter One titled "Introduction—The Importance of Acoustics," begins by questioning a conventional wisdom about sound and silence and then introducing acoustics in relation to sound and silence, as well as the diverse sub-fields of acoustics—together with the theoretical debate in the literature, my multilateral theory of acoustics, the relationship between theory and meta-theory, existential dialectics, sophisticated methodological holism, and some further clarifications.

Chapter Two titled "Sound and Its Doubleness" explores the nature of sound from the four perspectives of the mind, nature, society, and culture, with a deconstructive analysis of each.

Chapter Three titled "Silence and Its Duality" examines the nature of silence from the four perspectives of the mind, nature, society, and culture, with a constructive critique of each.

The last chapter titled "Conclusion—The Future of Acoustics" summarizes the analysis in the book in the dialectic context of sound and silence, with seventeen major theses, namely, (a) the first thesis: the absoluteness-relativeness principle, (b) the second thesis: the predictability-unpredictability principle, (c) the third thesis: the explicability-inexplicability principle, (d) the fourth thesis: the preciseness-vagueness principle, (e) the fifth thesis: the simpleness-complicatedness principle, (f) the sixth thesis: the openness-hiddenness principle, (g) the seventh thesis: the denseness-emptiness principle, (h) the eighth thesis: the slowness-quickness principle, (i) the nineth thesis: the expansion-contraction principle, (j) the tenth thesis: the theory-praxis principle, (k) the eleventh thesis: the convention-novelty principle, (l) the twelveth thesis: the evolution-transformation principle, (m) the thirteenth thesis: the symmetry-asymmetry principle, (n) the fourteenth thesis: the softness-hardness principle, (o) the fifteenth thesis: the regression-progression principle, (p) the sixteenth thesis: the sameness-difference principle, and (q) the seventeenth thesis: the post-human creativeness—to be elaborated in the rest of the book, with a summary in the concluding chapter.

This seminal project is to fundamentally alter the way that we think about sound and silence in acoustics, from the combined perspectives of the mind, nature, society, and culture, with enormous implications for the human future and what I originally called its "post-human" fate.

Some Clarifications

But some additional clarifications are needed, before a formal analysis of the nature of acoustics is to begin in Chapter Two.

These clarifications, lest any potential misunderstanding by the readers occurs, refer to (a) the conversation with my previous books, (b) the illustrative selection of case studies and examples, (c) detailed analysis versus overall analysis, (d) two distinctive features of using quotations, and (e) the use of neologisms.

These clarifications are something that I also regularly repeat in my previous books too, so these points are repeated hereafter, *often verbatim*—albeit in a different context this time.

In Conversation with My Previous Books

Firstly, my multilateral theory of acoustics is constructed on the theoretical foundation of my previous books (i.e., *FHC*, *FCD*, *FPHC*, *BDPD*, *BCPC*, *BCIV*, *FPHST*, *BNN*, *BWT*, *FC*, *FAE*, *ALD*, *FIA*, *FPHU*, *FPHK*, *FPHML*, *FPHE*, *FPHMM*, *FPHCT*, *FPHG*, *FPHUP*, *FPHL*, *FPHO*, *FPHMA*, *FPHS*, *FPHLAW*, *FPHFS*, *FPHWP*, *FPHP*, *BCOS*,

BEPE, *FPHGEOL*, *FPHCHESS*, *FPHR*, *FPHEDU*, and *FPHH*—as explicated in “The List of Abbreviations”).

For this reason, many of my books have some familiar words in their titles like “the future of post-human…” or “beyond…to,” as well as some tables for background information—as they constitute a series of books with the final aim to converge all of my original theories on numerous fields into a unified theory of everything (as already summarized in the section on theory and meta-theory in Chapter One).

Just consider the following two illustrations, namely, (a) in relation to other related theoretical debates and (b) in relation to different visions as worked out in my previous books.

(a) The first illustration is that the theoretical debate here also involves a few other related theoretical issues.

These other related theoretical issues have to do with some deeply contested theoretical debates, which, however, I had already gone to great lengths to analyze in my numerous (previous) books, as summarized below. It should also be stressed, lest misunderstanding occurs, that each of the theoretical debates (as listed below) does not exhaust the issues covered in the books, as each book covers more issues than what is listed below (for illustration only).

- On Communication
  –Ex: *The Future of Post-Human Mass Media*
  –Ex: *The Future of Post-Human Organization*
- On Science and Engineering
  –Ex: *The Future of Post-Human Engineering*
  –Ex: *The Future of Post-Human Urban Planning*
- On Uncertainty
  –Ex: *The Future of Complexity*
  –Ex: 2 volumes, *The Future of Human Civilization*
- On Genes and Memes
  –Ex: *Beyond Nature and Nurture*
- On Earthquake and Seismology
  –Ex: *The Future of Post-Human Geology*
- On Aesthetics and Music
  –Ex: *The Future of Aesthetic Experience*
  –Ex: 2 volumes, *The Future of Human Civilization*
- On Technology and Warfare
  –Ex: *The Future of Capitalism and Democracy*
  –Ex: *The Future of Post-Human War and Peace*

- On Spirituality
  - –Ex: 2 volumes, *The Future of Human Civilization*
  - –Ex: *The Future of Capitalism and Democracy*
  - –Ex: *The Future of Post-Human Religion*
  - –Ex: *Beyond Civilization to Post-Civilization*
  - –Ex: *The Future of Post-Human Sexuality*
  - –Ex: *The Future of Post-Human Martial Arts*
  - –Ex: *The Future of Post-Human Unconsciousness*
  - –Ex: *The Future of Post-Human Consciousness*
  - –Ex: *Beyond Ethics to Post-Ethics*
- On Creativity
  - –Ex: *The Future of Post-Human Creative Thinking*
  - –Ex: *The Future of Aesthetic Experience*
- On Learning
  - –Ex: *The Future of Post-Human Language*
  - –Ex: *The Future of Post-Human Education*
- On Personality
  - –Ex: *The Future of Post-Human Consciousness*
  - –Ex: *The Future of Post-Human Personality*
- On Cognition
  - –Ex: *The Future of Post-Human Unconsciousness*
  - –Ex: *The Future of Post-Human Consciousness*
  - –Ex: *The Future of Capitalism and Democracy*
  - –Ex: *The Future of Post-Human Creative Thinking*
  - –Ex: *The Future of Post-Human Personality*
- On Intuition
  - –Ex: *The Future of Post-Human Unconsciousness*
  - –Ex: *The Future of Post-Human Consciousness*
  - –Ex: *Beyond Capitalism to Post-Capitalism*
  - –Ex: *The Future of Capitalism and Democracy*
- On Emotion
  - –Ex: *The Future of Post-Human Unconsciousness*
  - –Ex: *The Future of Post-Human Consciousness*
  - –Ex: *The Future of Aesthetic Experience*
  - –Ex: *Beyond Capitalism to Post-Capitalism*
  - –Ex: *The Future of Capitalism and Democracy*
  - –Ex: *The Future of Post-Human Personality*
- On Behavior
  - –Ex: *The Future of Post-Human Language*
  - –Ex: *The Future of Post-Human Unconsciousness*
  - –Ex: *The Future of Post-Human Consciousness*

–Ex: *Beyond Capitalism to Post-Capitalism*
–Ex: *The Future of Capitalism and Democracy*
–Ex: *The Future of Post-Human Sexuality*
–Ex: *The Future of Post-Human Personality*
–Ex: *The Future of Post-Human Martial Arts*

- On No Morality Without Immoralitiy
  –Ex: 2 volumes, *The Future of Human Civilization*
  –Ex: *The Future of Capitalism and Democracy*
  –Ex: *Beyond Democracy to Post-Democracy*
  –Ex: *Beyond Ethics to Post-Ethics*
- On Being and Belonging
  –Ex: *The Future of Post-Human Consciousness*
  –Ex: 2 volumes, *The Future of Human Civilization*
- On Different Dimensions of Human Existence
  –Ex: 2 volumes, *The Future of Human Civilization*
- On No Freedom Without Unfreedom
  –Ex: 2 volumes, *The Future of Human Civilization*
  –Ex: *The Future of Capitalism and Democracy*
  –Ex: *Beyond Capitalism to Post-Capitalism*
  –Ex: *Beyond Democracy to Post-Democracy*
  –Ex: *Beyond Civilization to Post-Civilization*
- On Modernity, Post-Modernity, and After-Postmodernity
  –Ex: 2 volumes, *The Future of Human Civilization*
  –Ex: *Beyond Civilization to Post-Civilization*
- On Pre-Modernity
  –Ex: *Beyond Civilization to Post-Civilization*

Therefore, those who are interested in these other theoretical issues (and debates) which are related to the current context can consult these other books of mine for more details.

(b) And the second illustration is that one of the major theses of this book concerns the future, be it human and later post-human, which is closely linked with the different visions that I had worked out in my previous books, in relation to the mind, nature, society, and culture.

Of course, for the convenience of the readers, a summary of some of these visions will be provided whenever necessary, as shown in the two sections on meta-theory above (*often verbatim*) and in different tables at the end of this chapter and the concluding chapter, for instance.

Yet, I still expect the readers to read my previous books directly for more details—as this book is written in conversation with my previous ones.

Case Studies and Examples

Secondly, the case studies and examples in the chapters are not exhaustive but solely illustrative. Nor are they necessarily mutually exclusive, as they can be reclassified in a different way. And exceptions are allowed.

One is tempted to assume, however, that, since the case studies and their examples are not exhaustive, more case studies and examples are needed. But the problem here is that just adding more of them does not necessarily change the conclusions to be drawn and may even lead to redundancy.

After all, the case studies and examples used in this book are chosen with care and deemed sufficient for the purpose at hand, even though they are solely illustrative.

Detailed Analysis versus Overall Synthesis

Thirdly, there is a distinction between detailed analysis and overall synthesis, in that the former is, especially though not exclusively, more for the specialized viewpoint of a specific discipline, whereas the latter is, especially though not exclusively, more for the broad horizon of an interdisciplinary approach—relatively speaking, of course.

Or, by analogy, there is a distinction between depicting a "tree" (with tiny details) in a "forest" and viewing the entire "forest" itself (with the broader horizon).

This book, like all previous ones of mine, tries to strike a balance between the two forms of understanding, although this does not imply that the two forms should be exactly equally offered, in an exact 50-50 balance.

Instead, the book, more often than not, does not allow the tiny details to obscure the larger picture in a given case study, because the entire book is inter-disciplinary in nature and tries to look at the entire "forest," not being bogged down with the detailed analysis of an individual "tree" in the forest, by analogy.

In this sense, although detailed analysis can be provided whenever needed for a given case study, the important point to remember is the overall horizon with its broadness of scope.

Two Distinctive Features of Using Quotations

Fourthly, the way that quotes are used in this book (and others of mine) has two distinctive features, as explained below.

The first distinctive feature is that, at the end of each quotation, sometimes there are more than one reference—in which case the first reference is for the original citation, and the rest of the references are for the convenience of the reader for more information about the issue under quotation.

And the second feature is that quotations are used as often as possible, so as to let others speak for themselves, without the nuisance of putting my own words into their mouths. Even more importantly, this allows me to use the original text as an evidence for critical analysis at the end of each section.

The Use of Neologisms

And finally, I use different neologisms in my books, mostly to introduce my original concepts and theories, and good instances include those here or elsewhere in my previous books (e.g., "the multilateral theory of acoustics," "post-democracy," "hyper-spatial consciousness," "post-capitalism," and whatnot). Needless to say, they are used here solely for our current intellectual convenience, as they will be renamed differently in different ways in future history.

As I thus wrote in *FCD* (508-9), "all these terms 'post-capitalism,' 'post-democracy'...and other ones as introduced in...[the] project (e.g.,...'posthuman elitists,' and 'posthuman counter-elitists,' just to cite a few of them) are more for our current intellectual convenience than to the liking of future humans and post-humans, who will surely invent more tasteful neologisms to call their own eras, entities, and everything else, for that matter. But the didactic point here is to use the terms to foretell what the future might be like, not that its eras and entities must be called so exactly and permanently. After all, William Shakespeare (1995: Act II, Scene II, Line 47) well said long ago: 'What is in a name? That which we call a rose by any other name would smell as sweet.'"

As I had also stressed time and again before, each of the neologisms can be re-written as a different "X," only to be re-named differently by the powers that be in different eras of future history.

With these clarifications for thought in mind—I now proceed to Chapter Two on sound and its doubleness.

Table 1.1. Acoustics in Relation to Sound and Silence

• **Acoustics**
—Firstly, acoustics is an interdisciplinary science.
—Secondly, acoustics deals with both audible and non-audible kinds of sound.
—Thirdly, acoustics deals with both sound and silence.

• **Sound**
—Ex: the propagation of sound waves
—Ex:the transduction of sound waves

• **Silence**
—Firstly, "the word silence may also refer to any absence of communication, even in media other than speech." (WK 2010c; AHD 2000)
—Secondly, silence, alternatively but in the opposite direction, can be understood as "total communication, in reference to non verbal communication and spiritual connection." (WK 2010c)
—Thirdly, silence can also be understood in the context of linguistic anaysis.
—And finally, silence can also be understood in the context of cultural norms.

Notes: The examples in the categories are solely illustrative (not exhaustive), and the comparison is relative (not absolute), nor are they necessarily mutually exclusive. And some can be easily re-classified elsewhere. As generalities, they allow exceptions.
Source: From *Sec. 1.2* of *FPHA*. See book for citations.

Table 1.2. Five Great Future Transformation of Acoustics

- **New Sound Detection**
  —Ex: in the post-huamn era, the post-humans will further transform acoustics for more sophisticated sound detection in a way that the human world has never known.

- **New Sound Control**
  —Ex: in the post-huamn era, the post-humans will further transform acoustics for more sophisticated sound control in a way that the human world has never known.

- **New Sound Creation**
  —Ex: in the post-huamn era, the post-humans will further transform acoustics for more sophisticated sound creation in a way that the human world has never known.

- **New Sound Horizon**
  —Ex: in the post-huamn era, the post-humans will further transform acoustics for a broader understanding of sound horizon in a way that the human world has never known.

- **New Sound Spirituality**
  —Ex: in the post-huamn era, the post-humans will further transform acoustics for a more enlightened sound spirituality in a way that the human world has never known.

Notes: The examples in the categories are solely illustrative (not exhaustive), and the comparison is relative (not absolute), nor are they necessarily mutually exclusive. And some can be easily re-classified elsewhere. As generalities, they allow exceptions.
Source: From *Ch.4* of *FPHA*. See text for citations.

Table 1.3. The Theoretical Debate on Space-Time (Part I)

- **Isaac Newton's Absolutist (Substantivist) Theory of Space-Time**
  —space and time are independent from each other. The structure of space-time is $E3 \times E1$ (with the structure of space, *P*, as the set of spatial locations in a three-dimensional Euclidean space, *E3*, and the structure of time as the set of temporal moments, *T*, in the one-dimensional real time, *E1*).
  —space and time are also independent from the effects of matter and events. The existence of space and time is possible even in a world absent of matter (and, for that matter, even in a world absent of events), as if they were material objects but with their total unchangingness thorough time.

- **Albert Einstein's Relativist Theory of Space-Time**
  —space and time are interchangeable (not absolute), just as matter and energy are equivalent (not independent) with the famous equation, $E = mc^2$ (as in the special theory of relativity in 1905).
  —space-time and matter-energy are also relative in a grand union (as in the general theory of relativity in 1915). Thus, each pair affects the other pair, as "matter 'tells' spacetime how to curve [in a non-Euclidean geometry] and curved spacetime 'tells' matter how to behave….Space contracts near mass and dilates away from it. Time dilates near mass and contracts away from it. Clocks positioned farther away from the mass of the earth run faster than clocks closer to the earth." (L. Shlain 1991: 328-330)

(continued on next page)

Table 1.3. The Theoretical Debate on Space-Time (Part II)

- **Peter Baofu's Perspectival Theory of Space-Time**
  —space and time can be understood from multiple perspectives, be they in relation to culture, society, nature, and the mind, with each perspective revealing something about the nature of space-time and simultaneously delimiting its view. This is subject to "the regression-progression principle" in existential dialectics.
  —each perspective of space and time exists in society and culture with good reasons, with some being more successful and hegemonic (dominant) than others. This is subject to "the symmetry-asymmetry principle" in existential dialectics.
  —space and time will not last, to be eventually superseded (altered) by post-humans in different forms (e.g.,stretching/shrinking space-time, engineering more dimensions of space-time, and manipulating multiverses), be they here in this universe or in multiverses. Thus, even the physical existence of space-time cannot last forever, with ever more transformations in the process. This is subject to "the change-constancy principle" in existential dialectics.
  —the conventional wisdom (especially by physicists) of treating the physical perspective of space and time as the foundation of all other perspectives (of space and time) and of regarding them as much less important is a form of reductionism,committing what I call *the foundation fallacy*, in misleadingly dismissing the multiple perspectives of space and time in relation to culture, society, nature, and the mind.

Notes: The examples in each category are solely illustrative (not exhaustive), and the comparison is relative (not absolute), nor are they necessarily mutually exclusive. Some can be easily re-classified elsewhere. As generalities, they allow exceptions.
Source: A summary of *Sec.1.1*, *Sec.1.2*, and *Sec.1.3* of *FPHST*—and, for that matter, the rest of *FPHST*

Table 1.4. Main Reasons for Altering Space-Time

- **The Need to Make New Energy-Matter**
  —Ex: manipulating molecular bonds for new materials
  —Ex: creating nanotechnologies on the atomic scale
  —Ex: engineering the atomic nucleus
  —Ex: restructuring most elementary particles
  —Ex: inventing new forms of matter and energy

- **The Need to Create New Space-Time**
  —Ex: creating "warp drive" (as in science fiction) for space travel
  —Ex: creating "pocket universes"

- **The Need to Conquer the Cosmos unto Multiverses**
  —Ex: spreading floating consciousness and hyper-spatial consciousness, besides other forms that humans have never known, in the cosmos and beyond unto multiverses for ultimate conquest

Notes: The examples in each category are solely illustrative (not exhaustive) nor necessarily mutually exclusive, and the comparison is relative (not absolute). As generalities, they allow exceptions. Also, it should be stressed that the three reasons are all related, in that they all contribute to the evolution of intelligent life in the cosmos unto multiverses in the most distant future beyond our current knowledge.
Sources: A summary of *Sec.6.2* of *FPHST*. See also *FHC*, *FCD*, *FPHC*, *FC*, *FPHU*, and *FPHG*, for example.

Table 1.5. The Technological Frontiers of the Micro-World

- **Type I-Minus**
  —Ex: building structures and mining

- **Type II-Minus**
  —Ex: playing with the genetic makeups of living things

- **Type III-Minus**
  —Ex: manipulating molecular bonds for new materials

- **Type IV-Minus**
  —Ex: creating nanotechnologies on the atomic scale

- **Type V-Minus**
  —Ex: engineering the atomic nucleus

- **Type VI-Minus**
  —Ex: restructuring most elementary particles

- **Type Ω-Minus**
  —Ex: altering the structure of space-time

Notes: As already indicated in *Sec.4.4.2.2* of *FPHC*, the problem with this micro-classification (from Barrow's work) is that the civilization types (with the exception of Type Ω-Minus, for example) are not quite distinct, since many of them can be achieved more or less in a civilization, to the extent that Type II-minus and Type III-minus, just to cite two plausible types, can be historically contemporaneous, relatively speaking, unlike the vast historical distance between, say, Type O and Type I (or Type I and Type II) civilizations. In other words, the micro-classification here is not very useful to understand civilization types but is revealing to see the technological frontiers of the micro-world.
Sources: A reconstruction from J.Barrow (1998:133), as originally shown in *Table 4.7* of *FPHC*. See *FPHC* for more info.

Table 1.6. Finity, Transfinity, and Infinity (Part I)

- **Infinity**
  —The term "infinity," which is "symbolically represented with ∞," derives from the Latin *infinitas* (or "unboundedness"), to refer to "several distinct concepts—usually linked to the idea of 'without end'—which arise in philosophy, mathematics, and theology." (WK 2008c)
  —In the context of mathematics, "infinity" is often "treated as if it were a number (i.e., it counts or measures things: 'an infinite number of terms') but it is a different type of 'number' from the real numbers. Infinity is related to limits,...large cardinals,...projective geometry, extended real numbers and the absolute Infinite," for instance. (WK 2008c)

- **Transfinity**
  —To avoid confusion, my usage of the word "transfinite" (as used here in the ontological principle) is not the same as the one which was originally coined by Georg Cantor to refer to "cardinal numbers or ordinal numbers that are larger than all finite numbers, yet not necessarily absolutely infinite" (in the mind of God), as opposed to "relative" infinity (in the mind of Man). (WK 2008d) In other words, for Cantor, there are two kinds of infinity, "absolute" and "relative"—with "relative" infinity or "transfinity" (in the mind of Man) being dependent on "absolute" infinity (in the mind of God) for its existence.
  —By contrast, my usage of the word "transfinite" here differs radically from the one used by Cantor (and other mathematicians) and is more limited, in light of the problems confronting any attempt to understand the idea of infinity, be it by intuition, imagination, and conception (as already shown in the prevous three sub-sections).
  —With this clarification in mind, I allow numbers which can be many times larger—or smaller, for that matter—than the finite things that we encounter in daily life, but they do not have to be related to the idea of infinity at all (which may not exist).

(continued on next page)

Table 1.6. Finity, Transfinity, and Infinity (Part II)

- **Transfinity** ***(cont'd)***
  —Of course, there may be some *borderline* cases, in which it is not clear whether the number in question is transfinite (in my usage) or simply a mathematical convenience. A good example of a borderline case is the Planck unit of length for "the smallest space possibly measured in nature," which is "less than billionths of trillionths of trillionths of an inch" (or something like 1.6×10^{-35} meters). (P. Baofu 2006a; N. McAleer 1987: 219; WKV 2008; D. Corbett 2008)
  —The ideas concerning symmetry and dimensions in the examples above were addressed in Chapter Three and Chapter Four of *FPHG*. For now, it suffices to further add that it is mind-boggling to even conceive or imagine extreme numbers like the Planck unit of length (or others). Does it really exist at all? Or is it merely a mathematical convenience to describe a physical situation that the human mind does not understand properly?
  —In any event, "transfinity" (in my usage) can be used as a further extension (of number) to the finite numbers in daily life—albeit with some borderline cases in mind.

- **Finity**
  —It refers to numbers which are bounded (that is, with an end), especially (though not exclusively) in relation to things in everyday life.

Source: From *Sec. 1.4* and *Sec. 2.2.3* of *FPHG*

Table 1.7. Theoretical Speculations of Multiverses

• **"Baby Universes" (Ex: Andre Linde and others)**
—Ex: In a flat universe theory, "even if our part of it eventually collapses,…some spots in the cosmos would suddenly start inflating on their own, creating brand-new 'baby universes.'" (P. Baofu 2000: 623)

• **"Parallel Universes" (Ex: Stephen Hawking and others)**
—Ex: In quantum cosmology, there allows the existence of infinite numbers of parallel universes, with tunneling among them. (M. Kaku 1994: 256) Hawking later revised his views on this.

• **"Pocket Universes" (Ex: Alan Guth)**
—Ex: "As the pocket universes live out their lives and recollapse or dwindle away, new universes are generated to take their place....While life in our pocket universe will presumably die out, life in the universe as a whole will thrive for eternity." (A. Guth 1997: 248; P. Baofu 2002: 482)

• **"Brane Worlds" (Ex: Warren Siegel, Lisa Randall, and others)**
—Ex: Our universe is stuck on a membrane of space-time embedded in a larger cosmos, with different brane worlds connecting and/or colliding with each other.

Notes: These examples are solely illustrative (not exhaustive), and some of the items can be reclassified somewhere else. Nor are they always mutually exclusive. Since they are generalities, exceptions are expected.
Source: From *Table 4.8* of *FPHC*

Table 1.8. The Confusion between "Many Worlds" and "Multiverse" (Part I)

- **First Problem**
  —The first problem in this second confusion concerns a debate between the "many-worlds" interpretation of quantum mechanics by Hugh Everett (1957) and others, and the Copenhagen interpretation by Niels Bohr and Werner Heisenberg on the effects of quantum states after interacting with an external environment (e.g., by a measuring observer).
  —On the one hand, the Copenhagen interpretation of quantum mechanics is more indeterministic in its concern with the probabilities of the outcomes of a wave function after interacting with an external environment (e.g., by a measuring observer using light to detect, for example, "an electron in a particular region around the nucleus at a particular time." (WK 2008e) Each wave function for the state of a particle refers to "a mathematical representation used to calculate the probability for it to be found in a location, or state of motion." (WK 2008f) And the "wavefunction collapse" means that "the act of measurement causes the calculated set of probabilities to 'collapse' to the value defined by the measurement." (WK 2008f)
  —On the other hand, the Everett's interrpetation is more deterministic, in rejecting "the objective reality of wavefunction collapse" and "instead explaining the subjective appearance of wavefunction collapse with the mechanism of quantum decoherence." (WK 2008g) The words "quantum decoherence" here refers to "the mechanism by which quantum systems interact with their environments to exhibit probabilistically additive behavior," in such a way that "the quantum nature of the system is simply 'leaked' into the environment." (WK 2008g) For example, "[a]s a result of an interaction, the wave functions of the system and the measuring device become entangled with each other. Decoherence happens when different portions of the system's wavefunction become entangled *in different ways* [italic added] with the measuring device...." (WK 2008g) Thus, for Everett, the appearance of "wavefunction collapse" can be explained by "quantum decoherence" for quantum states to branch out in different independent worlds after interacting with an external environment.

(continued on next page)

Table 1.8. The Confusion between "Many Worlds" and "Multiverse" (Part II)

- **First Problem** ***(cont'd)***
  —Yet, a major criticism against Everett's interpretation is that these different independent worlds in Everett's sense "will never be accessible to us," because of the interference by the measuring observer in an external environment. (WK 2008g)
  —For illustration, "once a measurement is done, the measured system becomes entangled with both the physicist who measured it and a huge number of other particles, some of which are photons flying away towards the other end of the universe; in order to prove that the wave function did not collapse one would have to bring all these particles back and measure them again, together with the system that was measured originally. This is completely impractical, but even if one can theoretically do this, it would destroy any evidence that the original measurement took place (including the physicist's memory)." (WK 2008g)
  —But then, the critics ask, If these independent worlds in Everett's sense are inaccessible to us, how do we know much of anything really about them?

- **Second Problem**
  —What is relevant at the quantum mechanical level may not necessarily apply to the world at the physicsl level.
  —In fact, the persistent inability among contemporoary physicists to unifiy the theory of relativity for large objects at the physical level and quantum mechanics for tiny objects at the quantum mechanical level should warn us about the danger of reductionism, by reducing the understanding of different universes ("parallel universes") at the physical level from a generalization of "many worlds" at the quantum mechanical level.

- **Third Problem**
  —There is no empirical evidence to suggest that there are infinite (or extremely many) possible worlds in the cosmos as the many-worlds model by Everett would have us to believe.

Source: From *Sec.4.4.1* of *FPHG*

Table 1.9. Hyperspace and Its Challenge (Part I)

- **First Problem**
  —Hyperspace for intra- and inter-universal travel is assumed to be possible, on the basis of four main assumptions; although there may be others, these are the often cited ones.
  —The first assumption concerns the existence of "entry" and "exit" points in hyperspace, which is easier said than done to enter, go through, and then exit hyperspace.
  —For instance, using hyperspace "requires complex calculations...of single or multiple hyperspace jumps and the control of the jump....Peturbations such as those experienced by ship in space from the gravitional field around an object such as a planet or even a star are exacerbated in hyperspatial travel, since mass in real space distorts hyperspace in an equal measure. 'Jumping' near to a gravitational mass is likely to make resulting exit from hyperspace to be highly uncertain, with the level of improbability *i* increasing with the square of the distance to the nearest gravitional 'well.'" (WK 2008h)
  —Besides, "[o]nly spaceships equipped with a special force field can enter hyperspace, because exposure to hyperspace even for short period of time is hazardous to unprotected humans." (WK 2008h)

- **Second Problem**
  —The second assumption concerns the faster-than-light speed. It is well known that Einstein argued for the speed of light as the ultimate limit, but nowadays phyiscists do allow the violation of this limit under certain conditions (like the hypothetical existence of tachyons, as will be explained further in a later sub-section on time travel).
  —Even then, there is a challenge to explain in science fiction "why ships can travel faster than light in hyperspace," and two good arguments are that "hyperspace may be smaller than real space and therefore a star ship's propulsion seems to be greatly multiplied, or else the speed of light in hyperspace is not a barrier as it is in real space. Whatever the reasoning, the general effect is that ships traveling in hyperspace seem to have broken the speed of light, appearing at their destinations much more quickly and without the shift in time that the Theory of Relativity would suggest." (WK 2008h)

(continued on next page)

Table 1.9. Hyperspace and Its Challenge (Part II)

• **Third Problem**

—The third assumption concerns the speed of time (to be measured, say, by a clock), but this is elaborated in a different section on the related issue of "time travel" (as shown in a different table on time travel).

• **Fourth Problem**

—The fourth assumption concerns spatial shortcut, as it is possible, because of the curvature of space, in that "the manifold of ordinary three-dimensional space is curved in four or more 'higher' spacial dimensions (a 'hyperspace' in the geometric sense…). This curvature causes certain widely separated points in three-dimensional space to nonetheless be 'adjacent' to each other four-dimensionally. Creating an aperture in 4D space (a wormhole) between these locations can allow instantaneous transit between the two locations; a common comparison is that of a folded piece of paper, where a hole punched through two folded sections is more direct than a line drawn between them on the sheet. This idea probably arose out of certain popular descriptions of General Relativity and/or Riemannian manifolds, and may be the original form from which later concepts of hyperspace arose." (WK 2008h)

—But this does not mean that hyperpsace is always safe. In fact, "in some science fiction, the danger of hyperspace travel is due to the chance that the route through hyperspace may take a ship too close to a celestial body with a large gravitational field, such as a star. In such scenarios, if a starship passes too close to a large gravitational field while in hyperspace, the ship is forcibly pulled out of hyperspace and reverts to normal space. Therefore, certain hyperspace 'routes' may be mapped out that are safe, not passing too close to stars or other dangers." (WJ 2008h)

Source: From *Sec.4.4.1* of *FPHG*

Table 1.10. The Problems of Time Travel into the Future (Part I)

• **First Problem**

—The first problem concerns "time dilation under the Theory of Special Relativity" (which separates space-time from mass-energy as special cases). Suppose you are on a moving train and an observer outside the train is watching you along the tracks. As your train moves along at the speed of light (or something close to it), "time, as measured by your watch, ticked along at a slower pace than time measured by the observer. Not only that, distance changed, too. For the observer, a one-foot ruler whizzing by on the train would have measured less than a foot." (NO 2000)

—In fact, you may not even notice that this slower clicking of the clock; in fact, everything looks normal to you inside the train as it was before: "The weird thing is that, for you on the train, time wouldn't seem to be moving slower and your ruler wouldn't be shorter—all would appear normal." (NO 2008) By contrast, you instead think that "time on the rest of the Earth would appear to be ticking along slower and its rulers would be shorter." (NO 2000) So, both of you insist that the other clock (not his) is clicking more slowly—assuming, of course, an inertial frame of reference.

—Now, change the story a bit, in that, this time, instead of an "inertial reference frame," acceleration is allowed, and you are on your way to board a spaceship. As an illustration, suppose "[y]ou board a spaceship and take off for deep space. The ship approaches the speed of light. Time for you seems to pass as it always has. It takes you about five seconds to tie your shoe. But to an observer on Earth (assuming he or she could watch you), you are moving at a snail's pace. It takes hours to tie your shoe." (NO 2000) In any event, "you continue on your journey. You slow down, stop, and accelerate back to Earth. You arrive home. You have aged two years during your flight. Two hundred years have passed on Earth. You have successfully travelled forward through time." (NO 2000) In this case, both the observer (if he is still alive) and you agree that you have aged less.

(continued on next page)

Table 1.10. The Problems of Time Travel into the Future (Part II)

• **First Problem** ***(cont'd)***

—In other words, when you are back to Earth, a lot of things have changed since, and you may find rather strange to be in a new environment which you hardly recognize, when compared with what you used to think as your city, or the Earth (two years ago, well, in accordance to your time of measurement). This means that you are in the future (two hundred years later), even though you are still relatively young (only two years older). But, can you go back? Well, no. The reason is that "[a]ccording to relativity, you can only move through time in one direction." (NO 2000)

• **Second Problem**

—The second problem concerns "time dilation under the Theory of General Relativity" (which unifies space-time with mass-energy in a general theory), the same result of time dilation occurs.

—For instance, when "one clock is deeper in a gravity well than the other," a result is that "the clock deeper in the well" will "tick…more slowly; this effect must be taken into account when calibrating the clocks on the satellites of the Global Positioning System, and it could lead to significant differences in rates of aging for observers at different distances from a black hole." (WK 2008i)

—More precisely, by way of another example, "it has been calculated that, under general relativity, a person could travel forward in time at a rate four times that of distant observers by residing inside a spherical shell with a diameter of 5 meters and the mass of Jupiter. For such a person, every one second of their 'personal' time would correspond to four seconds for distant observers. Of course, squeezing the mass of a large planet into such a structure is not expected to be within our technological capabilities in the near future." (WK 2008i; R. Gott 2002)

—The point here, however, is that time travel into the future, in this sense, is allowed, theoretically speaking, but in a very limited way, from the theory of relativity. But one can also ask, What if the theory of relativity were wrong? Well, in that case, there would be a need to find a better theory to explain time travel into the future.

Source: From *Sec.4.4.2* of *FPHG*

Table 1.11. The Problems of Time Travel into the Past (Part I)

- **First Problem**
  - —The first problem concerns the faster-than-light-speed, since, in accordance to the theory of special relativity, when an object goes faster than the speed of light, something weird can happen, in that, instead of having the clock clicking more slowly, it is moving backward, in relation to some inertial frame of reference—which then raisies the issue of whether time travel into the past is possible at all.
  - —But the problem here is about the violation of the law of casuality. In everyday language, an excellent illustration of causality violation is that, suppose you travel back into the past and then kill your parents, this then means that you could not have been born. But this is not true, since you are alive.
  - —Yet, "in the case of a hypothetical signal moving faster than light, there would always be some frames in which the signal was received before it was sent, so that the signal could be said to have moved backwards in time." The violation of casuality exists in this case.
  - —The first part of the problem is that, in accordance to the theory of special relativity, "it would take an infinite amount of energy to accelerate a slower-than-light object to the speed of light"—let alone the energy needed to propel an object to go faster than the speed of light. (WK 2008i) Although some suggest the possibility of "negative energy," it remains to be seen to what extent this is true.
  - —And the second part of the problem is that, "although relativity does not forbid the theoretical possibility of tachyons which move faster than light at all times, when analyzed using quantum field theory it seems that it would not actually be possible to use them to transmit information faster than light and there is no evidence for their existence." (WK 2008i; S. Chase 1993)

(continued on next page)

Table 1.11. The Problems of Time Travel into the Past (Part II)

- **Second Problem**
  —The second problem concerns cosmic string and black holes, since it is the theory of general relativity which "extends the special theory to cover gravity, illustrating it in terms of curvature in spacetime caused by mass-energy and the flow of momentum." (WK 2008i)
  —On the other hand, "[g]eneral relativity describes the universe under a system of field equations…and there exist solutions to these equations that permit what are called 'closed time-like curves,' and hence time travel into the past….The first of these was proposed by Kurt Gödel, a solution known as the Gödel metric, but his (and many others') example requires the universe to have physical characteristics that it does not appear to have." (WK 2008i; K. Thorne 1994)
  —But then, the problem now becomes: "Whether general relativity forbids closed time-like curves for all realistic conditions is unknown." (WK 2008i)

- **Third Problem**
  —The third problem concerns wormholes and related ones, and "[w]ormholes are a hypothetical warped spacetime which are also permitted by the Einstein field equations of general relativity, although it would be impossible to travel through a wormhole unless it was what is known as a traversable wormhole." (WK 2008i; M. Visser 1995) In other words, "a wormhole is a hypothetical topological feature of spacetime that is fundamentally a 'shortcut' through space and time. Spacetime can be viewed as a 2D surface, and when 'folded' over, a wormhole bridge can be formed. A wormhole has at least two mouths which are connected to a single throat or tube. If the wormhole is traversable, matter can 'travel' from one mouth to the other by passing through the throat." (WK 2008j)
  —But the problem here is three-fold, as shown in what follows.

(continued on next page)

Table 1.11. The Problems of Time Travel into the Past (Part III)

- **Third Problem** ***(cont'd)***
  —The first part of the problem is that "there is no observational evidence for wormholes." (WK 2008j)
  —The second part of the problem is that it is not clear how exactly wormholes (and black holes, for that matter) could allow time travel into the past at all.
  —And the third part of the problem is that if using wormholes (and black holes, for that matter) to travel into the past requires something to go faster than light too, then this raises another issue concerning the two problems as previously indicated in the sub-section on the faster-than-light-speed.

- **Fourth Problem**
  —The fourth problem is that some physicists have argued that, even should time travel into the past be possible, by way of these three methods, the person still would not return to the same past history that he exactly experienced before, but in a different one. (WK 2008i)
  —In other words, "strictly speaking," time travel into the past is not "really" possible, in light of these criticisms, so there is no need to worry that one could travel into the past and kill one's parent, as a way to violate the law of casuality.

Source: From *Sec.4.4.2* of *FPHG*

Table 1.12. The Conceptual Dimensions of Consciousness (and Other Mental States)

- **On Heredity and Time**
  —Heredity and the Environment
  —The Past and the Present

- **On Layers of Mental States and Abnormality**
  —Consciousness, Unconsciousness, and Preconsciousness
  —Normality and Paranormality

- **On Organicity and Motivation**
  —Mechanicity and Organicity
  —Primary Motivations and Multiple Motivations

- **On Other (Mostly Epistemic) Considerations**
  —Synthesis and Analysis
  —Situation and the Subject
  —Process and Outcome
  —Reasoning and Other Modes of Thinking
  —Meta-Conceptual Nominalism and Realism

Notes:: These examples are solely illustrative (not exhaustive), and some of the items can be reclassified somewhere else. Nor are they always mutually exclusive. Since they are generalities, exceptions are expected.
Sources: From *FPHC*. A re-construction, but with my own contribution, originally from G.Lindzey & C.Hall, *Introduction to Theories of Personality* (NY: John Wiley & Sons, 1985).

Table 1.13. The Theoretical Levels of Consciousness (and Other Mental States) (Part I)

- **At the Micro-Physical Theoretical Level**
  —Quantum-Mechanics
  —Electromagnetism

- **At the Chemical Theoretical Level**
  —Biochemistry

- **At the Biological Theoretical Level**
  —Evolutionary Biology
  —Neuroscience
  —Artificial Intelligence

- **At the Psychological Theoretical Level**
  —Psychodynamic Psychology
    - Psychoanalytical Psychology
    - Analytical Psychology
    - Socially Oriented Psychology

  —Experimental (Behavioral) Psychology
    - Operant Reinforcement Theory
    - Stimulus-Response Theory
    - Social Learning Theory

  —Cognitive (Gestalt) Psychology
    - Humanist Psychology
    - Existential Psychology
    - Field Theory

  —Psychometric Psychology
    - Idiographics
    - Constitutional Psychology

  —Social Psychology
    - Symbolic Interactive Theory
    - Social Exchange Theory

(continued on next page)

Table 1.13. The Theoretical Levels of Consciousness (and Other Mental States) (Part II)

- **At the Organizational Theoretical Level**
  —Managerial-Bureaucratic Theory
  —Oligarchic Theory
  —Network Theory

- **At the Institutional Theoretical Level**
  —Functionalist Theory
  —Anomic Theory

- **At the Structural Theoretical Level**
  —Conflict Theory
    - Marxian Theory
    - Critical Theory
    - Weberian Theory

  —Games Theory (in Formal Theory)
  —Feminist Theory
    - Feminist Concerns
    - Feminist Hopes

- **At the Systemic Theoretical Level**
  —Equilibrium Theory
  —System Theory
  —Chaos Theory

- **At the Cosmological Theoretical Level**
  —Superluminal Model
  —The Theory of Floating Consciousness

(continued on next page)

Table 1.13. The Theoretical Levels of Consciousness (and Other Mental States) (Part III)

- **At the Cultural Theoretical Level**
  - —Substantive Theories
    - • Structuralist Theory
    - • Post-Structuralist Theory (in Postmodernism)
  - —Meta-Theories
    - • Epistemic Objectivism vs. Epistemic Historicism
    - • Epistemic Subjectivism vs. Epistemic Non-Subjectivism
      - –Phenomenology
      - –Ethnomethodology
      - –Hermeneutics
    - • Epistemic Relativism vs. Epistemic Absolutism
    - • Epistemic Reductionism vs. Epistemic Emergencism

- **At Other Levels**
  - —Structuration Theory
  - —Reflexive Socioanalysis

Notes: These examples are solely illustrative (not exhaustive), and some of the items can be reclassified somewhere else. Nor are they always mutually exclusive. Since they are generalities, exceptions are expected.
Source: From many different sources as indicated in *FPHC*

Table 1.14. The Thematic Issues of Consciousness (and Other Mental States)

- **The Factor of History**
  —Pre-Modernity
  —Modernity
  —Postmodernity
  —After-Postmodernity

- **The Importance of Needs**
  —Having (e.g., the everyday, the technological)
  —Belonging (e.g., the just and the good)
  —Being (e.g., the true, the holy, the sublime/beautiful, and the good)

Notes: These examples are solely illustrative (not exhaustive). Nor are they always mutually exclusive. Since they are generalities, exceptions are expected. Also, some categories overlap with others; for instance, the good is also related with being-ness, just as the everyday can be connected with belong-ness.
Sources: From *FPHC*, based on *FHC* and *FCD*

Table 1.15. Having, Belonging, and Being in Consciousness

- **Having (e.g., About the Everyday and Technological)**
  —Ex: Physiological needs (e.g., thirst, hunger, sex)
  —Ex: Safety (e.g., stability, freedom from fear and anxiety)

- **Belonging (e.g., About the Good and Just)**
  —Ex: Social interaction
  —Ex: Friendship, acquaintance
  —Ex: Love, family
  —Ex: Self-respect, respect from others

- **Being (e.g., About the True, Holy, Beautiful/Sublime, and Good)**
  —Ex: Understanding of reality about self and world
  —Ex: Spiritual quest for holiness, beauty, sublimity, and goodness

Notes: These examples are solely illustrative (not exhaustive), and some of the items can be reclassified somewhere else. Nor are they always mutually exclusive. Since they are generalities, exceptions are expected.
Sources: From *FPHC*, based on *FHC* and *FCD*

Table 1.16. The Having-Ness of Consciousness (and Other Mental States)

- **At the Micro-Physical Level**
  —Ex: The finer physics of nutrition

- **At the Chemical Level**
  —Ex: Food guide pyramid

- **At the Bio-Psychological Level**
  —Ex: The hierarchy of needs

- **At the Institutional Level**
  —Ex: The production of commodity fetishism

- **At the Organizational Level**
  —Ex: Mass standardization (and customization); bureaucratic rationality

- **At the Structural Level**
  —Ex: The divide between the haves and the have-nots

- **At the Cultural Level**
  —Ex: The bondage of greater expectations

- **At the Systemic Level**
  —Ex: The technological transformation of poverty with transvaluations

- **At the Cosmological Level**
  —Ex: Floating consciousness in deep space without the human physiological needs

Notes: These examples are solely illustrative (not exhaustive), and some of the items can be reclassified somewhere else. Nor are they always mutually exclusive. Since they are generalities, exceptions are expected.
Sources: From *FPHC*. See also *FCD* and *FHC*.

Table 1.17. The Belonging-Ness of Consciousness (and Other Mental States)

- **At the Micro-Physical Level**
  —Ex: Statistical physics and social networks

- **At the Chemical Level**
  —Ex: Olfaction and social relations

- **At the Bio-Psychological Level**
  —Ex: Dual human nature on relationships

- **At the Institutional Level**
  —Ex: Private property and social cleavage

- **At the Organizational Level**
  —Ex: Legal formalism and iron cage

- **At the Structural Level**
  —Ex: Class conflicts and social alienation

- **At the Cultural Level**
  —Ex: Ruthless society and the culture of law

- **At the Systemic Level**
  —Ex: Human extinction and post-human relations

- **At the Cosmological Level**
  —Ex: The transcendence of equality in the cosmos and beyond

Notes: These examples are solely illustrative (not exhaustive), and some of the items can be reclassified somewhere else. Nor are they always mutually exclusive. Since they are generalities, exceptions are expected.
Sources: From *FPHC*. See also *FCD* and *FHC*.

Table 1.18. The Being-Ness of Consciousness (and Other Mental States) (Part I)

- **At the Micro-Physical Level**
  —Ex: Space-time and matter-energy in classical mechanics, quantum mechanics, and the theory of relativity

- **At the Chemical Level**
  —Ex: Drugs and well being

- **At the Biological Level**
  —Ex: Neuroscience and mystical experiences

- **At the Psychological Level**
  —Ex: Reductionism and emergencism

- **At the Institutional Level**
  —Ex: The proliferation of cults and sects, and the corporate-art industry

- **At the Organizational Level**
  —Ex: The organization of ideas in the scheme of things

- **At the Structural Level**
  —Ex: Disciplinary power and control in the order of things

- **At the Systemic Level**
  —Ex: Mutualism and ecological psychology

- **At the Cultural Level**
  —Ex: The recurrence of competing spirits, and floating consciousness as a climax of evolution

(continued on next page)

Table 1.18. The Being-Ness of Consciousness (and Other Mental States) (Part II)

- **At the Cosmological Level**
  —Ex: The search for elsewherewhen, and the evolution of other consciousnesses (e.g., the hyper-spatial consciousness)

Notes: These examples are solely illustrative (not exhaustive), and some of the items can be reclassified somewhere else. Nor are they always mutually exclusive. Since they are generalities, exceptions are expected.
Sources: From *FPHC*. See also *FCD* and *FHC*.

Table 1.19. Cognitive Partiality in Different Mental States

- **The Conscious**
  - —*Biased*
    - Ex: Self-overratingness
    - Ex: Self-modesty
    - Ex: Self-fulfilling prophesy
  - —*Shallow*
    - Ex: Attribution error
    - Ex: Heuristics
    - Ex: Illusory thinking
  - —*Short-term*
    - Ex: Deindividuation
  - —*Materialistic*
    - Ex: Hierarchy of needs

- **The Subconscious**
    - Ex: mental reconstruction
    - Ex: stereotyping

- **The Unconscious**
    - Ex: automatic processing
    - Ex: cognitive intuition

Notes: These examples are solely illustrative (not exhaustive), and some of the items can be reclassified somewhere else. Nor are they always mutually exclusive. Since they are generalities, exceptions are expected. And the comparison is relative, not absolute.
Source: From *BCPC*

Table 1.20. Emotional Non-Neutrality and Behavioral Alteration in Different Mental States

- **The Emotional**
  —*Conscious*
    Ex: Feelings
  —*Subconscious*
    Ex: Moods
  —*Unconscious*
    Ex: Empathy
    Ex: Instincts

- **The Behavioral**
  —*Conscious*
    Ex: Foot-in-the-door
    Ex: Role playing
    Ex: Saying-is-believing
  —*Subconscious*
    Ex: Discrimination
  —*Unconscious*
    Ex: Non-Verbal Communication

Notes: These examples are solely illustrative (not exhaustive), and some of the items can be reclassified somewhere else. Nor are they always mutually exclusive. Since they are generalities, exceptions are expected. And the comparison is relative, not absolute.
Source: From *BCPC*

Table 1.21. The Limits of Intuition in Unconsciousness

• **Cognitive Intuition (or Intuitive Thinking)**
—*Features*
Ex: mental schemas
Ex: expertise
Ex: blindsight
—*Problems*
Ex: inaccurate
Ex: non-explanatory

• **Emotional Intuition**
—*Features*
Ex: empathy
Ex: instincts
—*Problems*
Ex: over-generalized
Ex: non-explanatory

• **Behavioral Intuition**
—*Features*
Ex: prosemics
Ex: kinesics
—*Problems*
Ex: unreliable
Ex: non-explanatory

Notes: These examples are solely illustrative (not exhaustive), and some of the items can be reclassified somewhere else. Nor are they always mutually exclusive. Since they are generalities, exceptions are expected. And the comparison is relative, not absolute.
Source: From *BCPC*

Table 1.22. The Wealth/Poverty Dialectics in Different Mental States: The Case of Cognition

- **The Conscious**
  - —*Biased*
    - Ex: If richer in self-serving bias, then poorer in understanding others
  - —*Shallow*
    - Ex: If richer in heuristic thinking, then poorer in comprehending reality
  - —*Short-term*
    - Ex: If richer in short-term tendency, then poorer in long-term planning
  - —*Materialistic*
    - Ex: If richer in physiological obsession, then poorer in spiritual enlightenment

- **The Subconscious**
  - Ex: If richer in mental reconstruction, then poorer in accuracy of recall
  - Ex: If richer in stereotyping, then poorer in understanding groups

- **The Unconscious**
  - Ex: If richer in automatic processing, then poorer in depth of analysis
  - Ex: If richer in cognitive intuition, then poorer in reliability of judgments

Notes: These examples are solely illustrative (not exhaustive), and some of the items can be reclassified somewhere else. Nor are they always mutually exclusive. Since they are generalities, exceptions are expected. And the comparison is relative, not absolute.
Source: From *BCPC*

Table 1.23. The Wealth/Poverty Dialectics in Different Mental States: The Case of Emotion and Behavior

- **Emotion**
  - —*Conscious*
    - Ex: If richer in relying on familiarity and similarity for social attraction, then poorer in having diverse social relations
  - —*Subconscious*
    - Ex: If richer in moodiness, then poorer in sober thinking
  - —*Unconscious*
    - Ex: If richer in empathy, then poorer in task-oriented efficiency
    - Ex: If richer in instincts, then poorer in critical thinking

- **Behavior**
  - —*Conscious*
    - Ex: If richer in foot-in-the-door technique, then poorer in instant Gratification
    - Ex: If richer in role playing, then poorer in identity stability
  - —*Subconscious*
    - Ex: If richer in discriminating, then poorer in social harmony
  - —*Unconscious*
    - Ex: If richer in kinesics and proxemics, then poorer in verbal skills, relatively speaking.

Notes: These examples are solely illustrative (not exhaustive), and some of the items can be reclassified somewhere else. Nor are they always mutually exclusive. Since they are generalities, exceptions are expected. And the comparison is relative, not absolute.
Source: From *BCPC*

Table 1.24. The Theoretical Debate on Nature vs. Nurture (Part I)

- **The Environmental Approach**
  - —*Thesis*: It focuses, relatively speaking, more on the environment (culture and society) in explaining the achievement gap among individuals and for that matter, countries or regions, when contrasted with the natural factors.
  - —*Discourse*: Examples include Jose Ortega y Gasset ("Man has no nature; what he has is history"), Ashley Montagu ("Man is man because he has no instincts, because everything he is and has become he has learned from his culture, from the man-made part of the environment, from other human beings"), Stephen Jay Gould ("[The] brain [is] capable of a full range of behaviors and predisposed to none"), and Jesse Jackson (who blames white racism for the failure of blacks to close the achievement gap between whites and blacks in America). The works on dependency theory in international political economy (with a Marxian influence) and on the Protestant work ethic (by Max Weber) also point to this environmental direction.

- **The Genetic Approach**
  - —*Thesis*: It focuses instead, relatively speaking again, on hereditory factors (e.g., genes and evolution) in explaining the achievement gap among individuals and for that matter, countries or regions, when contrasted with the envrionmental factors.
  - —*Discourse*: Examples are Hans Eysenck and William Sheldon (in constitutional psychology), Konrad Lorenz (in his work on innate aggressive human nature), Gary Marcus (on the complexities of human thought by a tiny number of genes) and Robert Plomin, together with Michael Owen and Peter McGuffin (on the genetic basis of complex human behaviors).

(continued on next page)

Table 1.24. The Theoretical Debate on Nature vs. Nurture (Part II)

- **The Compromise Approach**
  —*Thesis*: It seeks the middle-of-the-road argument in regard to nature and nurture and regards all differences among individuals and groups as the result of the mixture of both nature and nurture, more or less equally.
  —*Discourse*: Examples include C. Murray and R. Herrnstein ("It seems highly likely to us that both genes and environment have something to do with this issue") and Dan Dennett ("Surely 'everyone knows' that the nature-nurture debate was resolved long ago, and neither side wins since everything – is – a – mixture – of – both – and - it's – all – very -complicated...."

- **The Transcendent Approach**
  —*Thesis*: It goes beyond both nature and nurture (without, however, committing the compromise fallacy) in showing their closely intertwined interactions in producing the behavioral differences as often seen in individual human endeavors on the micro scale, and for that matter, in country (or regional) endeavors on the macro one—in the context of my five theses, namely, (a) the compromise fallacy, (b) no oppression without self-oppression, (c) no success without failure, (d) the factor of randomness, and (e) the post-human vision, to be elaborated in Chapter Six.
  —*Discourse*: Peter Baofu proposed this approach on the basis of his "theory of contrastive advantages" (as an original theoretical contribution to the debate, which was first proposed and analyzed in *The Future of Capitalism and Democracy*). In the end, the human genes will not last, to be eventually superseded by post-human life forms, so the debate between genes and memes has obscured something profoundly important about the future that the world has never known. And the debate is also misleading and faulty in its dichotomy.

Source: From *BNN*. See the book for citations and details.

Table 1.25. Physical Challenges to Hyper-Spatial Consciousness

- **The Understanding of a Higher-Dimensional World of Space-Time**
  —Ex: 4 for traditional aspects of space-time (e.g., length, width, breadth and time) plus 6 more new dimensions in theory of hyper-space, with profound implications for practical applications to new forms of consciousness.

- **The Mastering of Dark Matter and Dark Energy**
  —Ex: "ordinary matter" (e.g., atoms, molecules) as a mere 4.4% of the universe, with 23% made of "cold dark matter" and the rest (about 73%) of mysterious "dark energy," with fundamental significance to questions about the limit of the speed of energy (or info), the availability of energy for use, and the nature of space-time, just to cite some examples.

- **The Exploration of Multiverses**
  —Ex: theoretical speculation of other universes (e.g., "baby universes," "gateways" in black holes, "wave function of the universe," "many worlds," "brane worlds"), with potentially seminary discoveries of different physical laws in relation to matter-energy and space-time, and vital differences to the future of post-human conquest of other universes (for the emergence of new forms of consciousness).

Notes: These examples are solely illustrative (not exhaustive), and some of the items can be reclassified somewhere else. Nor are they always mutually exclusive. Since they are generalities, exceptions are expected. The point here is to give a rough picture of the evolution of consciousness to the hyper-spatial consciousness and others totally unknown to current earthlings. As a note of clarification, it makes no difference to my argument as to whether or not the hyper-spatial consciousness may emerge before, during, and after floating consciousness.
Source: From *Table 4.5* of *FPHC*

Table 1.26. The Theory of Floating Consciousness (Part I)

- **At the Micro-Physical Level**
  —Ex: intelligent life without the human physical-chemical system

- **At the Chemical Level**
  —Ex: space radiation and toxins

- **At the Bio-Psychological Level**
  —Ex: exo-biological evolution in deep space
  —Ex: genetic engineering of new beings

- **At the Institutional Level**
  —Ex: post-capitalism
  —Ex: post-democracy

- **At the Organizational Level**
  —Ex: less legal-formalistic routines

- **At the Structural Level**
  —Ex: alien forms of violence

- **At the Cultural Level**
  —Ex: transcending freedom
  —Ex: transcending equality

(continued on next page)

Table 1.26. The Theory of Floating Consciousness (Part II)

• **At the Cosmological Level**
—Ex: parallel universes
—Ex: pocket universes

• **At the Systemic Level**
—Ex: space habitats (in zero-gravity)

Notes: Each example draws from the works of different scholars in the field. For instance, at the cosmological level, the idea of parallel universes is from the theoretical speculation in quantum cosmology by Stephen Hawking and others, while the one of pocket universes comes from the theoretical work of Allan Guth at MIT. And at the institutional level, I proposed post-capitalism and post-democracy in *FCD* (and later, from *BDPD* and *BCPC*). In addition, the examples are solely illustrative (not exhaustive), and some of the items can be reclassified somewhere else. Nor are they always mutually exclusive. Since they are generalities, exceptions are expected.
Sources: From *FPHC*—and, originally, from *FCD*

Table 1.27. The Potential of Unfolding Unconsciousness

- **Superior Senses**
  - —*Through space*
    - Ex: clair-sentience (in feeling)
    - Ex: clair-voyance (in seeing)
    - Ex: clair-audience (in hearing)
    - Ex: clair-austance (in tasting)
    - Ex: clair-alience (in smelling)
    - Ex: clair-cognizance (in knowing)
  - —*Through time*
    - Ex: pre-cognition (in knowing), pre-sentiment (in feeling)
    - Ex: retro-cognition (in knowing), retro-sentiment (in feeling)

- **Superior Contacts**
  - —*With living minds*
    - Ex: telepathy
  - —*With the dead*
    - Ex: mediumship
  - —*With other-living entities*
    - Ex: telekinesis

Notes: The categories and examples are solely illustrative (not exhaustive), and the comparison is relative (not absolute), nor are they necessarily mutually exclusive. And some can be easily re-classified elsewhere. As generalities, they allow exceptions.
Source: A summary of Ch.5 of *FPHU*

Table 1.28. The Future Exploration of Unfolding Unconsciousness

- **In the Context of Nature**
  - —Ex: the higher-dimensional space-time in hyperspace theory
  - —Ex: the particle-wave interactions in quantum mechanics
  - —Ex: the study of brain waves in electromagnetic theory

- **In the Context of the Mind**
  - —Ex: the unconscious fantasies in psychoanalysis
  - —Ex: the mind-altering drugs in biochemistry
  - —Ex: the manipulation of neural activity in neurobiology

- **In the Context of Culture**
  - —Ex: the correlation between anomalous experience and moral/religious/aesthetic interests in culture studies
  - —Ex: the role of myths and language in epistemology

- **In the Context of Society**
  - —Ex: different organizational agendas in anomalous research
  - —Ex: the scholarly divide on anomalous study
  - —Ex: interests and propaganda in institutioanl research
  - —Ex: the transformative power of technology in research on social systems
  - —Ex: the natural and non-natural selection of other worlds in exo-biology
  - —Ex: the role of the post-human conscious mind in qualitative demography

Notes: The examples in each category are solely illustrative (not exhaustive), and the comparison is relative (not absolute), nor are they necessarily mutually exclusive. And some can be easily re-classified elsewhere. As generalities, they allow exceptions.
Source: A summary of Chs.2-5 of *FPHU*, in conjunction with the rest of the book

Table 1.29. Creative Techniques and Traits

- **Creative Techniques**
  - —*Comprehensiveness*
    - Ex: Broadness of Scope
    - Ex: In-Depth Analysis
    - Ex: Multifaceted Taxonomy
    - Ex: Numerous Theories
  - —*Arrangement*
    - Ex: Separation (or Subtraction)
    - Ex: Combination (or Addition)
    - Ex: Replacement
    - Ex: Accommodation
    - Ex: Reversal
    - Ex: Stretching
  - —*Serendipity*
    - Ex: Play
    - Ex: Chance

- **Creative Traits**
  - —*Openness / Risk*
    - Ex: Ability to welcome new ideas or to do new things, with risk
  - —*Discipline*
    - Ex: Ability to sacrifice whatever it takes to succeed
  - —*Resilience / Confidence*
    - Ex: Ability to take failures to mind, not to heart
  - —*Trust*
    - Ex: Ability to cooperate with others for team work, if necessary

Notes: The examples in the categories are solely illustrative (not exhaustive), and the comparison is relative (not absolute), nor are they necessarily mutually exclusive. And some can be easily re-classified elsewhere. As generalities, they allow exceptions.
Source: From *Sec.4.2* of *FPHCT*. For more details, also read the rest of the book.

Table 1.30. The Desirability of Creativity, and Its Dark Sides (Part I)

- **In Relation to Instrumental Rationality**
  —*Desirability*
  - Ex: For the sake of efficiency and effectiveness in this capitalist era of our time.

  —*Undesirability*
  - Ex: What Max Weber (1930) called the "iron cage" of capitalism, where live the "sensualists without heart, specialists without spirit"—or something which the Frankfurt School has forcefully asked us to be freed from.

- **In Relation to Substantive Rationality**
  —*Desirability*
  - Ex: For the sake of God, the King, Motherland, or other comparable ideals.

  —*Undesirability*
  - Ex: Domination and oppression of various forms—something which the moderns had struggled to free themselves from in the first place, since the dawn of modernity.

- **In Relation to Autonomous Rationality**
  —*Desirability*
  - Ex: For the sake of itself, the autonomy of creative endeavor.

  —*Undesirability*
  - Ex: The decadence of its own degeneration (e.g., the donwsides of postmodernism in our postmodern times, or what Nietzsche once called the unbearable "unreality" and "falseness" of the autonomous artist's "innermost existence"—and, in other cases, the painful suffering from different mental illnesses for those highly creative individuals).

(continued on next page)

Table 1.30. The Desirability of Creativity, and Its Dark Sides (Part II)

Notes: The examples in the categories are solely illustrative (not exhaustive), and the comparison is relative (not absolute), nor are they necessarily mutually exclusive. And some can be easily re-classified elsewhere. As generalities, they allow exceptions.
Also, I already went to great lenghts to explain the desirability and its dark sides of all these three rationalities (although there may be others too) in the two volumes of *FHC*—and, for that matter, *FCD*.
Source: From *Sec.4.9* of *FPHCT*. For more details, also read the rest of the book.

Table 1.31. Posthuman-Ism, Post-Humanism, and Trans-Humanism

- **Post-Humanism**
  —The neologism "post-human" used in my books should *not* be confused with another term which looks similar but has a totally different meaning in the literature of postmodernism, namely, "post-humanism"—which constitutes a critique of "humanism" as tradionally understood (especially, though not exclusively, in relation to the idea of progress in science and reason in the Enlightenment project). (WK 2008)
  —My works reject the project of "postmodernism" and propose the future world of what I orignally called "after-postmodernity" in *FHC* and *FCD*, for instance.

- **Trans-Humanism**
  —Also, the neologism "post-human" used in my books should *not* be confused with a similar term which is used to champion the ideology of technology for the fturue co-existence between humans and cyborgs in "trans-humanism." (WK 2008a)
  —Instead, my term "post-human" in relation to "posthuman-ism" also rejects "trans-humanism" (especially, though not exclusively, in relation to the promises of technology) and refers to something else altogether, that is, the future extinction of humans and its post-human successors in deep space and beyond unto multiverses.
  —Chapter Two of my book on ethics (*BEPE*) provides a more comprehensive critique of trans-humanism.

- **Posthuman-Ism**
  —With these two clarifications in mind, the "post-humans" (as envisioned in my books) can take different forms, and I proposed, in my different books, some of them, such as "unfolding unconsciousness," "floating consciousness," "hyper-spatial consciousness," "thinking machines," "thinking robots," "genetically altered superior beings," "cyborgs," and others (as already analyzed in my previous works).
  —They will evolve in conjunction with other visions of mine in relation to nature, society, and culture.

Sources: From *Sec.1.6* of *FPHE* (and also *BEPE*). For more details, also read the rest of the books (and other books of mine).

Table 1.32. Three Great Future Transformations of the Martial Body (Part I)

• **Virtual Battle**
 —Ex: virtual fight
 —Ex: virtual memory

• **The Hyper-Martial Body**
 —*In relation to the physical dimension*
 • Ex: striking (e.g., punching, kicking, trapping, acupressure-striking, and so on in an amazing way)
 • Ex: grapping (e.g., throwing, pinning, joint-locking, and so on in an amazing way)
 • Ex: running and walking (e.g., on different surfaces)
 • Ex: jumping (e.g., through walls, trees, etc.)
 • Ex: flying (e.g., through space)
 • Ex: enduring pain (e.g., absorbing punches, kicks, and so on in an amazing way)
 • Ex: using extremely advanced hi-tech weapons
 • Ex: healing (e.g., with special energy, etc.)
 • Ex: entering (e.g., into different dimensions of space)
 • Ex: sensing (e.g., in extraordinary ways through space and time, etc.)
 • Ex: wearing hi-tech suits for defensive and offensive purposes (e.g., like a hi-tech armor)
 —*In relation to the mental dimension*
 • Ex: the ability to relax the martial body in a way that current humans cannot.
 • Ex: the ability to concentrate the martial mind in a way that current humans are not accustomed to.
 • Ex: the ability to breathe and be timely in preparing the martial body for fighting in a way that current humans are not good at.

• **The Post-Martial Floating Mind**
 —Ex: the mind without the human body
 —Ex: the quest for spirituality beyond martial arts

(continued on next page)

Table 1.32. Three Great Future Transformations of the Martial Body (Part II)

Notes: The examples in the categories are solely illustrative (not exhaustive), and the comparison is relative (not absolute), nor are they necessarily mutually exclusive. And some can be easily re-classified elsewhere. As generalities, they allow exceptions.
Source: From *Sec. 2.5.1* of *FPHMA*. See text for citations.

Table 1.33. Three Great Future Transformations of the Sexual Body

- **Virtual Pleasure**
  —Ex: virtual Sex
  —Ex: virtual Recall

- **The Hyper-Sexual Body**
  —*In relation to the physical dimension*
  - Ex: the ability to perform differenet sexual positions without the various physical limitations that humans currently have.
  - Ex: the ability to produce unusual amount of sexual energy to endure in sexual acts in a way that humans currently cannot.
  - Ex: the ability to engage in multiple sexual experiences comparable to (but different from) non-human sexuality (e.g., hermaphrodite, monoecious, sex-switching, and many other types) that current humans cannot.

  —*In relation to the mental dimension*
  - Ex: the ability to relax the sexual body in a way that current humans are not accustomed to.
  - Ex: the ability to concentrate the sexual mind in a way that current humans are not good at.
  - Ex: the ability to breathe and be patient in preparing the sexual body for sexuality in a way that current humans are not able to.

- **The Post-Sexual Floating Mind**
  —Ex: the mind without the human body
  —Ex: the quest for spirituality beyond sexuality

Notes: The examples in the categories are solely illustrative (not exhaustive), and the comparison is relative (not absolute), nor are they necessarily mutually exclusive. And some can be easily re-classified elsewhere. As generalities, they allow exceptions.
Source: From *Sec. 3.4.2* of *FPHS*. See text for more info.

Table 1.34. The Trinity of Pre-Modernity

- **Pre-Free-Spirited Pre-Modernity (Pre-Modernism) and Its Internal Split**
  —Competing worldviews and values both within and between linear centric (e.g., Islamic, Christian, Judaic, Imperial Roman) and cyclical-centric (e.g., Confucian, Taoist, Hindu, and Buddhist) orientations
  —Compare modernism with pre-modernism here in relation to the seven dimensions of human existence like the true and the holy (e.g., different versions of epistemic dogmas and religious superstitions), the everyday and the technological (e.g.,different versions of non-technophilism and non-consumerism), the beautiful/sublime (e.g., different versions of aesthetic non-autonomy), and the good and the just (e.g., different versions of moral particularism).

- **Pre-Capitalist Pre-Modernity (Pre-Modernization) and Its Own Discontents**
  —Competing versions of societal arrangements (e.g., feudalism, monarchism, and the holy order)

- **Hegemonic Pre-Modernity and Its Countervailing Forces**
  —Different power centers and their enemies (e.g., the Roman Empire and the "barbarian hordes," the "Holy Crusades" and the Muslims, the Middle Kingdom and the invading tribes, different social castes in India, and warring Greek city-states)

Notes: The examples in each category are solely illustrative (not exhaustive) nor necessarily mutually exclusive, and the comparison is relative (not absolute). As generalities, they allow exceptions. Also, it does not matter what the "base" era is in the analysis of any trinity. And in the present context, the "base" era is modernity (for instance, with its "free-spirited modernity" and the other two parts). So, for pre-modernity, the trinity takes the form of, say, "pre-free-spirited pre-modernity," together with the other two parts.
Sources: From Ch.2 of *BCIV* and also the 2 volumes of *FHC*

Table 1.35. The Trinity of Modernity (Part I)

- **Free-Spirited Modernity (Modernism) and Its Internal Split**
  —*On the True and the Holy*
  - The freedom from the dogmas of the past to the better understanding of, and union with, the world and self (Ch.3 of *FHC*)
  - Alternative discourses: about the true (e.g., anti-science discourses) and the holy (non-mainstream theologies) (Ch.3 of *FHC*)

  —*On the Technological and the Everyday*
  - The freedom from life harshness to the higher technophilic, consumeristic lifeform (Ch.2 of *FHC*)
  - Alternative discourses: about the everyday (e.g., transcendental mindsets) and the technological (e.g., Arcadianism) (Ch.2 of *FHC*)

  —*On the Good and the Just*
  - The freedom from the theo-aristocratic tyranny to the moral universality for a just society (Ch.5 of *FHC*)
  - Alternative discourses: about the just (e.g., Communism, Anarchism) and the good (e.g., Nazism/Fascism, and Zarathustrianism) (Chs.5-6 of *FHC*)

  —*On the Beautiful and the Sublime*
  - The freedom from the external distortion of aesthetic pleasure to the boundless infinity of totality in artistic autonomy (Ch.4 of *FHC*)
  - Alternative discourses: about the beautiful/sublime (e.g., kitsch and historical avant-gardism) (Ch.4 of *FHC*)

- **Capitalist Modernity (Modernization) and Its Own Discontents**
  —*During the Industrial Revolution*
  - Ex: Marx on the institution of inequality (Ch.1 of *FHC*)

  —*During the Modern Rational-Instrumental Epoch*
  - Ex: Weber on the politics of soft liberal institutions (Ch.5 of *FHC*)

  —*During the Great Depression*
  - Ex: Keynes on the myth of the free market (Chs.1-3 of *FHC*)

  —*During the Cold War*
  - Ex: Lasch on the narcissistic culture industry (Chs.2-3 of *FHC*)

(continued on next page)

Table 1.35. The Trinity of Modernity (Part II)

- **Hegemonic Modernity and Its Countervailing Forces**
  —*The Legacies of Colonialism and Imperialism*
    - Ex: European colonization of most of the modern world (Ch.1 of *FHC*)

  —*The Struggle for Decolonialization*
    - Ex: The countervailing forces of resentment, rechantment, and regionalism (Chs.1 & 6 of *FHC*)

Notes: The examples in each category are solely illustrative (not exhaustive) nor necessarily mutually exclusive, and the comparison is relative (not absolute). As generalities, they allow exceptions.
Sources: From the 2 volumes of *FHC*—and also from *FCD*

Table 1.36. The Trinity of Post-Modernity

- **Free-Spirited Postmodernity (Postmodernism) and Its Internal Split**
  —*On the True and the Holy*
    - Postmodern performative turn for knowing and its enemies (Ch.3 of *FHC*)
    - Postmodern comparative theology and its opponents (Ch.3 of *FHC*)
  —*On the Technological and the Everyday*
    - Postmodern corporate technological mindset and its adversaries (Ch.2 of *FHC*)
    - Postmodern postmaterialism and its critics (Ch.2 of *FHC*)
  —*On the Good and the Just*
    - Postmodern politics of difference and its foes (Ch.5 of *FHC*)
  —*On the Beautiful and the Sublime*
    - Postmodern deconstruction and its dissenters (Ch.4 of *FHC*)

- **Capitalist Postmodernity (Postmodernization) and Its Own Discontents**
  —*During the Post-Cold War and Beyond*
    - Ex: post-Fordism and its shortcomings (Ch.6 of *FHC*; Chs.6-7 of *FCD*)

- **Hegemonic Postmodernity and Its Countervailing Forces**
  —*The Debate on the Global Village*
    - Ex: uni-civilizationalism vs. multi-civilizationalism (Ch.6 of *FHC*)
  —*The Resistance Movement*
    - Ex: rechantment and the politics of civilizational claims (e.g., Islamic, Confucian and other ethos in relation to the Same) (Ch.6 of *FHC*; Ch.10 of *FCD*)
    - Ex: resentment and the politics of resurgence (e.g., the rising Chinese superpower, the growing EU, and other players in relation to the U.S. and her allies) (Ch.6 of *FHC*; Ch.8 of *FCD*)
    - Ex: regionalism and the politics of inequality (e.g., trans- or international blocs, the North-South divide, NGO's) (Ch.6 of *FHC*; Ch.5 of *FCD*)

Notes: The examples in each category are solely illustrative (not exhaustive) nor necessarily mutually exclusive, and the comparison is relative (not absolute). As generalities, they allow exceptions.
Sources: From *FCD* and the 2 volumes of *FHC*

Table 1.37. The Trinity of After-Postmodernity

- **Free-Spirited After-Postmodernity (After-Postmodernism) and Its Internal Split**
  —The discourse of naked contingency (Ch.10 of *FCD*; Ch.4 of *FPHC*)

- **Post-Capitalist After-Postmodernity (After-Postmodernization) and Its Own Discontents**
  —Different versions of post-capitalism and post-democracy, and their enemies (Ch.10 of *FCD*; Chs.3-4 of *FPHC*)

- **Hegemonic After-Postmodernity and Its Countervailing Forces**
  —The Cyclical Progression of Hegemony in Multiverses (Chs.9-10 of *FCD*; Ch.4 of *FPHC*)

Notes: The examples in each category are solely illustrative (not exhaustive) nor necessarily mutually exclusive, and the comparison is relative (not absolute). As generalities, they allow exceptions.
Sources: From *FCD* and also *FHC*

• PART TWO •

Sound

CHAPTER 2
SOUND AND ITS DOUBLENESS

> Life 's but a walking shadow, a poor player That struts and frets his hour upon the stage And then is heard no more: it is a tale Told by an idiot, full of sound and fury, Signifying nothing.
> —W. Shakespeare (BT 2010b)

The Orderliness of Sound

Acoustics deals with both sound and silence, as already explained in *Sec. 1.2*.

This distinction between sound and silence is true, even if it is more a question of degree than of kind. This then warns us that there is a close relationship between sound and silence, as one cannot be completely separated from the other.

With this disitnction in mind, this book offers a dialectic treatment of sound and silence in that, if sound has its order, it has its disturbance too—just as, if silence has its serenity, it likewise has its turmoil.

The inquiry concernng sound is the subject matter of this chapter (Chapter Two), whereas that of silence will be examined in the next chapter (Chapter Three).

However, the close relationship between sound and silence poses a scholarly challenge here, in that one cannot analyze sound (or silence, for that matter) completely separately.

This then suggests that the issues concerning sound in this chapter (Chapter Two) and silence in the next chapter (Chapter Three) can overlap, as each issue on one can be relevant to the other too.

With this caveat in mind—a good way to explore the doubleness of sound is by way of the evaluation of the extent to which it (sound) is in fact both possible and desirabe.

This chapter thus accepts the challenging task to provide a comprehensive analysis of silence in the context of acoustics from the four perspectives of the mind, nature, society, and culture (in accordance to my sophisticated methodological holism, as explained in *Sec. 1.8*).

Thus, this chapter is organized in four main sections, in relation to (2.2) sound and the mind, (2.3) sound and nature, (2.4) sound and society, and (2.5) sound and culture—to be examined hereafter, respectively (and summarized in *Table 2.1*).

Sound and the Mind

Sound, when examined from the perspective of the mind, can reveal something intriguing about its possibility and desirability, and this can be shown by way of two case studies, namely, (2.2.1) sound, perception limits, and the psychoacoustic model and (2.2.2) sound, noise sensibility, and extroversion-introversion—to be addressed in what follows, in that order.

Sound, Perception Limits, and the Psychoacoustic Model

The psychoacoustic model, as already introduced in *Sec. 1.4*, is "the branch of science studying the psychological and physiological responses associated with sound (including speech and music)." (WK 2010d)

Four Limits of the Human Perception of Sound

One of the most fascinating aspects of human perception in relation to sound has to do with "the limits of perception." Over time, psychoacousticians had identified four major limits of the human perception of sound, as described below (and summarized in *Table 2.2*).

(a) High Frequency Limit

The first limit of the human perception of sound has to do with "the highest frequency limit," which, for the human ear, is about 20,000 Hz (or 20 kHz). (WK 2010d)

"High frequencies" means "high pitches" and "lower frequencies have low pitches." (WK 2010j)

However, this upper frequencey limit is not constant, because it can decrease over time, "due to either hearing damage (e.g. from listening to loud music) or aging." (WK 2001j)

The lower frequency limit, for the human ear, is "below 20 Hz," but "these can be perceived via the body's sense of touch." (WK 2010d)

Besides frequency limit—there is also "frequency resolution" (about changes in pitch) which can be detected by the human ear, which "is 3.6 Hz within the octave of 1,000–2,000 Hz. That is, changes in pitch larger than 3.6 Hz can be perceived in a clinical setting. However, even smaller pitch differences can be perceived through other means. For example, the interference of two pitches can often be heard as a (low-)frequency difference pitch. This effect of phase variance upon the resultant sound is known as beating." (WK 2010d;H. Olson 1967)

The human ear can also pay attention to sound pressure (in relation to loudness); for instance, it "can detect pressure changes as small as 2×10−10 atm and as great as or greater than 1 atm. For this reason, sound pressure level is also measured logarithmically [non-linearly], with all pressures referenced to 1.97385×10−10 atm. The lower limit of audibility is therefore defined as 0 dB, but the upper limit is not as clearly defined. The upper limit is more a question of the limit where the ear will be physically harmed or with the potential to cause noise-induced hearing loss." (WK 2010d)

(b) Absolute Threshold of Hearing

The second limit of the human perception of sound has to do with "the absolute threshold of hearing," which refers to "the lower limits of audibility" or "the minimum threshold at which a sound can be heard" by the human ear, which is "frequency dependent." (Wk 2010d)

For instance, "[b]y measuring this minimum intensity for testing tones of various frequencies, a frequency dependent absolute threshold of hearing (ATH) curve may be derived. Typically, the ear shows a peak of sensitivity (i.e., its lowest ATH) between 1,000 Hz and 5,000 Hz, though the threshold changes with age, with older ears showing decreased sensitivity above 2,000 Hz." (WK 2010d)

The frequency-dependent ATH here refers to "the lowest of the equal-loudness contours. Equal-loudness contours indicate the sound pressure level (dB), over the range of audible frequencies, which are perceived as being of equal loudness. Equal-loudness contours were first measured by [H.] Fletcher and [W.] Munson at Bell Labs in 1933 using pure tones reproduced via headphones, and the data they collected are called Fletcher-Munson curves. Because subjective loudness was difficult to measure, the Fletcher-Munson curves were averaged over many subjects." (WK 2010d)

Later, "[D. W.] Robinson and [R. S.] Dadson refined the process in 1956 to obtain a new set of equal-loudness curves for a frontal sound source measured in an anechoic chamber. The Robinson-Dadson curves were standardized as ISO 226 in 1986. In 2003, ISO 226 was revised as equal-loudness contour using data collected from 12 international studies." (WK 2010d)

(c) Temporal Masking

The third limit of the human perception of sound has to do "temporal masking" (or "non-simultaneous masking"), which "occurs when a sudden stimulus sound makes inaudible other sounds which are present immediately preceding or following the stimulus." (WK 2010l)

And "masking which obscures a sound immediately preceding the masker is called *backward masking* or *pre-masking* and masking which obscures a sound immediately following the masker is called *forward masking* or *post-masking*. Temporal masking's effectiveness attenuates exponentially from the onset and offset of the masker, with the onset attenuation lasting approximately 20 ms and the offset attenuation lasting approximately 100 ms." (WK 2010l)

(d) Simultaneous Masking

And the fourth limit of the human perception of sound has to do "simultaneous masking," which refers to "masking between two concurrent sounds. Sometimes called frequency masking or spectral masking since it is often observed when the sounds share a frequency band; e.g., two sine tones at 440 Hz and 450Hz can be perceived clearly when separated. They cannot be perceived clearly when simultaneous." (WK 2010m)

For instance, S. Gelfand once gave a good example to illustrate this phenomenon, in that, "let us say that for a given individual, the sound of a cat scratching a post in an otherwise quiet environment is first audible at a level of 10 decibels (dB). However, in the presence of a 'masking' noise—for example, a vacuum cleaner that is running simultaneously—that same individual cannot detect the sound of the cat scratching unless the level of the scratching sound is at least 26 dB. We would say that the unmasked threshold for that individual for the target sound (i.e., the cat scratching) is 10 dB, while the masked threshold is 26 dB. The amount of masking is simply the difference between these two thresholds—in our example, 16 dB." (WK 2010m)

A Response from the Critics

However, the psychological analysis of the limits of human hearing is not without its own problems. Consider a few of them below, for illustration, as summarized in *Table 2.2*.

Firstly, one major criticism of the psychological analysis of the limits of human hearing is that psychoacoustics focuses primarily on the "psychological and physiological responses associated with sound"—but not on other aspects of sound which fall outside the domains of psychology and physiology, for example, in relation to the societal, cultural, and natural contexts. (WK 2010d)

Secondly, another major criticism of the psychological analysis of the limits of human hearing is that, even among psychoacousticians, there are disagreements on different issues.

For instance, psychoacousticians disagree over the best methods to measure absolute threshold of hearing (ATH), since, over the years, there have been different "psychophysical methods" of measuring ATH, and good examples include (i) "classical methods," (ii) "modified classical methods," (iii) "forced-choice methods," and (iv) "adaptive methods," just to cite four main competing ones. (WK 2010k; I. Hirsh 1952; S. Gelfand 2004; H. Levitt 1971; J. Miller 2002)

And thirdly, still another major criticism of the psychological analysis of the limits of human hearing is that much work remains to be done to understand better the limits of human hearing.

For instance, although "[t]he lower limit of audibility is [clearly] defined as 0 dB," there is uncertainty over "the upper limit," which "is not as clearly defined," because "[t]he upper limit is more a question of the limit where the ear will be physically harmed or with the potential to cause noise-induced hearing loss." (WK 2010d)

A More Detached Perspective

These criticisms of psychoacoustics about the limits of human hearing do not mean to dismiss its usefulness but to help us learn from the two opposing sides of the debate, such that the possibility and desirability of sound (from the perspective of the mind with sound, perception limits, and the psychoacoustic model as a case study here) are not to the extent that the respective defenders on both sides would like us to believe.

In addition, the analysis of psychoacoustics about the limits of human hearing can teach us something valuable about the ontological principles in existential dialectics, and good examples include the absoluteness-relativeness principle, the predictability-unpredictability principle, the explicability-inexplicability principle, the preciseness-vagueness principle, the simpleness-complicatedness principle, the openness-hiddenness principle, the denseness-emptiness principle, the change-constancy principle, the slowness-quickness principle, the expansion-contraction principle, the theory-praxis principle, the convention-novelty principle, the evolution-transformation principle, the symmetry-asymmetry principle, the regression-progression principle, and the same-difference principle.

For instance, in relation to the absoluteness-relativeness principle, if there is absoluteness (e.g., the measurement of the absolute threshold of hearing by "the method of limits," with "several series of descending and ascending runs"), there is relativeness (e.g., what is true for "the method of limits" in its use of "descending and ascending runs" is not so for "the method of constant stimuli" with "no ascending or descending trials"). And the reverse direction also holds true.

In relation to the predictability-unpredictability principle, if there is predictability (e.g., the predictable tendency of a subject in "the method of constant stimuli" to respond either "yes" or "no" after each presentation of a sound stimuli), there is unpredictabiltiy (e.g., the unpredictability of the presentation of the sound stimuli, which is given by the tester "at completely random order" in "the method of constant stimuli"). And the reverse direction also holds true.

In relation to the explicability-inexplicability principle, if there is explicability (e.g, the explanation by psychoacousticians that humans have the audio range "between 20 Hz and 20,000 Hz"), there is inexplicability (e.g., the lack of sufficient explanation by psychoacousticians of why humans have this kind of audio range but other animals like "dogs can perceive vibrations higher than 20,000 Hz"). (WK 2010a) And the reverse direction also holds true.

In relation to the preciseness-vagueness principle, if there is preciseness (e.g., the precise identification by psychoacousticians that "[t]he lower limit of audibility is [clearly] defined as 0 dB"), there is vagueness (e.g., the vagueness in the identification, since psychoacousticians do not define clearly "the upper limit," which is treated instead as "a question of the limit where the ear will be physically harmed or with the potential to cause noise-induced hearing loss," which varies from person to person). (WK 2010d) And the reverse direction also holds true.

In relation to the simpleness-complicatedness principle, if there is simpleness (e.g., the relatively simple analysis of the limits of human hearing by psychoacousticians), there is complicatedness (e.g., the relatively more complicated analysis of the limits of human hearing by questioning the claims and assumptions used in psychoacoustics, as shown in the criticisms). And the reverse direction also holds true.

In relation to the openness-hiddenness principle, if there is openness (e.g., the open exploration of the relationship between the measurement of the absolute hearing threshold and the understanding of human "auditory system"), there is hiddenness (e.g., the hidden bias in the measurement by each method, since each has its own critieria, assumptions, and disadvantages). And the reverse direction also holds true.

In relation to the denseness-emptiness principle, if there is denseness (e.g., the relatively denser concentration of sound waves at higher frequenccies), there is emptiness (e.g., the relatively less dense, or more empty, concentration of sound waves at lower frequencies). And the reverse direction also holds true.

In relation to the change-constancy principle, if there is change (e.g., the ever new ways to measure absolute threshold of hearing, as shown by the use of different methods over time like "classical methods," "modified classical methods," "forced-choice methods," and "adaptive methods"), there is constancy (e.g., the ever constancy of the attempt to measure absolute threshold of hearing over time). And the reverse direction also holds true.

In relation to the order-chaos principle, if there is order (e.g., the relatively orderly arrangement of either "yes" or "no after each presentation of a sound stimuli in "the method of constant stimuli" to measure ATH), there is chaos (e.g., the relatively disorderly or random presentation of the sound stimuli, since they are presented "at completely random order" by the tester in "the method of constant stimuli"). And the reverse direction also holds true.

In relation to the slowness-quickness principle, if there is slowness (e.g., the relatively slower speed of someone to hear a sound when there is a masking by other sounds to obscure it), there is quickness (e.g., the relatively quicker speed of someone to hear a sound if there is no masking by other sounds to obscure it). And the reverse direction also holds true.

In relation to the expansion-contraction principle, if there is expansion (e.g., the relatively more developed ability of humans to hear the audio range "between 20 Hz and 20,000 Hz"), there is contraction (e.g., the relatively less developed ability of humans to hear sounds outside the audio range "between 20 Hz and 20,000 Hz"). And the reverse direction also holds true.

In relation to the theory-praxis principle, if there is theory (e.g., the theoretical construction of the limits of human hearing in psychoacoustics), there is praxis (e.g., the practical application of the theoretical construction of the limits of human hearing by acoustic engineers, who, "when compressing a digital signal,...can safely assume that any frequency beyond approximately 20 kHz will not have any effect on the perceived sound of the finished product, and thus use a high-shelving filter to cut everything outside this range"). (WK 2010j) And the reverse direction also holds true.

In relation to the convention-novelty principle, if there is convention (e.g., the conventional wisdom about measuring ATH by the "classical methods" as first proposed by Gustav Theodor Fechner in his work *Elements of Psychophysics* in the 19th century), there is novelty (e.g., the alternative novel challenge to the conventional wisdom about the "classical methods," by new ways of measuring ATH nowadays, like "forced-choice methods," and "adaptive methods," for instance). (WK 2010k; Hirsh 1952; S. Gelfand 2004) And the reverse direction also holds true.

In relation to the evolution-transformation principle, if there is evolution (e.g., the natural evolution of humans to hear the audio range "between 20 Hz and 20,000 Hz"), there is transformation (e.g., the technical transformation of human ability to hear sounds outside the audio range with the invention of instruments for ultrasonography and elastography to study "the ultrasonic frequency range," for instance). (WK 2010) And the reverse direction also holds true.

In relation to the symmetry-asymmetry principle, if there is symmetry (e.g., the co-existence of different ways to measure ATH, like "the method of constant stimuli" and "the method of limits"), there is asymmetry (e.g., "the method of constant stimuli" is better in one way, because its "catch trials" can "help to detect the amount of a listener's guessing"—but "the method of limits" is better in a different way, because it requires only a smaller number of trials, which can save time "required to complete the testing"). And the reverse direction also holds true.

In relation to the regression-progression principle, if there is regression (e.g., the regression made by the psychoacoustic model to understand sounds, as shown in the criticisms by the critics), there is progression (e.g., the progress made by the psychoacoustic model to help us understand the limits of human hearing). And the reverse direction also holds true.

And in relation to the same-difference principle, if there is similarity in outcome (e.g., the contribution to the molding and regulation of beliefs and values, regardless of whether this be done by way of understanding sounds from the scientific perspective of psychoacoustics or from the pre-scientific perspective of magic, shamanism, etc.), there is difference in outcome (e.g., the contribution to the molding and regulation of beliefs and values by way of understanding sounds from the scientific perspective of psychoacoustics for a more logical-rational worldview—but the contribution to the molding and regulation of beliefs and values by way of understanding sounds from the pre-scientific perspective of magic, shamanism, etc for a more religious worldview). And the reverse direction also holds true.

Sound, Noise Sensibility, and Extroversion-Introversion

Another way to evaluate the possibility and desirability of sound from the perspective of the mind is to explore sound, noise sensibility, and extroversion-introversion.

I suggest this relationship between noise sensibility and extroversion-introversion for analysis, because there has not been any discussion in the literature about this relationship (although introversion and extroversion are central traits in the study of human personality).

Extroversion, Introversion, and Ambiversion

In the literature on human personality, "[t]he terms introversion and extroversion were first popularized by Carl Jung [1921]. Virtually all comprehensive models of personality include these concepts. Examples include Jung's analytical psychology, Eysenck's three-factor model, Cattell's 16 personality factors, the Big Five personality traits, the four temperaments, the Minnesota Multiphasic Personality Inventory, the Myers Briggs Type Indicator, and Socionics." (WK 2010n)

It is important to note that "[e]xtroversion and introversion are typically viewed as a single continuum. Thus, to be high on one is necessarily to be low on the other. Carl Jung and the authors of the Myers-Briggs provide a different perspective and suggest that everyone has both an extroverted side and an introverted side, with one being more dominant than the other. In any case, people fluctuate in their behavior all the time, and even extreme introverts and extroverts do not always act according to their type." (WK 2010n)

With this note in mind—the single continuum has three traits for consideration, namely, "extroversion," "introversion," and "ambiversion," as shown below: (WK 2010n)

- "Extroversion"—refers to "the act, state, or habit of being predominantly concerned with and obtaining gratification from what is outside the self." Thus, "[e]xtroverts tend to enjoy human interactions and are generally enthusiastic, talkative, assertive, and gregarious. They take pleasure in activities that involve large social gatherings, such as parties, community activities, public demonstrations, and business or political groups. Politics, teaching, sales, managing and brokering are fields that favor extroversion. An extroverted person is likely to enjoy time spent with people and find less reward in time spent alone. They tend to be energized when around other people, and they are more prone to boredom when they are by themselves." (WK 2010n)

- "Introversion"—refers to "the state of or tendency toward being wholly or predominantly concerned with and interested in one's own mental life." In this sense, "[i]ntroverts are people whose energy tends to expand through reflection and dwindle during interaction. Introverts tend to be more reserved and less outspoken in large groups. They often take pleasure in solitary activities such as reading, writing, music, drawing, tinkering, playing video games, watching movies and plays, and using computers, along with some more reserved outdoor activities such as fishing….The archetypal artist, writer, sculptor, engineer, composer, and inventor are all highly introverted. An introvert is likely to enjoy time spent alone and find less reward in time spent with large groups of people, though he or she may enjoy interactions with close friends. Trust is usually an issue of significance: a virtue of utmost importance to an introvert choosing a worthy companion. They prefer to concentrate on a single activity at a time and like to observe situations before they participate, especially observed in developing children and adolescents. Introverts are easily overwhelmed by too much stimulation from social gatherings and engagement. They are more analytical before speaking. Introversion is not the same as shyness or the social outcast. Introverts choose solitary over social activities by preference, whereas shy people (who may be extroverts at heart) avoid social encounters out of fear and whereas the social outcast faces solitude, but not by choice." (WK 2010n; L. Helgoe 2008; GE 1998; M. Laney 2002; M. Whitten 2001)
- "Ambiversion"—refers to "people who fall more or less directly in the middle and exhibit tendencies of both groups. An ambivert is normally comfortable with groups and enjoys social interaction, but also relishes time alone and away from the crowd." (WK 2010n; D. Cohen 1979)

Of course, because of the continuum in the study, all three types differ from each other in terms of degree, not of kind.

The Question about Causes

But then, the next question to ask is about the causes of these three personality traits. There is no lack of theories to speculate on them. Consider a few below, for illustration.

Firstly, Carl Jung argued that "introversion and extroversion refer to the direction of psychic energy. If a person's psychic energy usually flows outwards then he or she is an extrovert, while if the energy usually flows inwards, the person is an introvert. Extroverts feel an increase of perceived energy when interacting with a large group of people, but a decrease of energy when left alone. Conversely, introverts feel an increase of energy when alone, but a decrease of energy when surrounded by a large group of people." (WK 2010n; TIME 1955)

Secondly, Hans Eysenck (1967) claimed that "extroversion-introversion" is "the degree to which a person is outgoing and interactive with other people. These behavioral differences are presumed to be the result of underlying differences in brain physiology. Extroverts seek excitement and social activity in an effort to heighten their arousal level, whereas introverts tend to avoid social situations in an effort to keep such arousal to a minimum." (WK 2010n)

Originally, Eysenck proposed that "extroversion was a combination of two major tendencies, impulsiveness and sociability. He later added several other more specific traits, namely liveliness, activity level, and excitability. These traits are further linked in his personality hierarchy to even more specific habitual responses, such as partying on the weekend." (WK 2010n)

And thirdly, some twin studies found "a genetic component of 39% to 58%. In terms of the environmental component, the shared family environment appears to be far less important than individual environmental factors that are not shared between siblings." (WK 2010n; A. Tellegen 1988)

Another study found that "introverts have more blood flow in the frontal lobes of their brain and the anterior or frontal thalamus, which are areas dealing with internal processing, such as planning and problem solving. Extroverts have more blood flow in the anterior cingulate gyrus, temporal lobes, and posterior thalamus, which are involved in sensory and emotional experience. This study and other research indicates that introversion-extroversion is related to individual differences in brain function." (WK 2010n; D. Johnson 1999)

The Relationship between Personality and Noise Sensibility

These findings on the differences between extroverts and introverts allow me to propose that they do not react to sound (or noise) in the same way.

For illustration, extroverts tends to be more "impulsive," "social," "lively," "active," "partying," engaging in "public demonstrations," etc.

People who enjoy "partying" a lot, for instance, are more likely to live a less quiet life and more easily get excited by loud music, as shown in the song "Party All the Time" by Eddie Murphy in 1985 or "La Vida Loca" by Ricky Martin in 1999. By the same logic, people who enjoy participation in "public demonstrations" are more likely to get thrilled by their boisterous activities.

By contrast, people who are introverts are less likely to seek out excitment in loud parties (or public demonstrations) and more likely to stay alone in a quiet environment (or in "solitary activities" like "reading," "reflection," etc.).

In this way, there is a difference, at least in degree of noise sensibility, between extroverts and introverts.

And this is something that the literature on acoustics does not really address.

Siding with Neither

This suggestion of mine about the correlation between personality and noise sensibility does not mean to favor extroversion over introversion (or vice versa) but is to show the two opposing sides of the debate, such that the possibility and desirability of sound (from the perspective of the mind with sound, noise sensibility, and extroversion-introversion as a case study here) are not to the extent that the respective defenders on both sides would like us to believe.

More importantly, the analysis of sound, noise sensibility, and extroversion-introversion reveals something valuable about the ontological principles in existential dialectics, and good examples include the absoluteness-relativeness principle, the partiality-totality principle, the predictability-unpredictability principle, the explicability-inexplicability principle, the finiteness-transfiniteness principle, the preciseness-vagueness principle, the simpleness-complicatedness principle, the openness-hiddenness principle, the denseness-emptiness principle, the slowness-quickness principle, the expansion-contraction principle, the theory-praxis principle, the convention-novelty principle, the evolution-transformation principle, the symmetry-asymmetry principle, the seriousness-playfulness principle, the regression-progression principle, and the same-difference principle.

For instance, in relation to the absoluteness-relativeness principle, if there is absoluteness (e.g., the absolute view of an extrovert for more "social interaction" overall), there is relativeness (e.g., what is good for an extrovert for more "social interaction" overall is not necessarily so for an introvert for more "private space" overall). And the reverse direction also holds true.

In relation to the partiality-totality principle, if there is partiality (e.g., the more partial perspective of an introvert or, alternatively, an extrovert), there is totality (e.g., the more holistic view of all the perspectives of both introverts and extroverts, such that the whole is not the sum of its parts, in that they do not all add up into a harmonious entirety). And the reverse direction also holds true.

In relation to the predictability-unpredictability principle, if there is predictability (e.g., the predictable tendency of an extrovert to enjoy "partying"), there is unpredictability (e.g., the more difficult task to predict exactly which specific point on the single continuum of extroversion-introversion an extrovert is at a given time). And the reverse direction also holds true.

In relation to the explicability-inexplicability principle, if there is explicability (e.g., the explanation by Jung that the differences between extroversion and introversion have more to do with "a person's psychic energy"), there is inexplicability (e.g., the lack of sufficient explanation, if viewed within the framework of Jung, of why there cannot be other explanations on the basis of brain physiology as suggested by Eysenck or biological causes as proposed by others, for example). And the reverse direction also holds true.

In relation to the finiteness-transfiniteness principle, if there is finiteness (e.g., the finite number of categories on the single continuum, like "extroversion," "introversion," and "ambiversion"), there is transfiniteness (e.g., the transfinite number of points on the single continuum that any person can choose from, so as to identify with). And the reverse direction also holds true.

In relation to the preciseness-vagueness principle, if there is preciseness (e.g., the precise identification of "a genetic component of 39% to 58%" in some twin studies), there is vagueness (e.g., the vagueness in the identification, since it is not clear what the exact degree of the genetic basis is, for a given twin). And the reverse direction also holds true.

In relation to the simpleness-complicatedness principle, if there is simpleness (e.g., the relatively simple analysis of introversion and extroversion), there is complicatedness (e.g., the relatively more complicated analysis of introversion and extroversion by questioning the claims about different causes, which remain in contention). And the reverse direction also holds true.

In relation to the openness-hiddenness principle, if there is openness (e.g., the open exploration, by Jung, of the correlation between psychic energy and the differences between extroverts and introverts), there is hiddenness (e.g., the hidden bias in Jung's view, because of its focus on psychic energy, but not on other causes). And the reverse direction also holds true.

In relation to the denseness-emptiness principle, if there is denseness (e.g., the relatively denser concentration of extroverts in America, which is regarded as "an extroverted society"), there is emptiness (e.g., the relatively less dense, or more empty, concentration of extroverts in "Japan or regions where Buddhism, Sufism etc. prevail" and "prize introversion"). (WK 2010n; M. Laney 2002; S. Diamond 2008) And the reverse direction also holds true.

In relation to the slowness-quickness principle, if there is slowness (e.g., the relatively slower speed for an extrovert to get turned off by loud parties), there is quickness (e.g., the relatively quicker speed for an introvert to get turned off by loud parties). And the reverse direction also holds true.

In relation to the expansion-contraction principle, if there is expansion (e.g., the relatively more developed ability of an extrovert to excel in social interaction), there is contraction (e.g., the relatively less developed ability of an extrovert to excel in contemplation or meditation). And the reverse direction also holds true.

In relation to the theory-praxis principle, if there is theory (e.g., the theoretical construction of the extroversion-introversion single continuum by psychologists like Jung and others), there is praxis (e.g., the practical application of the extroversion-introversion theoretical construct to the work on the geographical variation of the traits in the United States by Peter Rentfrow [2008] and others). (WK 2010n) And the reverse direction also holds true.

In relation to the convention-novelty principle, if there is convention (e.g., the conventional wisdom about the differences between extroverts and introverts), there is novelty (e.g., the alternative novel view by some scholars nowadays, who argued that "[a]cknowledging that introversion and extroversion are normal variants of behavior can help in self-acceptance and understanding of others," as "an extrovert can accept his or her introverted partner's need for space, while an introvert can acknowledge his or her extroverted partner's need for social interaction"). (WK 2010n) And the reverse direction also holds true.

In relation to the evolution-transformation principle, if there is evolution (e.g., the natural evolution of humans to understand each other in terms of their different personalities in the state of nature), there is transformation (e.g,. the technical transformation of human understanding of each other by the technical invention of such personality models as "Eysenck's three-factor model, Cattell's 16 personality factors, the Big Five personality traits, the four temperaments, the Minnesota Multiphasic Personality Inventory, the Myers Briggs Type Indicator, and Socionics"). (WK 2010n) And the reverse direction also holds true.

In relation to the symmetry-asymmetry principle, if there is symmetry (e.g., the co-existence of different personality traits like extroversion and introversion), there is asymmetry (e.g., "an extroverted person is likely to enjoy time spent with people and find less reward in time spent alone"—but "[i]ntroverts are people whose energy tends to expand through reflection and dwindle during interaction"). And the reverse direction also holds true.

In relation to the seriousness-playfulness principle, if there is seriousness (e.g., an extrovert can be serious in political public demonstrations), there is playfulness (e.g., an extrovert can enjoy himself by having fun in loud parties, like in political public demonstrations). And the reverse direction also holds true.

In relation to the regression-progression principle, if there is regression (e.g., the regression made by each personality trait in relation to things which it may not excel in, like engaging in solitary activities by an extrovert), there is progression (e.g,. the progress made by each personality trait in relation to things which it may excel in instead, like involving in large social gatherings by an extrovert). And the reverse direction also holds true.

And in relation to the same-difference principle, if there is similarity in outcome (e.g., the contribution to the molding and regulation of beliefs, values and behaviors, regardless of whether this be done by way of extroversion or introversion), there is difference in outcome (e.g., the contribution to the molding and regulation of beliefs, values and behaviors by way of extroversion for a more social lifeworld—but the contribution to the molding and regulation of beliefs, values and behaviors by way of introversion for a more solitary lifeworld). And the reverse direction also holds true.

Sound and Nature

Sound, when examined from the perspective of nature, can cast an interesting light on its possibility and desirability, and this can be shown by way of two case studies, namely, (2.3.1) sound, animals, and survival and (2.3.2) sound, earthquakes, and seismology—to be addressed hereafter, respectively.

Sound, Animals, and Survival

The use of sounds is important for animals in different ways, especially in relation to their survival in the state of nature. Consider, for illustration, the amazing story of caterpillars when they whistle.

The Whistles of Caterpillars

It has been known "for more than 100 years that many caterpillars can generate clicking or squeaking noises," but "researchers have only recently begun to experimentally investigate how these noises are made and what roles they might play." (C. Choi 2010)

For instance, researchers at Queen's University in Kingston, Canada and Carleton University in Ottawa used "high-speed video" and found that caterpillars "pulled their heads back to compress the body cavity while they whistled. Unlike reptiles, birds and mammals, insects don't breathe using their mouths, but with holes in their sides known as spiracles, and the scientists reasoned they were forcing air out these holes to whistle, generating squeaking noises." (C. Choi 2010)

So, in order to confirm this hypothesis, "researcher Veronica Bura at Carleton University gently applied latex over all eight pairs of the caterpillars' abdominal spiracles and then uncovered each pair systematically while pinching the larva. The whistles definitely came from the eighth pair, generating trains of whistles lasting up to four seconds each, and spanning frequencies that ranged from those audible to birds and humans up to ultrasound." (C. Choi 2010)

But a good question to ask is, Why should caterpillars whistle, since making sound would attract predators instead? To answer this question, the researchers "studied captive yellow warblers (Dendroica petechia), a bird that is known to frequently eat caterpillars and lives where the walnut sphinx caterpillar does. The scientists put walnut sphinx caterpillars on twigs in cages with yellow warblers and patiently filmed the encounter. Surprisingly, when the birds attacked, the caterpillars whistled and the bird typically flinched and hopped or flew away. In tests with three warblers and two attacks each, the caterpillars got away completely unscathed." (C. Choi 2010)

The trick of whistling worked for the caterpillars to foil the attacks by the birds in the experiments, because, as Jayne Yack at Carleton University said, "[t]hese birds are clearly startled by the unexpected sounds coming out of this caterpillar. They dove for cover." (C. Choi 2010)

So, the researchers came to the conclusion that "[t]he sounds are probably not advertising that the walnut sphinx caterpillars are distasteful. The birds simply appear to be startled, 'because these sounds are unexpected,' Yack said." (C. Choi 2010)

The Sounds of Animals

But one should not have the wrong impression that only insects like caterpillars make sounds for survival. More advanced animals in the evolutionary tree do the same for their survival. Consider, for illustration, a few examples of some animals making sounds, in what follows.

(a) Bats

Bats are well-known to have amazing sound-detecting abilities. For instance, they "use a variety of ultrasonic ranging (echolocation) techniques to detect their prey. They can detect frequencies as high as 100,000 Hz, although there is some disagreement on the upper limit." (WK 2010o; J. Cancel 1998)

(b) Rodents/Insects

Rodents and insects also have their own remarkable abilities. For instance, "[t]here is evidence that ultrasound in the range emitted by bats causes flying moths to make evasive manoeuvres because bats eat moths. Ultrasonic frequencies trigger a reflex action in the noctuid moth that causes it to drop a few inches in its flight to evade attack. Tiger moths also emit clicks which jam bats' echolocation." (WK 2010o; G. Jones 2000; Y. Hui 2003)

(c) Dogs

Dogs, by contrast, "can hear sound at higher frequencies than humans can. A dog whistle exploits this by emitting a high frequency sound to call to a dog. Many dog whistles emit sound in the upper audible range of humans, but some, such as the silent whistle, emit ultrasound at a frequency in the range 18,000–22,000 Hz." (WK 2010o)

(d) Dolphins and whales

Fish can do some amazing tricks too. For instance, "[i]t is well known that some whales can hear ultrasound and have their own natural sonar system. Some whales use the ultrasound as a hunting tool (for both detection of prey and as an attack)." (WK 2010o; VQ 2010)

Other types of fish "can detect ultrasound. In the order Clupeiformes, members of the subfamily Alosinae (shad), have been shown to be able to detect sounds up to 180,000 Hz, while the other subfamilies (e.g. herrings) can hear only up to 4,000 Hz." (WK 2010o; D. Mann 2001)

A Note of Caution

But these findings about the sounds of animals should be treated with a note of caution, as they have their own problems. Consider, for illustration, a few criticisms below.

Firstly, one major criticism against these findings about the sounds of animals is that the analysis is a bit simplistic. For instance, while it is fascinating to know that some insects like noctuid moths can take some counter-measures like dropping a few inches in their flight to evade attack by bats, it is not clear if the bats can learn over time to take their own counter-measures to defeat the counter-measures of the moths; so the cycle of detection and counter-detection continues to the next round

Secondly, another major criticism against these findings about the sounds of animals is that a better understanding can also allow humans to control the animals better for exploitation. For instance, the better understanding of ultrasound now allows "the beef cattle industry to improve…the yield of cattle operations," by using ultrasound technology "to evaluate fat thickness, rib eye area, and intramuscular fat in living animals," for instance. (WK 2010o; D. Bennett 2005; W. Wagner 2010; A. McKinnon 1993)

And thirdly, still another major criticism against these findings about the sounds of animals is that it is also not clear to what extent the use of either ultrasound or audio sound by animals is successful for their own purposes. For instance, in the case about the caterpillars, the study showed that the birds "dove for cover" in the experiments when they heard the whistles of the caterpillars, but it is not clear to what extent the caterpillars are successful in real life. After all, bats need to eat, and moths are known to be one major source of their diets.

A Redemptive Lesson

But these criticisms against these findings about the sounds of animals do not mean to deny their usefulness but are to teach us the two opposing sides of the debate, such that the possibility and desirability of sound (from the perspective of nature with sound, animals, and survival as a case study here) are not to the extent that the respective defenders on both sides would like us to believe.

Moreover, the analysis of sound, animals, and survival reveals something relevant about the ontological principles in existential dialectics, and good examples include the absoluteness-relativeness principle, the predictability-unpredictability principle, the explicability-inexplicability principle, the finiteness-transfiniteness principle, the preciseness-vagueness principle, the simpleness-complicatedness principle, the openness-hiddenness principle, the denseness-emptiness principle, the slowness-quickness principle, the expansion-contraction principle, the theory-praxis principle, the convention-novelty principle, the evolution-transformation principle, the symmetry-asymmetry principle, the softness-hardness principle, the seriousness-playfulness principle, the regression-progression principle, and the same-difference principle.

For instance, in relation to the absoluteness-relativeness principle, if there is absoluteness (e.g., the absolute orientation of bats to eat moths), there is relativeness (e.g., what is good for bats to eat moths is not necessarily so for moths to be eaten by bats). And the reverse direction also holds true.

In relation to the predictability-unpredictability principle, if there is predictability (e.g., the predictable tendency of flying moths to take counter-measures like dropping a few inches in their flight to evade attack by bats), there is unpredictability (e.g., the more difficult task to predict when a particular counter-measure by a flying moth will necessarily succeed in fighting off the attack by a bat). And the reverse direction also holds true.

In relation to the explicability-inexplicability principle, if there is explicability (e.g., the explanation by the researchers in the two universities in Canada that "[t]he sounds are probably not advertising that the walnut sphinx caterpillars are distasteful" but that "[t]he birds simply appear to be startled"), there is inexplicability (e.g., the lack of sufficient explanation by the researchers of why this explanation must be true, since there are alternative explanations, for example, in that the birds may get scared by the sounds, that the sounds may disorient their flight, and so on). And the reverse direction also holds true.

In relation to the finiteness-transfiniteness principle, if there is finiteness (e.g., the finite number of the kinds of animals like bats and others which eat moths), there is transfiniteness (e.g., the transfinite number of individual moths which have been eaten by animals like bats and others, in the history of the earth). And the reverse direction also holds true.

In relation to the preciseness-vagueness principle, if there is preciseness (e.g., the precise identification of "a few inches in their flight" that a moth may drop to evade attack by a bat), there is vagueness (e.g., the vagueness in the identification, since it is not clear how many inches exactly are involved, or why only a few inches are necessary). And the reverse direction also holds true.

In relation to the simpleness-complicatedness principle, if there is simpleness (e.g., the relatively simple analysis of the sounds of animals for different purposes), there is complicatedness (e.g., the relatively more complicated analysis of the sounds of animals by asking some critical questions about the claims by researchers, as shown in the criticisms). And the reverse direction also holds true.

In relation to the openness-hiddenness principle, if there is openness (e.g., the open, or detectable, nature of moths which can be detected by the ultrasound in the range emitted by bats during their attempt to attack moths), there is hiddenness (e.g., the hidden, or evasive, nature of moths, like dropping a few inches in their flight to evade attack). And the reverse direction also holds true.

In relation to the denseness-emptiness principle, if there is denseness (e.g., the relatively denser concentration of moths to be the targets of being eaten by bats), there is emptiness (e.g., the relatively less dense, or more empty, concentration of lions to be the targets of being eaten by bats). And the reverse direction also holds true.

In relation to the slowness-quickness principle, if there is slowness (e.g., the relatively slower readiness of a flying noctuid moth to emit sounds to evade attack by a bat), there is quickness (e.g., the relatively quicker readiness of a flying noctuid moth to exercise "a reflex action…that cause[s] it to drop a few inches in its flight to evade attack"). And the reverse direction also holds true.

In relation to the expansion-contraction principle, if there is expansion (e.g., the relatively more developed ability of a flying noctuid moth to exercise "a reflex action…that cause[s] it to drop a few inches in its flight to evade attack"), there is contraction (e.g., the relatively less developed ability of a flying noctuid moth to emit sounds to evade attack by a bat). And the reverse direction also holds true.

In relation to the theory-praxis principle, if there is theory (e.g., the theoretical construction of the nature of ultrasound by acousticians), there is praxis (e.g., the practical application of the understanding of ultrasound to the beef cattle industry to improve…the yield of cattle operations"). And the reverse direction also holds true.

In relation to the convention-novelty principle, if there is convention (e.g., the conventional wisdom that many caterpillars make sounds), there is novelty (e.g., the alternative novel challenge to the conventional wisdom by the researchers in two Canadian universities who "experimentally investigate[d] how these noises are made and what roles they might play"). And the reverse direction also holds true.

In relation to the evolution-transformation principle, if there is evolution (e.g., the natural evolution of cattles to live in the state of nature, with some in good health and others in bad one), there is transformation (e.g., the technical transformation of cattles to survive, by the human invention of ultrasound technology "to improve…the yield of cattle operations"). And the reverse direction also holds true.

In relation to the symmetry-asymmetry principle, if there is symmetry (e.g., the co-existence of different counter-measures by moths to avoid being eaten by bats), there is asymmetry (e.g., the counter-measure by a flying noctuid moth to exercise "a reflex action…that cause[s] it to drop a few inches in its flight to evade attack"—but the counter-measure by a Tiger moth to "emit clicks which jam bats' echolocation"). And the reverse direction also holds true.

In relation to the softness-hardness principle, if there is softness (e.g., the use of ultrasound by whales to communicate with each other for mating), there is hardness (e.g., the use of ultrasound by whales to hunt and kill). And the reverse direction also holds true.

In relation to the seriousness-playfulness principle, if there is seriousness (e.g., the serious business of a male whale to make sounds to a female whale for reproduction), there is playfulness (e.g., the playful nature of a male whale to make sounds to a female whale for seduction). And the reverse direction also holds true.

In relation to the regression-progression principle, if there is regression (e.g., the regression made by the study of animal sounds, as shown by the criticisms about some major problems in the findings), there is progression (e.g., the regression made by the study of animal sounds, for a better understanding of animal communication and survival). And the reverse direction also holds true.

And in relation to the same-difference principle, if there is similarity in outcome (e.g., the contribution to the molding and regulation of behaviors, regardless of whether this be done by way of evasive manoeuvres or by way of charming manoeuvres), there is difference in outcome (e.g., the contribution to the molding and regulation of behaviors by way of evasive manoeuvres by a flying moth vis-à-vis a bat in the violent battle for survival—but the contribution to the molding and regulation of behaviors by way of charming manoeuvres by a male whale to a female whale in the seductive battle for reproduction). And the reverse direction also holds true.

Sound, Earthquakes, and Seismology

Another way to evaluate the possibility and desirability of sound from the perspective of nature is about sound, earthquakes, and seismology.

The word "seismology" refers to "the scientific study of earthquakes and the propagation of elastic waves through the Earth. The field also includes studies of earthquake effects, such as tsunamis as well as diverse seismic sources such as volcanic, tectonic, oceanic, atmospheric, and artificial processes (such as explosions). A related field that uses geology to infer information regarding past earthquakes is paleoseismology." (WK 2010p)

So, a good way to study sound and earthquakes is to understand the nature of the waves propagated through the Earth and its surface, namely, "seismic waves."

Seismic Waves

The term "seismic waves" refers to "elastic waves that propagate in solid or fluid materials. Because seismic waves commonly propagate efficiently and interact with internal structure, they provide the highest resolution noninvasive methods for studying Earth's interior." (WK 2010p)

There are two kinds of seismic waves through solids, namely, (a) "P-waves" (or "primary waves") and (b) "S-waves" (or "secondary waves"). (WK 2010p) The two kinds of seismic waves are also known as "body waves" through solids (as opposed to "interface waves" through solid-air, as explained below).

And a third kind of waves is known as "surface waves" (also known as "interface waves") along a solid-air interface, which have two kinds, namely, "Rayleigh and Love" waves; but "[i]nterface waves can be theoretically explained in terms of interacting P- and/or S-waves." (WK 2010p)

Let us consider these different kinds of waves more closely in what follows.

(a) Body Waves and Surface Waves

The first major type of seismic waves is "pressure waves" (or "primary waves," also known simply as P-waves"), which "are longitudinal waves that travel at maximum velocity within solids and are therefore the first waves to appear on a seismogram." (WK 2010p)

By contrast, the second major type of seismic waves is "S-waves" (or "transverse waves," "shear waves," or simply known as "secondary waves" instead), which are "transverse waves that travel more slowly than

P-waves and thus appear later than P-waves on a seismogram. Particle motion is perpendicular to the direction of wave propagation. Shear waves do not exist in fluids with essentially no shear strength, such as air or water." (WK 2010p)

Both "P-waves" and "S-waves" are "body waves," because they travel through solids under the Earth.

And the third type of seismic waves is "[s]urface waves" (also known as "interface waves") which "travel more slowly than P-waves and S-waves, but because they are guided by the surface of the Earth (and their energy is thus trapped near the Earth's surface) they can be much larger in amplitude than body waves, and can be the largest signals seen in earthquake seismograms. They are particularly strongly excited when the seismic source is close to the surface of the Earth, such as the case of a shallow earthquake or explosion." (WK 2010p)

(b) Seismic Waves and Earthquakes

Seismic waves are important to understand the nature of earthquakes, and they do not behave in the same way.

For instance, "[o]ne of the earliest important discoveries (suggested by Richard Dixon Oldham in 1906 and definitively shown by Harold Jeffreys in 1926) was that the outer core of the Earth is liquid." (WK 2010p)

This layer of liquid has implications for P-waves and S-waves. The reason is that "S-waves" which "shake side-to-side" need "rigid material so they do not pass through the outer core. Thus, the liquid core causes a 'shadow' on the side of the planet opposite of the earthquake where no direct S-waves are observed." (WK 2010p)

In addition, when "P-waves...pass through the core," the subsequent "reduction in P-wave velocity of the outer core also causes a substantial delay for P waves penetrating the core from the (seismically faster velocity) mantle." (WK 2010p)

Yet, "[u]sing seismic tomography with earthquake waves, the interior of the Earth has been completely mapped to a resolution of several hundred kilometers. This process has enabled scientists to identify convection cells, mantle plumes and other large-scale features of the inner Earth." (WK 2010p)

Earthquake Prediction and Its Problems

All these findings are useful for earthquake prediction, but there are also unresolved problems.

(a) Earthquake Prediction

Nowadays, "seismographs are instruments that sense and record the motion of the Earth. Networks of seismographs today continuously monitor the seismic environment of the planet, allowing for the monitoring and analysis of global earthquakes and tsunami warnings, as well as recording a variety of seismic signals arising from non-earthquake sources ranging from explosions (nuclear and chemical), to pressure variations on the ocean floor induced by ocean waves (the global microseism), to cryospheric events associated with large icebergs and glaciers." (WK 2010p)

In fact, "[o]ne of the first attempts at the scientific study of earthquakes followed the 1755 Lisbon earthquake. Other especially notable earthquakes that spurred major developments in the science of seismology include the 1857 Basilicata earthquake, 1906 San Francisco earthquake, the 1964 Alaska earthquake and the 2004 Sumatra-Andaman earthquake." (WK 2010p)

(b) Prediction Problems

However, there are also problems with "earthquake prediction," which refers, in a technical sense, to the "forecasting of a probable timing, location, magnitude and other important features of a forthcoming seismic event." (WK 2010p)

Consider, for illustration, a few problems below.

Firstly, one major problem with earthquake prediction is that it is so difficult to offer timely warnings well in advance for preparation. In fact, "[m]ost seismologists do not believe that a system to provide timely warnings for individual earthquakes has yet been developed, and many believe that such a system would be unlikely to give significant warning of impending seismic events." (WK 2010p)

Secondly, another major problem with earthquake prediction is that there are emergent phenomena in nature which are not entirely predictable, as uncertainty has a role to play too, and this is something that I already went to great lengths to explain in *The Future of Complexity* (2007) and *The Future of Post-Human Geology* (2010), for instance.

This problem remains, even though "[m]ore general forecasts, however, are routinely used to establish seismic hazard." (WK 2010p)

And thirdly, still another major problem with earthquake prediction is that they are not very precise. For instance, "various attempts have been made by seismologists and others to create effective systems for precise earthquake predictions, including the VAN method. Such methods have yet to be generally accepted in the seismology community," because they are not very accurate. (WK 2010p)

From a Broader Perspective

But these criticisms against earthquake prediction should not be misleadingly misconstrued to dismiss its usefulness but are to show us the two opposing sides of the debate, such that the possibility and desirability of sound (from the perspective of nature with sound, earthquakes, and seismology as a case study here) are not to the extent that the respective defenders would on both sides like us to believe.

More fundamentally, the analysis of sound, earthquakes, and seismology can teach us something refreshing about the ontological principles in existential dialectics, and good examples include the absoluteness-relativeness principle, the predictability-unpredictability principle, the explicability-inexplicability principle, the preciseness-vagueness principle, the simpleness-complicatedness principle, the openness-hiddenness principle, the denseness-emptiness principle, the slowness-quickness principle, the expansion-contraction principle, the theory-praxis principle, the convention-novelty principle, the evolution-transformation principle, the symmetry-asymmetry principle, the regression-progression principle, and the same-difference principle.

For instance, in relation to the absoluteness-relativeness principle, if there is absoluteness (e.g., the absolute tendency of "S-waves" which do not travel "in fluids with essentially no shear strength, such as air or water"), there is relativeness (e.g., what is true for "S-waves" which travel only in solids is not so for "interface waves" which travel through solid-air). And the reverse direction also holds true.

In relation to the predictability-unpredictability principle, if there is predictability (e.g., the predictable tendency of seismologists to engage in "[m]ore general forecasts" about "seismic hazard"), there is unpredictability (e.g., the more difficult task for seismologists to make forecasts about "seismic hazard" which are very accurate and very timely). And the reverse direction also holds true.

In relation to the explicability-inexplicability principle, if there is explicability (e.g., the explanation by seismologists that they understand the interior of the Earth better nowadays, because of "seismic tomography with earthquake waves," which allowed them, for instance, "to identify convection cells, mantle plumes and other large-scale features of the inner Earth"), there is inexplicability (e.g., the lack of sufficient explanation by seismologists of why they are still unable to predict earthquakes very accurately and very timely, if they think that they understand the interior of the Earth so well—since, for the critics, there can still be a lot of things that the sesimologists do not quite understand). And the reverse direction also holds true.

In relation to the preciseness-vagueness principle, if there is preciseness (e.g., the precise identification by earthquake engineers of some forecasts which are to estimate "the probability of…a particular size affecting a particular location within a particular time span"), there is vagueness (e.g., the vagueness in this identification, since "[s]uch methods have yet to be generally accepted in the seismology community," as they are not very accurate). (WK 2010p) And the reverse direction also holds true.

In relation to the simpleness-complicatedness principle, if there is simpleness (e.g., the relatively simple analysis of earthquake prediction by earthquake engineers), there is complicatedness (e.g., the relatively more complicated analysis of earthquake prediction by the critics who question the assumptions and claims by earthquake engineers, as shown in the criticisms). And the reverse direction also holds true.

In relation to the openness-hiddenness principle, if there is openness (e.g., the open exploration of the correlation between earthquake prediction and the understanding of the interior of the Earth), there is hiddenness (e.g., the hidden bias in this supposed correlation, because of the assumptions and claims which have been questioned by the critics, such that earthquake prediction is not what the proponents would like us to believe). And the reverse direction also holds true.

In relation to the denseness-emptiness principle, if there is denseness (e.g., the relatively denser concentration of S-waves in solids), there is emptiness (e.g., the relatively less dense, or more emtpy, concentration of "S-waves" in liquids). And the reverse direction also holds true.

In relation to the slowness-quickness principle, if there is slowness (e.g., the relatively slower speed of S-waves to travel through solids, when compared with P-waves), there is quickness (e.g., the relatively faster speed of P-waves to travel through solids, when compared with S-waves). And the reverse direction also holds true.

In relation to the expansion-contraction principle, if there is expansion (e.g., the relatively more developed ability of S-waves to travel through solids), there is contraction (e.g., the relatively less developed ability of S-waves to travel through liquids). And the reverse direction also holds true.

In relation to the theory-praxis principle, if there is theory (e.g., the theoretical construction of body waves and surface waves in seismology), there is praxis (e.g., the practical application of the knowledge of body waves and surface waves to the field of earthquake engineering for earthquake prediction). And the reverse direction also holds true.

In relation to the convention-novelty principle, if there is convention (e.g,. the conventional wisdom about the core of the Earth as solids), there is novelty (e.g., the alternative novel challenge to this conventional wisdom when "Richard Dixon Oldham in 1906 and…Harold Jeffreys in 1926" discovered that "the outer core of the Earth is liquid"). And the reverse direction also holds true.

In relation to the evolution-transformation principle, if there is evolution (e.g., the natural evolution of humans in the state of nature to sense earthquakes when they occurred), there is transformation (e.g., the technical transformation of human understanding of earthquakes by the invention of seismographs for earthquake prediction before they even occurred). And the reverse direction also holds true.

In relation to the symmetry-asymmetry principle, if there is symmetry (e.g., the existence of body waves and surface waves), there is asymmetry (e.g., body waves travel through solids—but surface waves travel through solid-air). And the reverse direction also holds true.

In relation to the regression-progression principle, if there is regression (e.g., the regression made by earthquake engineers, as shown by the problems pointed out by the critics), there is progression (e.g., the progress made by earthquake engineers for general earthquake forecasts). And the reverse direction also holds true.

And in relation to the same-difference principle, if there is similarity in outcome (e.g., the contribution to the molding and regulation of beliefs and values, regardless of whether this be done by way of earthquake engineering or by way of shamanism), there is difference in outcome (e.g., the contribution to the molding and regulation of beliefs and values by way of earthquake engineering for a more secular lifeworld—but the contribution to the molding and regulation of beliefs and values by way of shamanism for a more sacred lifeworld). And the reverse direction also holds true.

Sound and Society

Sound, when examined from the perspective of society, reveals to us some powerful features in relation to its possibility and desirability, and this can be shown by way of two case studies, namely, (2.4.1) sound, underwater acoustics, and naval warfare, and (2.4.2) sound, small talk, and social skills—to be addressed in what follows, in that order.

Sound, Underwater Acoustics, and Naval Warfare

An excellent case study to understand sound from the perspective of society has to do with sound, underwater acoustics, and naval warfare.

As a start, the word "underwater acoustics" (or "hydroacoustics") refers to "the study of the propagation of sound in water and the interaction of the mechanical waves that constitute sound with the water and its boundaries. The water may be in the ocean, a lake or a tank." (WK 2010q)

Generally speaking, "frequencies associated with underwater acoustics are between 10 Hz and 1,000,000 Hz. The propagation of sound in the ocean at frequencies lower than 10 Hz is usually not possible without penetrating deep into the seabed, whereas frequencies above 1,000,000 Hz are rarely used because they are absorbed very quickly." (WK 2010q)

As a field of study, "underwater acoustics is closely related to a number of other fields of acoustic study, including sonar, transduction, acoustic signal processing, acoustical oceanography, bioacoustics, and physical acoustics." (WK 2010q)

Sound in Water

Sound in water has some distinctive features. Consider, for illustration, a few of them below, as summarized in *Table 2.3*.

(a) Sound Waves

The first feature of sound in water is that its "sound wave propagating underwater consists of alternating compressions and rarefactions of the water. These compressions and rarefactions are detected by a receiver, such as the human ear or a hydrophone, as changes in pressure. These waves may be man-made or naturally generated." (WK 2010q)

(b) Sound Absorption

The second feature of sound in water is that sound absorption (or attenuation) at low frequency is "weak," while "additional contributions at lower frequency in seawater are associated with the ionic relaxation of boric acid (up to c. 10,000 Hz) and magnesium sulfate (c. 10,000 Hz-500,000 Hz)." (WK 2010q; R. Francois 1982)

On the other hand, "[t]he main cause of sound attenuation in fresh water, and at high frequency in sea water (above 100,000 Hz) is viscosity." (WK 2010q)

In addition, "[s]ound may be absorbed by losses at the fluid boundaries. Near the surface of the sea losses can occur in a bubble layer or in ice, while at the bottom sound can penetrate into the sediment and be absorbed." (WK 2010q; R. Francois 1982)

(c) Sound Reflection/Scattering

The third feature of sound in water is that sound can reflect and scatter, both (c.1) in boundary interactions and (c.2) at target.

(c.1) In relation to boundary interactions, "the water surface and bottom are reflecting and scattering boundaries." (WK 2010q)

At the water surface—"the sea-air surface can be thought of as a perfect reflector. The impedance contrast is so great that little energy is able to cross this boundary….At high frequency (above about 1,000 Hz) or when the sea is rough, some of the incident sound is scattered." (WK 2010q) Another contribution is "the presence of wind generated bubbles or fish close to the sea surface. The bubbles can also form plumes that absorb some of the incident and scattered sound, and scatter some of the sound themselves." (WK 2010q; D. Weston 1989; G. Norton 2001)

And on the seabed—"[t]he acoustic impedance mismatch between water and the bottom is generally much less than at the surface and is more complex. It depends on the bottom material types and depth of the layers. Theories have been developed for predicting the sound propagation in the bottom" by N. Chotiros (1995) and M. Buckingham (2000). (WK 2010q)

(c.2) In relation to target, "[t]he reflection of sound at a target whose dimensions are large compared with the acoustic wavelength depends on its size and shape as well as the impedance of the target relative to that of water. Formulae have been developed for the target strength of various simple shapes as a function of angle of sound incidence. More complex shapes may be approximated by combining these simple ones." (WK 2010q; R. Urick 1983)

(d) Sound Propagation

The fourth feature of sound in water is that "[u]nderwater acoustic propagation depends on many factors," like speed, depth, temperature, range, etc. (WK 2010q)

In relation to speed—"[t]he direction of sound propagation is determined by the sound speed gradients in the water. In the sea the vertical gradients are generally much larger than the horizontal ones." (WK 2010q)

In relation to temperature—"[a]t equatorial and temperate latitudes in the ocean the surface temperature is high enough to reverse the pressure effect, such that a sound speed minimum occurs at depth of a few hundred metres. The presence of this minimum creates a special channel known as Deep Sound Channel, previously known as the SOFAR (sound fixing and ranging) channel, permitting guided propagation of underwater sound for thousands of kilometres without interaction with the sea surface or the seabed." (WK 2010q)

And in relation to range—"as sound propagates underwater there is a reduction in the sound intensity over increasing ranges, though in some circumstances a gain can be obtained due to focussing. Propagation loss (sometimes referred to as transmission loss) is a quantitative measure of the reduction in sound intensity between two points, normally the sound source and a distant receiver….At short range the propagation loss is dominated by spreading while at long range it is dominated by absorption and/or scattering losses." (WK 2010q)

Sound for Naval Warfare

The understanding of the main features of sound in water is not an idle exercise, since it has been used by the military in different countries for naval warfare, with the use of submarines as an excellent case in point. (WK 2010r)

As countries increasingly use submarines for their navies, "submarine navigation undercover" needs to make good use of undercover acoustics, and "[t]he challenges of underwater navigation have become more important as submarines spend more time underwater, travelling greater distances and at higher speed." (WK 2010r)

After all, "[m]ilitary submarines travel underwater in an environment of total darkness with neither windows nor lights. Operating in stealth mode, they cannot use their active sonar systems to ping ahead for underwater hazards such as undersea mountains, drilling rigs or other submarines. Surfacing to obtain navigational fixes is precluded by pervasive anti-submarine warfare detection systems such as radar and satellite surveillance. Antenna masts and antenna-equipped periscopes can be raised to obtain navigational signals but in areas of heavy surveillance, only for a few seconds or minutes; current radar technology can detect even a slender periscope while submarine shadows may be plainly visible from the air." (WK 2010r; A. Bivens 2004)

Submarine navigation can be classified into two types for consideration, namely, (a) "surface and near-surface navigation, and (b) "deep water navigation," as shown below and summarized in *Table 2.3*): (WK 2010r)

(a) Surface and Near-Surface Navigation

The surface and near-surface navigtion refers to "the surface" or the "periscope depth," where submarines travel, with the use of the following methods to find out their position: (WK 2010r)

- "Satellite navigation"
  - "Global positioning system (GPS)—by entering waypoints internally, able to navigate at a more precise level."
  - "NAVSAT"
- "Terrestrial radio-based navigation systems; largely superseded by satellite systems"
  - "LORAN—seldom if rarely used anymore."
  - "CHAYKA, the Russian counterpart of LORAN"
  - "OMEGA, the Western counterpart of the Alpha Navigation System, no longer in use"
  - "Alpha, the Russian counterpart of the Omega Navigation System"
- "Celestial navigation using the periscope, or sextant—seldom used anymore due to advancement in technology"
- "Radar navigation"—"radar signals are easily detected so radar is normally only used in friendly waters entering and exiting ports. With the implementation a more advanced radar system, many new techniques have been implemented in this process."
- "Active sonar"—"like radar, active sonar systems are readily detected, so active sonar is usually used only entering and exiting ports."
- "Pilotage"—"in coastal and internal waters, surfaced submarines rely on the standard system of navigational aids (buoys, navigational markers, lighthouses, etc.), utilizing the periscopes for obtaining lines of position to plot a triangulation fix."
- "Voyage Management System"—"referred to as the VMS, utilizes digital charts with other external sources fed in, to establish the ship's position. Other information may also be entered in manually in establishing a high quality fix or position."

Of course, there can be other methods, but these ones as cited above are the most important.

(b) Deep Water Navigation

By contrast, deep water navigation refers to "below periscope depth," where "submarines determine their position" by using the following methods: (WK 2010r; UK 2010; JCS 2003; S. Hamn 1995)

- "Dead reckoning course information obtained from the ship's gyrocompass" provides "measured speed and estimates of local ocean currents," so "this could also be considered an estimated position as long as the ocean current is computed in."

- "Inertial navigation system" is "an estimated position source, utilizes acceleration, deceleration, and pitch and roll in computing."
- "Bottom contour navigation may be used in areas where detailed hydrographic data has been charted and there is adequate variation in sea floor topography. Fathometer depth measurements are compared to charted depth patterns."

Again, there can be other methods, but these ones here are the most important.

Problems with Submarine Navigation

But submarine navigation is not without its own problems. Consider, for illustration, a few of them below, as summarized in *Table 2.3*.

Firstly, one major problem with submarine navigation is that it does not always work, because submarines collide underwater from time to time with other submarines, drilling rigs, undersea mountains, and so on.

For instance, even tthe USS San Francisco (SSN-711), known for its quietness underwater, "suffered substantial damage after colliding at high-speed with an undersea mountain" in 2005. (WK 2010r)

Secondly, another major problem with submarine navigation is that submarines also become the target of "anti-submarine warfare detection systems such as radar and satellite surveillance." (WK 2010r)

And thirdly, still another major problem with submarine navigation is that submarines can be sunk by anti-submarine weapons and can also suffer substantial damage from internal explosion or from fire onboard sometimes.

Favoring No One

But these criticisms against submarine navigation do not mean to totally reject its effectiveness for naval warfare but are to show us the two opposing sides of the debate, such that the possibility and desirability of sound (from the perspective of society with sound, underwater acoustics, and naval warfare as a case study here) are not to the extent that the respective defenders on both sides would like us to believe.

Moreover, the analysis of sound, underwater acoustics, and naval warfare reveals something important about the ontological principles in existential dialectics, and good examples include the absoluteness-relativeness principle, the predictability-unpredictability principle, the explicability-inexplicability principle, the preciseness-vagueness principle, the simpleness-complicatedness principle, the openness-hiddenness principle, the denseness-emptiness principle, the change-constancy principle, the slowness-quickness principle, the expansion-contraction principle, the theory-praxis principle, the convention-novelty principle, the evolution-transformation principle, the symmetry-asymmetry principle, the softness-hardness principle, the regression-progression principle, and the same-difference principle.

For instance, in relation to the absoluteness-relativeness principle, if there is absoluteness (e.g., the absolute tendency of submarines to not use active sonar systems and radar navigation when in non-friendly waters), there is relativeness (e.g., what is true in non-friendly waters when submarines do not use active sonar systems and radar navigation in order to avoid detection is not necessarily so in friendly waters when submarines can use active sonar systems and radar navigation to enter and exit ports). And the reverse direction also holds true.

In relation to the predictability-unpredictability principle, if there is predictability (e.g., the predictable tendency of submarines to operate in stealth mode when in non-friendly waters, in order to avoid detection), there is unpredictability (e.g., the more difficult task to predict exactly which particular trip that a submarine takes in non-friendly waters will necessarily be successful, since it can collide underwater from time to time with other submarines, drilling rigs, undersea mountains, and so on). And the reverse direction also holds true.

In relation to the explicability-inexplicability principle, if there is explicability (e.g., the explanation that sound reflection/scattering on the seabed "depends on the bottom material types and depth of the layers," as proposed by the theories developed by N. Chotiros and M. Buckingham, for instance), there is inexplicability (e.g., the lack of sufficient explanation in the theories of why there are not other important factors in determining sound reflection/scattering on the seabed, except material types and depth of the layers). And the reverse direction also holds true.

In relation to the preciseness-vagueness principle, if there is preciseness (e.g., the precise identification of the seven important methods used by submarines for surface and near-surface navgiation), there is vagueness (e.g., the vagueness in the identification of the seven important methods, since it is not clear why there are seven of them, not nine, ten, and so on—or which one is the best overall, etc.). And the reverse direction also holds true.

In relation to the simpleness-complicatedness principle, if there is simpleness (e.g., the relatively simple formulae which "have been developed for the target strength of various simple shapes as a function of angle of sound incidence"), there is complicatedness (e.g., the relatively more complex formulae for more complex shapes which "may be approximated by combining these simple ones"). And the reverse direction also holds true.

In relation to the openness-hiddenness principle, if there is openness (e.g., the open operation by submarines in friendly waters, by using active sonar systems and radar navigation, which can be easily detected), there is hiddenness (e.g., the stealth mode in which submarines travel when in non-friendly waters, in order to avoid detection). And the reverse direction also holds true.

In relation to the denseness-emptiness principle, if there is denseness (e.g., the relatively denser concentration of submarines using active sonar systems and radar navigation in friendly waters), there is emptiness (e.g., the relatively less dense, or more empty, concentration of submarines using active sonar systems and radar navigation in non-friendly waters). And the reverse direction also holds true.

In relation to the change-constancy principle, if there is change (e.g., the ever new methods used by submarines for navigation over time), there is change (e.g., the ever constancy of the availability of methods to be used by submarines for navigation). And the reverse direction also holds true.

In relation to the slowness-quickness principle, if there is slowness (e.g., the relatively slower speed of old submarines in the early days when traveling in deep water), there is quickness (e.g., the relatively quicker speed of new submarines nowadays when traveling in deep water). And the reverse direction also holds true.

In relation to the expansion-contraction principle, if there is expansion (e.g., the relative expansion of the influence of "global positioning system" or GPS for submarine navigation nowadays), there is contraction (e.g., the relative contraction or decline of the influence of "terrestrial radio-based navigation systems" for submarine navigation nowadays). And the reverse direction also holds true.

In relation to the theory-praxis principle, if there is theory (e.g., the theoretical construction of global positioning system or GPS), there is praxis (e.g., the practical application of GPS to submarine navigation for naval warfare). And the reverse direction also holds true.

In relation to the convention-novelty principle, if there is convention (e.g., the conventional wisdom for ships in coastal and internal waters to rely on "the standard system of navigational aids...[like] buoys, navigational markers, lighthouses, etc."), there is novelty (e.g., the alternative novel challenge to this standard system by the new way of satellite navigation). And the reverse direction also holds true.

In relation to the the evolution-transformation principle, if there is evolution (e.g., the natural evolution of humans to live on land, unlike fish living in water), there is transformation (e.g., the technical transformation of human ability to live in water by the invention of submarines). And the reverse direction also holds true.

In relation to the symmetry-asymmetry principle, if there is symmetry (e.g., the co-existence of different navigation systems, like OMEGA and Alpha), there is asymmetry (e.g., the preferred usage of OMEGA in the West—but the preferred usage of Alpha in Russia). And the reverse direction also holds true.

In relation to the softness-hardness principle, if there is softness (e.g., the use of GPS for civilian communication), there is hardness (e.g., the use of GPS for navel warfare). And the reverse direction also holds true.

In relation to the regression-progression principle, if there is regression (e.g., the regression made by submarine navigation, as shown in the problems pointed out by the critics), there is progression (e.g., the progress made by submarine navigation for naval warfare). And the reverse direction also holds true.

And in relation to the same-difference principle, if there is similarity in outcome (e.g., the contribution to the molding and regulation of beliefs and behaviors, regardless of whether this be done by way of stealthy submarine navigation or by way of open commercial ship navigation), there is difference in outcome (e.g., the contribution to the molding and regulation of beliefs and behaviors by way of stealthy submarine navigation for a more secretive lifeworld—but the contribution to the molding and regulation of beliefs and behaviors by way of open commercial ship navigation for a more transparent lifeworld). And the reverse direction also holds true.

Sound, Small Talk, and Social Skills

Another way to evaluate the possibility and desirability of sound from the perspective of society has to do with sound, small talk, and social skills.

The term "small talk" here refers to a way to break the "awkward silence" by making some conversational noise about trivia (which is not to mean much of anything), that is, a kind of "conversation for its own sake" (or "comments on what is perfectly obvious."), in that "[i]t is an informal type of discourse that does not cover any functional topics of conversation or any transactions that need to be addressed." (WK 2010s & 2010t; B. Malinowski 1923)

Th discourse on small talk "was initially studied in 1923 by Bronisław Malinowski, who coined the term *phatic communion* to describe it. The ability to conduct small talk is a social skill." (WK 2010s)

Social Functions of Small Talk

Small talk, as a way to break "the awkward silence" in making some conversational noise so understood, serves some social functions, as summarized in *Table 2.4*. (WK 2010t)

In other words, "small talk is a bonding ritual and a strategy for managing interpersonal distance. It serves many functions in helping to define the relationships between friends, work colleagues, and new acquaintances." (WK 2010s; T. Bickmore 1999)

Hereafter are three major examples of the social functions of small talk: (WK 2010s)

- "In particular, it helps new acquaintances to explore and categorize each other's social position." (J. Laver 1975)
- "Small talk is closely related to the need for people to maintain positive face—to feel approved-of by those who are listening to them."
- "It lubricates social interactions in a very flexible way, although the desired function is often dependent on the point in the conversation at which the small talk occurs." (J. Holmes 2000)

There are three different points in a conversation at which small talk occurs, as shown below. (WK 2010s)

- "At the Beginning of a Conversation"—this is "[w]here there is already a relationship between the two talkers, their small talk serves as a gentle introduction before engaging in more functional topics of conversation. It allows them to signal their own mood and to sense the mood of the other person. In a business meeting, it enables people to establish each other's reputation and level of expertise. When the talkers do not know each other, it allows them to show that they have friendly intentions and desire some sort of positive interaction."
- "At the End of a Conversation"—this is where "[s]uddenly ending an exchange may risk appearing to reject the other person. Small talk can be used to mitigate that rejection, affirm the relationship between the two people, and soften the parting."
- "As a Space Filler"—this is where "[in] many cultures, silences between two people are usually considered uncomfortable. Tension

can be reduced by starting phatic talk until a more substantial subject arises."

Surely, there are exceptions, and a good example concerns "couples in an intimate relationship," which "can signal their level of closeness by a lack of small talk. They can comfortably accept silence in circumstances that would be uncomfortable for two people who were only casual friends." (WK 2010s; A. Jaworski 2000)

Topics of Small Talk

Besides certain social functions, small talk has certain topics, which "usually depend…on any pre-existing relationship between the two people, and the circumstances of the conversation. In either case, someone initiating small talk will tend to choose a topic for which they can assume a shared background knowledge, to prevent the conversation being too one-sided." (WK 2010s; J. Holmes 2000a)

There are two kinds of topics here, namely, "either direct or indirect. Direct topics include personal observations such as health or looks. Indirect topics refer to a situational context such as the latest news, or the conditions of the communicative situation." (WK 2010s; E. Ventola 1979)

In any event, hereafter are four "safe" topics under "most circumstances": (WK 2010s; J. Holmes 2000)

- Ex: "The weather"
- Ex: "Recent shared experiences, for example *'Good party last night, wasn't it?'*"
- Ex: "Television and films"
- Ex: "Sports"

However, there is a caveat here, in that "[t]he level of detail offered should not overstep the bounds of interpersonal space. When asked, *'How are you?'* by an acquaintance they do not know well, a person is likely to choose a simple, generalised reply such as, *'Fine, thank you.'* In this circumstance it would probably not be appropriate for them to reply with a list of symptoms of any medical conditions they were suffering from. To do so would assume a greater degree of familiarity between the two people than is actually the case, and this may create an uncomfortable situation." (WK 2010s; J. Holmes 2000a)

Patterns of Small Talk

In addition, small talk follows some recognizable patterns. For instance, "Klaus Schneider [1988]...theorises that such a conversation consists of a number of fairly predictable segments, or 'moves,'" as shown below: (WK 2010s)

- "The First Move"—"is usually phrased so that it is easy for the other person to agree. It may be either a question, or a statement of opinion with a tag question. For example, an opening line such as *'Lovely weather, isn't it?'* is a clear invitation for agreement.
- "The Second Move"—"is the other person's response. In functional conversations that address a particular topic, [H.] Grice's Maxim of Quantity [1975] suggests that responses should contain no more information than was explicitly asked for. Schneider claims that one of the principles of small talk contradicts the Maxim of Quantity. He suggests that politeness in small talk is maximised by responding with a more substantial answer. Going back to the example of *'Lovely weather, isn't it?'*, to respond by just saying *'Yes'* is less polite than saying, *'Yes, very mild for the time of year.'*"
- "Subsequent Moves"—"Schneider describes that subsequent moves may involve an acknowledgement such as *'I see,'* a positive evaluation such as *'That's nice,'* or what's called *'idling behaviour,'* such as *'Mmm,'* or *'Really?'*."

However, even here, there is certain social structure governing small talk, as explained below.

Social Structure of Small Talk

The social structure of small talk can be illustrated by way of its gender differences.

For instance, "speech patterns between women tend to be more collaborative than those of men, and tend to support each other's involvement in the conversation. Topics for small talk are more likely to include compliments about some aspect of personal appearance. For example, *'That dress really suits you.'* Small talk between women who are friends may also involve a greater degree of self disclosure. Topics may cover more personal aspects of their life, their troubles, and their secrets. This self-disclosure both generates a closer relationship between them, and also is a signal of that closeness." (WK 2010s; D. Tannen 1992)

On the other hand, "men's small talk tends to be more competitive. It may feature verbal sparring matches, playful insults, and putdowns. However, in a way these are also both creators and signals of solidarity; the men are signalling that they are comfortable enough with each other's company to be able to say these things without their being taken as insults." (WK 2010s)

Problems with the Discourse on Small Talk

But there are problems with this discourse on small talk. Consider, for illustration, some criticisms below, as summarized in *Table 2.4*.

Firstly, one major criticism of the discourse on small talk is that, as Robert Merton (1968) once pointed out in his critique of functionalism in sociology, social functons can be "undesirable," relative to different moral evaluations. (P. Baofu 2004: 69)

In the case of small talk, it can degenerate into a waste of time for much gossiping about nothing.

Secondly, another major criticism of the discourse on small talk is that, as Robert Merton (1968) also pointed out in his critique of functionalism in sociology, social functons "are not necessarily indispensable (since there can be 'functional alternatives,' 'functional equivalents or substitutes')." (P. Baofu 2004: 69)

In the case of small talk, there can be alternative social functions not listed above.

Thirdly, still another major criticism of the discourse on small talk is that, as Robert Merton (1968) also pointed out in his critique of functionalism in sociology, social functons do not necessarily exist, because "there exist 'nonfunctions' which are consequences...left over from previous historical epochs but no longer relevant to a current system. (G. Ritzer 2003:93-4)" (P. Baofu 2004: 69)

In the case of small talk, it may well be much ado about nothing, especially when it ends up wasting much time in some cases.

And fourthly, one more major criticism of the discourse on small talk is that whether it is appropriate or not to engage in small talk about a certain topic can be culturally relative.

For instance, "[w]eather is a common topic in regions where the climate has great variation and can be unpredictable. Questions about the family are usual in some Asian and Arabic countries, although male followers of the Islamic faith avoid asking about another man's wife. In cultures that are highly status-oriented such as China and Japan, small talk between new acquaintances may feature numerous questions that enable social categorization of each other. In many European cultures it is common to discuss the weather, politics or the economy, although in some countries personal finance issues such as salary are considered taboo." (WK 2010s; G. Hofstede 2000; J. Grzega 2008)

In Retrospect

But these criticisms against the discourse on small talk do not mean to dismiss its validity outright but are to show us the two opposing sides of the debate, such that the possibility and desirability of sound (from the perspective of society with sound, small talk, and social skills as a case study here) are not to the extent that the respective defenders on both sides would like us to believe.

Moreover, the analysis of sound, small talk, and social skills can teach us something insightful about the ontological principles in existential dialectics, and good examples include the absoluteness-relativeness principle, the predictability-unpredictability principle, the explicability-inexplicability principle, the preciseness-vagueness principle, the simpleness-complicatedness principle, the openness-hiddenness principle, the denseness-emptiness principle, the change-constancy principle, the slowness-quickness principle, the expansion-contraction principle, the convention-novelty principle, the evolution-transformation principle, the symmetry-asymmetry principle, the softness-hardness principle, the seriousness-playfulness principle, the regression-progression principle, and the same-difference principle.

For instance, in relation to the absoluteness-relativeness principle, if there is absoluteness (e.g., the absolute norms of "male followers of the Islamic faith [to] avoid asking about another man's wife" in small talk), there is relativeness (e.g., what is not appropriate for "male followers of the Islamic faith [to]...ask...about another man's wife" in small talk" is not necessarily so for members of other societies, say, in the U.S., who may not mind). And the reverse direction also holds true.

In relation to the predictability-unpredictability principle, if there is predictability (e.g., the predictable tendency of small talk to address "[w]eather...in regions where the climate has great variation and can be unpredictable"), there is unpredictability (e.g., the more difficult task to predict exactly what particular aspects of weather a small talk at a given point will concern with). And the reverse direction also holds true.

In relation to the explicability-inexplicability principle, if there is explicability (e.g., the explanation that small talk is not needed in situations involving "couples in an intimate relationship," which "can signal their level of closeness by a lack of small talk"), there is inexplicability (e.g., the lack of sufficient explanation, within this view, of why the absence of small talk among couples in an intimate relationship cannot mean something else instead, like a relationship in crisis, and so on—not that this must necessarily imply good intimate relatinship). And the reverse direction also holds true.

In relation to the preciseness-vagueness principle, if there is preciseness (e.g., the precise identification of four major "safe" topics under "most circumstances"), there is vagueness (e.g., the vagueness in the identification, since it is not clear why there must be these four major "safe" topics, not others, whether more or less). And the reverse direction also holds true.

In relation to the simpleness-complicatedness principle, if there is simpleness (e.g., the relatively simple analysis of small talk in terms of its social functions), there is complicatedness (e.g., the relatively more complicated analysis of small talk by questioning its assumptions and claims, as shown in the criticisms). And the reverse direction also holds true.

In relation to the openness-hiddenness principle, if there is openness (e.g., the open exploration of the relationship between small talk and social skills), there is hiddenness (e.g., the hidden bias in the discourse on small talk and social skills, because it dogmatically assumes that small talk is necessarily functional for better social skills). And the reverse direction also holds true.

In relation to the denseness-emptiness principle, if there is denseness (e.g., the relatively denser concentration of people who engage in small talk if they are not in an intimate relationship), there is emptiness (e.g., the relatively less dense, or more empty, concentration of people who engage in small talk if they are in an intimate relationship). And the reverse direction also holds true.

In relation to the change-constancy principle, if there is change (e.g., the ever new topics of small talk over time), there is constancy (e.g., the ever constant existence of topics of small talk). And the reverse direction also holds true.

In relation to the slowness-quickness principle, if there is slowness (e.g., the relatively slower speed of someone like H. Grice to give more info than was explicitly asked for in a small talk), there is quickness (e.g., the relatively quicker speed of someone like Schneider to give more info than was explicitly asked, because "politeness in small talk is maximised by responding with a more substantial answer"). And the reverse direction also holds true.

In relation to the expansion-contraction principle, if there is expansion (e.g., the relatively more developed ability of those who indulge in small talk over time to improve their social skills), there is contraction (e.g., the relatively less developed ability of those who indulge in small talk over time to focus on serious conversation). And the reverse direction also holds true.

In relation to the convention-novelty principle, if there is convention (e.g., the conventional wisdom about Grice's Maxim of Quantity, which suggests that responses should contain no more info than was explicitly asked for in small talk), there is novelty (e.g., the alternative novel challenge to this conventional wisdom about the Maxim of Quantity, by Schneider for a new view which suggested instead that it is polite to give more info than was explicitly asked for). And the reverse direction also holds true.

In relation to the evolution-transformation principle, if there is evolution (e.g., the natural evolution of humans to engage in small talk over time in the state of nature), there is transformation (e.g., the technical transformation of human tendency to engage in small talk by the invention of some etiquettes like Grice's Maxim of Quantity and Schneider's counter-proposal). And the reverse direction also holds true.

In relation to the symmetry-asymmetry principle, if there is symmetry (e.g., the co-existence of different etiquettes about small talk), there is asymmetry (e.g., Grice's Maxim of Quantity in the 1970's—but Schneider's counter-proposal in the 1980's). And the reverse direction also holds true.

In relation to the softness-hardness principle, if there is softness (e.g., the use of "compliments about some aspect of personal appearance" among women in small talk), there is hardness (e.g., the use of "verbal sparring matches, playful insults, and putdowns" among men in small talk). And the reverse direction also holds true.

In relation to the seriousness-playfulness principle, if there is seriousness (e.g.,the serious business of social functions in small talk), there is playfulness (e.g.,the playful nature of small talk, when men engage in "playful insults"). And the reverse direction also holds true.

In relation to the regression-progression principle, if there is regression (e.g., the regression made by small talk, as shown in the criticisms by the critics), there is progression (e.g., the progress made by small talk for the enhancement of social skills). And the reverse direction also holds true.

And in relation to the same-difference principle, if there is similarity in outcome (e.g., the contribution to the molding and regulation of beliefs and behaviors, regardless of whether this be done by way of small talk or substantial talk), there is difference in outcome (e.g., the contribution to the molding and regulation of beliefs and behaviors by way of small talk for a relatively smoother social world—but the contribution to the molding and regulation of beliefs and behaviors by way of substantial talk for a relatively blunter social world). And the reverse direction also holds true.

Sound and Culture

Sound, when examined from the perspective of culture, can cast an illuminating light on its possibility and desirability, and this can be shown by way of two case studies, namely, (2.5.1) sound, harmonic series, and musical culture, and (2.5.2) sound, and noise aesthetics—to be addressed in what follows, respectively.

Sound, Harmonic Series, and Musical Culture

An excellent case study here to understand sound from the perspective of culture has to do with sound, harmonic series, and musical culture.

In Western culture, "musical acoustics or music acoustics is the branch of acoustics concerned with researching and describing the physics of music—how sounds employed as music work." (WK 2010u)

The Sound of Music

To understand how sounds are employed as music work, it is important to be familiar with some important aspects of the sound of music, namely, (a) the physical aspect, (b) the psychological aspect, and (c) the musical aspect, as summarized in *Table 2.5*.

(a) The Physical Aspect

The first important aspect of the sound of music is physical, in relation to the physics of sound.

To start, "[w]henever two different pitches are played at the same time, their sound waves interact with each other—the highs and lows in the air pressure reinforce each other to produce a different sound wave. As a result, any given sound wave which is more complicated than a sine [basic] wave can be modelled by many different sine waves of the appropriate frequencies and amplitudes (a frequency spectrum)." (WK 2010u)

In the context of music, "[w]hen the original sound sources are perfectly periodic, the note consists of several related sine waves (which mathematically add to each other) called the fundamental and the harmonics, partials, or overtones." (WK 2010u)

To clarify these words (like "fundamental," "overtones," "partials," and "harmonics") further—it is important to remember that "[t]he fundamental is the frequency at which the entire wave vibrates. Overtones are other sinusoidal components present at frequencies above the fundamental. All of the frequency components that make up the total waveform, including the fundamental and the overtones, are called partials. Together they form the harmonic series." (WK 2010u)

(a) The Psychological Aspect

The second important aspect of the sound of music is psychological, in relation to the psychology of sound.

As a reminder, what is perceived as "sound" has to do with "[v]ariations in air pressure against the ear drum, and the subsequent physical and neurological processing and interpretation," which then "give rise to the subjective experience called sound." (WK 2010u)

So, the subjective experience of the sound of music has to do with "periodic or regular vibrations rather than non-periodic ones" in the sound perceived, because "musical sounds typically have a definite pitch." (WK 2010u)

Then, "[t]he transmission of these variations through air is via a sound wave. In a very simple case, the sound of a sine wave, which is considered to be the most basic model of a sound waveform, causes the air pressure to increase and decrease in a regular fashion, and is heard as a very pure tone. Pure tones can be produced by tuning forks or whistling. The rate at which the air pressure oscillates is the frequency of the tone, which is measured in oscillations per second, called hertz. Frequency is the primary determinant of the perceived pitch. Frequency of musical instruments can change with altitude due to changes in air pressure." (WK 2010u)

(c) The Musical Aspect

The third important aspect of the sound of music is musical, in relation to musical acoustics.

Recall what is described above about the differences among different terms like "fundamental," "overtones," "partials," and "harmonics"—which are important to understand the sound of music, like the nature of harmonic series.

For instance, "[o]vertones which are perfect integer multiples of the fundamental are called harmonics. When an overtone is near to being harmonic, but not exact, it is sometimes called a harmonic partial, although they are often referred to simply as harmonics. Sometimes overtones are created that are not anywhere near a harmonic, and are just called partials or inharmonic overtones." (WK 2010u)

There are different harmonics of course, because "[t]he fundamental frequency is considered the first harmonic and the first partial. The numbering of the partials and harmonics is then usually the same; the second partial is the second harmonic, etc." (WK 2010u)

However, "if there are inharmonic partials, the numbering no longer coincides. Overtones are numbered as they appear above the fundamental. So strictly speaking, the first overtone is the second partial (and usually the second harmonic). As this can result in confusion, only harmonics are usually referred to by their numbers, and overtones and partials are described by their relationships to those harmonics." (WK 2010u)

This caveat aside—"[if] two notes are simultaneously played, with frequency ratios that are simple fractions (e.g. 2/1, 3/2 or 5/4), then the composite wave will still be periodic with a short period, and the combination will sound consonant." (WK 2010u)

A good example is "a note vibrating at 200 Hz and a note vibrating at 300 Hz (a perfect fifth, or 3/2 ratio, above 200 Hz) will add together to make a wave that repeats at 100 Hz: every 1/100 of a second, the 300 Hz wave will repeat thrice and the 200 Hz wave will repeat twice. Note that the total wave repeats at 100 Hz, but there is not actually a 100 Hz sinusoidal component present." (WK 2010u)

Moreover, "the two notes will have many of the same partials. For instance, a note with a fundamental frequency of 200 Hz will have harmonics at": (WK 2010u)

(200,) 400, 600, 800, 1000, 1200,...

And "a note with fundamental frequency of 300 Hz will have harmonics at": (WK 2010u)

(300,) 600, 900, 1200, 1500,...

So, these "two notes have the harmonics 600 and 1200 in common, and more will coincide further up the series." (WK 2010u)

What is perceived as "harmony" in music is precisely this "combination of composite waves with short fundamental frequencies and shared or closely related partials." (WK 2010u)

Conversely, "[w]hen two frequencies are near to a simple fraction, but not exact, the composite wave cycles slowly enough to hear the cancellation of the waves as a steady pulsing instead of a tone. This is called beating, and is considered to be unpleasant, or dissonant." (WK 2010u)

Problems with the Sound of Music

However, there are problems with this mainstream analysis of the sound of music. Consider, for illustration, some criticisms below, as summarized in *Table 2.5*.

Firstly, one major criticism of this mainstream analysis of the sound of music is that what is perceived as "harmonic" also depends on "hearing sensitivity." (WK 2010q; W. Richardson 1995)

For instance, it is known that there are "differences in hearing sensitivity," be they among humans or animals.

And secondly, another major criticism of this mainstream analysis of the sound of music is that whether or not harmonic music is beautiful or not also depends on differences in cultural tradition.

For instance, the American experimental composer John Cage is known for his fun with chaotic music—which is drastically different from harmonic music as portrayed above. Cage even made use of chance by using *I Ching* (and sometimes dice and coin too) in his musical composition. (O.Hardison 1989:171; P. Baofu 2000) Some like it, but others dislike it.

Above the Fray

Yet, these criticisms against the mainstream analysis of the sound of music should not be mistakenly regarded as a total rejection of its usefulness but serve to reveal to us the two opposing sides of the debate, such that the possibility and desirability of sound (from the perspective of culture with sound, harmonic series, and musical culture as a case study here) are not to the extent that the respective defenders on both sides would like us to believe.

In addition, this case study of sound, harmonic series, and musical culture can teach us something valuable about the ontological principles in existential dialectics, and good examples include the absoluteness-relativeness principle, the predictability-unpredictability principle, the explicability-inexplicability principle, the preciseness-vagueness principle, the simpleness-complicatedness principle, the openness-hiddenness principle, the denseness-emptiness principle, the order-chaos principle, the expansion-contraction principle, the theory-praxis principle, the convention-novelty principle, the evolution-transformation principle, the symmetry-asymmetry principle, the softness-hardness principle, the seriousness-playfulness principle, the regression-progression principle, and the same-difference principle.

For instance, in relation to the absoluteness-relativeness principle, if there is absoluteness (e.g., the absolute tendency of those who are fun with harmonic series in music to focus on the combination of sound waves which are "consonant"), there is relativeness (e.g., what is beautiful for the fans of harmonic music to focus on consonant sounds is not necessarily so for the fans of John Cage, who prefer chaos and chance in music instead). And the reverse direction also holds true.

In relation to the predictability-unpredictability principle, if there is predictability (e.g., the predictable tendency of consonant sounds to be produced when they are "composite waves with short fundamental frequencies and shared or closely related partials"), there is unpredictability (e.g., the more difficult task to predict exactly which particular cultures in future eras will appreciate consonant sounds, under what occassions, etc.). And the reverse direction also holds true.

In relation to the explicability-inexplicability principle, if there is explicability (e.g., the explanation that sound waves which are "consonant" are therefore "pleasant," as opposed to those dissonant ones which are "unpleasant" to hear), there is inexplicability (e.g., the lack of sufficient explanation, within this view, of why "consonant" sounds in music are necessarily "pleasant" and "dissonant" ones are necessarily "unpleasant," as some people, for instance, enjoy chaotic music). And the reverse direction also holds true.

In relation to the preciseness-vagueness principle, if there is preciseness (e.g., the identification of the "[v]ariations in air pressure against the ear drum, and the subsequent physical and neurological processing and interpretation," as the origination of "the subjective experience called sound"), there is vagueness (e.g., the vagueness in the identification, since it is not clear how and why exactly a given sound is to be "interpreted" in different cultures and "sensitive" to different individuals). And the reverse direction also holds true.

In relation to the simpleness-complicatedness principle, if there is simpleness (e.g., the relative simple nature of sine waves), there is complicatedness (e.g., the relatively more complicated nature of other waves which are "modelled by many different sine waves" of different "appropriate frequencies and amplitudes"). And the reverse direction also holds true.

In relation to the openness-hiddenness principle, if there is openness (e.g., the open exploration of the correlation between consonant sound waves and pleasant sounds), there is hiddneness (e.g., the hidden bias in the exploration of this correlation between consonant sound waves and pleasant sounds, because it downgrades "dissonant" sound waves as "unpleasant"). And the reverse direction also holds true.

In relation to the denseness-emptiness principle, if there is denseness (e.g., the relatively denser concentration of people who like harmonic series to find consonant sounds pleasant), there is emptiness (e.g., the relatively less dense, or more empty, concentration of people who, like Cage, prefers chaotic music to find consonant sounds pleasant). And the reverse direction also holds true.

In relation to the order-chaos principle, if there is order (e.g., the relatively orderly patterns of harmonic series, like the "combination of composite waves with short fundamental frequencies and shared or closely related partials"), there is chaos (e.g., the relatively disorderly nature of chaotic music by Cage, with the use of chance in *I Ching* and sometimes dice and coin too). And the reverse direction also holds true.

In relation to the expansion-contraction principle, if there is expansion (e.g., the relatively more developed ability of a musical composer of harmonic series to make use of consonant sounds), there is contraction (e.g., the relatively less developed ability of a musical composer of harmonic series to make use of dissonant sounds). And the reverse direction also holds true.

In relation to the theory-praxis principle, if there is theory (e.g., the theoretical construction of harmonic series), there is praxis (e.g,. the practical application of harmonic series to the clinical use of music for musical therapy). (WK 2010u) And the reverse direction also holds true.

In relation to the convention-novelty principle, if there is convention (e.g., the conventional wisdom about consonant sounds as pleasant for beautiful music), there is novelty (e.g., the alternative novel challenge to this conventional wisdom about pleasant consonant sounds, by Cage with his new view that chaotic music can be beautiful too). And the reverse direction also holds true.

In relation to the evolution-transformation principle, if there is evolution (e.g,. the natural evolution of humans to be sensitive to different sounds in the state of nature), there is transformation (e.g., the technical transformation of human ability to listen to sounds by the invention of chaotic music, like the one by Cage). And the reverse direction also holds true.

In relation to the symmetry-asymmetry principle, if there is symmetry (e.g., the co-existence of different overtones), there is asymmetry (e.g., some overtones are harmonic—whereas others are inharmonic, because they are "created that are not anywhere near a harmonic"). And the reverse direction also holds true.

In relation to the softness-hardness principle, if there is softness (e.g., consonant sounds), there is hardness (e.g., dissonant sounds). And the reverse direction also holds true.

In relation to the seriousness-playfulness principle, if there is seriousness (e.g., the serious nature of chaotic music for a different musical sensibility), there is playfulness (e.g., the playful nature of chaotic music when it plays with chance in the composition). And the reverse direction also holds true.

In relation to the regression-progression principle, if there is regression (e.g., the regression made by harmonic music, as shown in the criticisms by the critics), there is progression (e.g., the progress made by harmonic music for the cultivation of pleasant consonant sounds). And the reverse direction also holds true.

And in relation to the same-difference principle, if there is similarity in outcome (e.g., the contribution to the molding and regulation of beliefs and values, regardless of whether this be done by way of harmonic music or chaotic music), there is difference in outcome (e.g., the contribution to the molding and regulation of beliefs and values by way of harmonic music for a more conservative aesthetic experience—but the contribution to the molding and regulation of beliefs and values by way of chaotic music for a more radical aesthetic experience). And the reverse direction also holds true.

Sound, and Noise Aesthetics

Another way to evaluate the possibility and desirability of sound from the perspective of culture has to do with sound and noise aesthetics.

Definitions

I use the term "noise aesthetics" here to relate it to the historical phenomenon of the exploitation of sound (especially noise) for music, also known as "noise music," which refers to the "varieties of avant-garde music and sound art that may use elements such as cacophony, dissonance, atonality, noise, indeterminacy, and repetition in their realization." (WK 2010ee)

Noise music so defined "can feature distortion, various types of acoustically or electronically generated noise, randomly produced electronic signals, and non-traditional musical instruments. Noise music may also incorporate manipulated recordings, static, hiss and hum, feedback, live machine sounds, custom noise software, circuit bent instruments, and non-musical vocal elements that push noise towards the ecstatic." (WK 2010ee; T. Sangild 2002; C. Kelly 2009)

However, the term "noise music" has been defined in different ways over time. For instance, "Ben Watson...points out that Ludwig van Beethoven's *Grosse Fuge* (1825) 'sounded like noise' to his audience at the time. Indeed, Beethoven's publishers persuaded him to remove it from its original setting as the last movement of a string quartet. He did so, replacing it with a sparkling Allegro. They subsequently published it separately." (2010ee; B. Watson 2009)

In a similar tone, Paul Hegarty (2007) "cites the work of noted cultural critics Jean Baudrillard, Georges Bataille and Theodor Adorno and through their work traces the history of 'noise.' He defines noise at different times as 'intrusive, unwanted,' 'lacking skill, not being appropriate' and 'a threatening emptiness.' He traces these trends starting with 18th century concert hall music. Hegarty contends that it is John Cage's composition *4'33"*, in which an audience sits through four and a half minutes of 'silence' (Cage 1973), that represents the beginning of noise music proper. For Hegarty, 'noise music,' as with *4'33"*, is that music made up of incidental sounds that represent perfectly the tension between 'desirable' sound (properly played musical notes) and undesirable 'noise' that make up all noise music from Erik Satie to NON to Glenn Branca." (WK 2010ee)

Douglas Kahn (1999), on the other hand, "discusses the use of noise as a medium and explores the ideas of Antonin Artaud, George Brecht, William Burroughs, Sergei Eisenstein, Fluxus, Allan Kaprow, Michael McClure, Yoko Ono, Jackson Pollock, Luigi Russolo, and Dziga Vertov." (WK 2010ee)

And Jacques Attali (1985) "explores the relationship between noise music and the future of society. He indicates that noise in music is a predictor of social change and demonstrates how noise acts as the subconscious of society—validating and testing new social and political realities." (WK 2010ee; A. Weiss 1995)

Characteristics

With these different definitions in mind—noise music has distinctive characteristics. For illustration, consider a few of them below, as summarized in *Table 2.6*.

Firstly, one major characteristic of noise music is that it gives noise a more positive evaluation in the context of aesthetics via music (than is the case in traditional melodic music).

For instance, "[i]n music, dissonance is the quality of sounds which seems 'unstable,' and has an aural 'need' to 'resolve' to a 'stable' consonance. Despite the fact that words like 'unpleasant' and 'grating' are often used to describe the sound of harsh dissonance, in fact all music with a harmonic or tonal basis—even music which is perceived as generally harmonious—incorporates some degree of dissonance." (WK 2010ee)

By contrast, relatively speaking of course, in traditional melodic music, noise has a negative connotation. For instance, "the word noise means unwanted sound or noise pollution. In electronics noise can refer to the electronic signal corresponding to acoustic noise (in an audio system) or the electronic signal corresponding to the (visual) noise commonly seen as 'snow' on a degraded television or video image. In signal processing or computing it can be considered data without meaning; that is, data that is not being used to transmit a signal, but is simply produced as an unwanted by-product of other activities. Noise can block, distort, or change the meaning of a message in both human and electronic communication. White noise is a random signal (or process) with a flat power spectral density. In other words, the signal contains equal power within a fixed bandwidth at any center frequency. White noise is considered analogous to white light which contains all frequencies." (WK 2010ee; E. Hecht 2001; UCB 2010)

And secondly, another major characteristic of noise music is that it makes good use of different technologies for noise creation.

For instance, "[i]n much the same way the early modernists were inspired by naïve art, some contemporary digital art noise musicians are excited by the archaic audio technologies such as wire-recorders, the 8-track cartridge, and vinyl records. Many artists not only build their own noise-generating devices, but even their own specialized recording equipment and custom software (for example, the C++ software used in creating the viral symphOny by Joseph Nechvatal)." (WK 2010ee; S. Pedersen 2007; A. Observatori 2008)

Historical Examples

These different definitions and characteristics of noise music give us a general idea of noise music, which can be further illustrated by some historical examples as they have been created over time, as shown below (and summarized in *Table 2.6*).

(a) The Art of Noises

The first historical example of noise music has to do with "the art of noises." (WK 2010ee)

For instance, "Luigi Russolo, a futurist artist of the very early 20th century, was perhaps the first noise artist. His 1913 manifesto, *L'Arte dei Rumori*, translated as *The Art of Noises*, stated that the industrial revolution had given modern men a greater capacity to appreciate more complex sounds. Russolo found traditional melodic music confining and envisioned noise music as its future replacement. He designed and constructed a number of noise-generating devices called Intonarumori and assembled a noise orchestra to perform with them. A performance of his *Gran Concerto Futuristico* (1917) was met with strong disapproval and violence from the audience, as Russolo himself had predicted. None of his intoning devices have survived, though recently some have been reconstructed and used in performances. Although Russolo's works bear little resemblance to modern noise music, his pioneering creations cannot be overlooked as an essential stage in the evolution of this genre." (WK 2020ee; L. Russolo 1986; P. Hegarty 2007)

At the beginning, for Russolo (1986), "the art of music sought purity, limpidity and sweetness of sound. Then different sounds were amalgamated, care being taken, however, to caress the ear with gentle harmonies. Today music, as it becomes continually more complicated, strives to amalgamate the most dissonant, strange and harsh sounds. In this way we come ever closer to noise-sound." (WK 2010ee)

(b) Found Sound

The second historical example of noise music has to do with "found sound." (WK 2010ee)

In the early 20th century, the "found sound" movement had different pioneers. For instance, there was "Parade, a performance produced at the Chatelet Theatre, Paris, on May 18, 1917, that was conceived by Jean Cocteau, with design by Pablo Picasso, choreography by Leonid Massine, and music by Eric Satie. The extra-musical materials used in the production were referred to as *trompe l'oreille* sounds by Coctueau and included a dynamo, Morse code machine, sirens, steam engine, airplane motor, and typewriters." (WK 2010ee; J. Chadabe 1996)

Then, "Arseny Avraamov's composition *Symphony of Factory Sirens* involved navy ship sirens and whistles, bus and car horns, factory sirens, cannons, foghorns, artillery guns, machine guns, hydro-airplanes, a specially designed steam-whistle machine creating noisy renderings of Internationale and Marseillaise for a piece conducted by a team using flags and pistols when performed in the city of Baku in 1922." (WK 2010ee; M. Callanan 2010)

Later, in 1923, "Arthur Honegger created *Pacific 231*, a modernist musical composition that imitates the sound of a steam locomotive.

Another example is Ottorino Respighi's 1924 orchestral piece *Pines of Rome*, which included the phonographic playback of a nightingale recording." (WK 2010ee; J. Chadabe 1996; D. Albright 2004)

At about the same time, in 1924, "George Antheil created a work entitled *Ballet Mécanique* with instrumentation that included 16 pianos, 3 airplane propellers, and 7 electric bells. The work was originally conceived as music for the Dada film of the same name, by Dudley Murphy and Fernand Léger, but in 1926 it premiered independently as a concert piece." (WK 2010ee; J. Chadabe 1996)

And in 1930, "Paul Hindemith and Ernst Toch recycled records to create sound montages and in 1936 Edgard Varese experimented with records, playing them backwards, and at varying speeds. Varese had earlier used sirens to create what he called a 'continuous flowing curve' of sound that he could not achieve with acoustic instruments. In 1931 Varese's Ionisation for 13 players featured 2 sirens, a lions's roar, and used 37 percussion instruments to create a repertoire of unpitched sounds making it the first musical work to be organized solely on the basis of noise. In remarking on Varese's contributions the American composer John Cage stated that Varese had 'established the present nature of music' and that he had 'moved into the field of sound itself while others were still discriminating 'musical tones' from noises.'" (WK 2010ee; R. Rice 2010; M. Nyman 1974)

(c) Experimental Music

The third historical example of noise music has to do with "experimental music." (WK 2010e)

For instance, "[i]n the 1940s,…[u]nder the influence of Henry Cowell in San Francisco, Lou Harrison and John Cage began composing music for 'junk' percussion ensembles, scouring junkyards and Chinatown antique shops for appropriately-tuned brake drums, flower pots, gongs, and more." (WK 2010ee; H. Cowell 2004)

Later, "[i]n late 1947 Antonin Artaud (1995) recorded *Pour en Finir avec le Jugement de Dieu* (*To Have Done with the Judgment of god*), an audio piece full of the seemingly random cacophony of xylophonic sounds mixed with various percussive elements, mixed with the noise of alarming human cries, screams, grunts, onomatopoeia, and glossolalia….Also in 1949, Boulez befriended John Cage, who was visiting Paris to do research on the music of Erik Satie. John Cage had been pushing music in even more startling directions during the war years, writing for prepared piano, junkyard percussion, and electronic gadgetry." (WK 2010ee; P. Hegarty 2007; A. Ross 2007)

(d) Postmodern Developments

The fourth historical example of noise music has to do with "postmodern developments." (WK 2010ee)

Perhaps, the art critic Rosalind Krauss said it best when he "argued that by 1968 artists such as Robert Morris, Robert Smithson, and Richard Serra had 'entered a situation the logical conditions of which can no longer be described as modernist.' Sound art found itself in the same condition, but with an added emphasis on distribution. Antiform process art became the terms used to describe this post-modern post-industrial culture and the process by which it is made. Serious art music responded to this conjuncture in terms of intense noise, for example the La Monte Young Fluxus composition *89 VI 8 C. 1:42–1:52 AM Paris Encore* from *Poem For Chairs, Tables, Benches, Etc.* Young's composition *Two Sounds* (1960) was composed for amplified percussion and window panes and his *Poem for Tables, Chairs and Benches* (1960) used the sounds of furniture scraping across the floor." (WK 2010ee; R. Krauss 1986)

(e) Industrial Music

The fifth historical example of noise music has to do with "industrial music." (WK 2010e)

For instance, "[i]n the 1970s, the concept of art itself expanded and groups like Survival Research Laboratories, Borbetomagus and Elliott Sharp embraced and extended the most dissonant and least approachable aspects of these musical/spatial concepts. Around the same time, the first postmodern wave of industrial noise music appeared with Throbbing Gristle, Cabaret Voltaire, and NON (aka Boyd Rice). These cassette culture releases often featured zany tape editing, stark percussion and repetitive loops distorted to the point where they may degrade into harsh noise." (WK 2010ee; R. Young 2009)

Then, both in the 1970s and 1980s, "industrial noise groups like Current 93, Hafler Trio, Throbbing Gristle, Coil, Laibach, Steven Stapleton, Thee Temple ov Psychick Youth, Smegma, Nurse with Wound, Einstürzende Neubauten, The Haters, and The New Blockaders performed industrial noise music mixing loud metal percussion, guitars, and unconventional 'instruments' (such as jackhammers and bones) in elaborate stage performances. These industrial artists experimented with varying degrees of noise production techniques." (WK 2010ee; B. Duguid 1995)

And other examples include "Conceptual Art and the Neo-Dada [which] use of techniques such as assemblage, montage, bricolage, and appropriation." (WK 2010ee)

(f) Noise Rock and No Wave Music

The sixth historical example of noise music has to do with "noise rock and no wave music." (WK 2010ee)

For instance, "[t]he aptly-named noise rock fuses rock to noise, usually with recognizable 'rock' instrumentation, but with greater use of distortion and electronic effects, varying degrees of atonality, improvisation, and white noise. One notable band of this genre is Sonic Youth who took inspiration from the No Wave composers Glenn Branca and Rhys Chatham (himself a student of LaMonte Young)." (WK 2010ee)

And "Marc Masters, in his book on the no wave, points out that aggressively innovative early dark noise groups like Mars and DNA drew on punk rock, avant-garde minimalism and performance art. Important in this noise trajectory are the nine nights of noise music called *Noise Fest* that was organized by Thurston Moore of Sonic Youth in the NYC art space White Columns in June 1981, followed by the *Speed Trials* noise rock series organized by Live Skull members in May 1983." (WK 2010ee; M. Masters 2007; R. Young 2009)

(g) Post-Industrial and Post-Digital Music

And the seventh historical example of noise music has to do with "post-industrial and post-digital music." (WK 2010ee)

For instance, "there has been a flood of noise musicians whose ambient, microsound, or glitch-based work is often subtler to the ear. Kim Cascone refers to this development as a postdigital movement and describes it as an 'aesthetic of failure'....Goodman characterizes this widespread outpouring of free noise based media as a 'noise virus.'" (WK 2010ee; C. Kelly 2009; K. Cascone 2002; S. Goodman 2009)

Others like "Richie Hawtin, Jan Jelinek, Ricardo Villalobos, Decomposed Subsonic, Trentemøller, and other Minimal techno and Microhouse DJs have been using noise elements such as buzz, hum, and clicks as sonic flavor since the early 1990s." (WK 2010ee; S. Goodman 2009)

Problems with Noise Aesthetics

Yet, noise aesthetics has alos been subject to a lot of criticisms. For illustration, consider a few examples below, as summarized in *Table 2.6*.

Firstly, one major criticism of noise aesthetics is that it celebrates a mixture of whatever techniques and tools available to make new kinds of noise, to the point that it can easily degenerate into the vice of what I originally called "irreverent eclecticism" in Chapter 4 of *The Future of Human Civilization*.

Charles Baudelaire aptly observed about the dark side of eclecticism, when he wrote that "[e]clecticism has at all periods and places held itself superior to past doctrines because, coming last onto the scene, it finds the remotest horizons already open to it; but this impartiality only goes to prove the impotence of the eclectics. People who are so lavish with their times for reflection are not complete man: they lack the element of passion. No matter how clever he may be, an eclectic is but a feeble man; for he is a man without love. Therefore he has no ideal...; neither star nor compass. Doubt has led certain artists to beg the aid of all the other arts. Experiment with contradictory means, the encroachment of one art upon another, the importation of poetry, wit, and sentiment into painting—all these modern miseries are vices peculiar to the eclectics." (P. Baofu 2000: 370; H. Janson 1995:888)

And secondly, another major criticism of noise aesthetics is that it does not have universal appeal.

For instance, those who favor hamornic music dislike noise aesthetics, as shown in the previous section on harmonic series. Here is a battle between the two camps in the history of music, the conservative for harmony and the radical for dissonance.

More on this issue was already analyzed in my other books like *The Future of Aesthetic Experience* (2007) and *The Future of Human Civilization* (2000).

A Note of Caution

Of course, these criticisms against noise aesthetics should not be misconstrued as a total dismissal of its usefulness but are to show us the two opposing sides of the debate, such that the possibility and desirability of sound (from the perspective of culture with sound and noise aesthetics as a case study here) are not to the extent that the respective defenders on both sides would like us to believe.

Furthermore, this case study of sound and noise aesthetics can cast an illuminating light on the nature of ontological principles in existential dialectics, and good examples include the absoluteness-relativeness principle, the predictability-unpredictability principle, the explicability-inexplicability principle, the preciseness-vagueness principle, the simpleness-complicatedness principle, the openness-hiddenness principle, the denseness-emptiness principle, the change-constancy principle, the order-chaos principle, the expansion-contraction principle, the theory-praxis principle, the convention-novelty principle, the evolution-transformation principle, the symmetry-asymmetry principle, the softness-hardness principle, the seriousness-playfulness principle, the regression-progression principle, and the same-difference principle.

For instance, in relation to the absoluteness-relativeness principle, if there is absoluteness (e.g., the absolute view of Russolo that "found traditional melodic music confining and envisioned noise music as its future replacement"), there is relativeness (e.g., what is appealing for Russolo in regard to noise music is not necessary so for the defenders of traditional melodic music who prefer harmonic series).

In relation to the predictability-unpredictability principle, if there is predictability (e.g., the predictability of those in post-industrial and post-digital music like "Richie Hawtin, Jan Jelinek, Ricardo Villalobos, Decomposed Subsonic, Trentemøller, and other Minimal techno and Microhouse DJs" to use "noise elements such as buzz, hum, and clicks as sonic flavor since the early 1990s"), there is unpredictability (e.g., the more difficult task to predict exactly how long and to what extent this use of "buzz, hum, and clicks as sonic flavor" will last in which particular future eras).

In relation to the explicability-inexplicability principle, if there is explicability (e.g., the explanation by those in noise music that dissonance is a desirable quality for music), there is inexplicability (e.g,. the lack of sufficient explanation, by those in noise music, of why dissonance is necessarily desirable for music, since, for the critics, its positive evaluation is not universal).

In relation to the preciseness-vagueness principle, if there is preciseness (e.g., the precise classification of some techniques which are central for industrial music, like "assemblage, montage, bricolage, and appropriation"), there is vagueness (e.g., the vagueness in the classification, since it is not clear why only techniques like these aforecited, not some others, are included).

In relation to the simpleness-complicatedness principle, if there is simpleness (e.g., at the beginning in the older days, the art of noises tended to be simple, pure, and sweet), there is complicatedness (e.g., nowadays, the art of noises "becomes continually more complicated, strives to amalgamate the most dissonant, strange and harsh sounds").

In relation to the openness-hiddenness principle, if there is openness (e.g., the open exploration of sound for music), there is hiddenness (e.g., the hidden bias in sound music, because of its obsession with noise and dissonance, for instance).

In relation to the denseness-emptiness principle, if there is denseness (e.g., the relatively denser concentration of noise musicians in the second half of the 20th century in the West), there is emptiness (e.g., the relatively less dense, or more empty, concentration of noise musicians in the second half of the 20th century in Communist China and the Soviet Union).

In relation to the change-constancy principle, if there is change (e.g., the ever new ways to do noise music, as shown in the seven historical examples), there is constancy (e.g., the ever constancy of the appeal of noise as a basis for music for some individuals).

In relation to the order-chaos principle, if there is order (e.g., the relatively orderly nature of harmonic series in traditional melodic music), there is chaos (e.g., the relatively chaotic nature of dissonance in noise music).

In relation to the expansion-contraction principle, if there is expansion (e.g., the relative rise of the influence of noise music in the West, especially though not exclusively during the 20th century), there is contraction (e.g., the relative decline of the influence of traditional melodic music in the West especially in the 20th century).

In relation to the theory-praxis principle, if there is theory (e.g., the theoretical construction of "airplane propellers" in aeronautics), there is praxis (e.g., the practical application of the sound of "airplane propellers" to the field of "found sound" in noise music).

In relation to the convention-novelty principle, if there is convention (e.g., the conventional wisdom in traditional music, about the distinction between "musical tones" and "noise"), there is novelty (e.g., the alternative novel challenge to this conventional wisdom about music by Varese in found music, who proposed a new way of doing music, that is, by incorporating noise into music itself).

In relation to the evolution-transformation principle, if there is evolution (e.g., the natural evolution of humans to hear sounds in the state of nature), there is transformation (e.g,. the technical transformation of human ability to hear sounds by the invention of different musical instruments, like those used by noise music as a modern addition).

In relation to the symmetry-asymmetry principle, if there is symmetry (e.g., the existence of different genres of music), there is asymmetry (e.g., traditional melodic music focuses on harmonic series—but noise music focuses on dssonance).

In relation to the softness-hardness principle, if there is softness (e.g., the pleasant sounds in harmonic series), there is hardness (e.g., "the most dissonant, strange and harsh sounds" in the contemporary art of noises).

In relation to the seriousness-playfulness principle, if there is seriousness (e.g., the serious business by those in noise music to create a new kind of music), there is playfulness (e.g., the playful nature by those in noise music to play with different techniques and instruments in the process of creating new music).

In relation to the regression-progression principle, if there is regression (e.g., the progression made by noise music, as shown in the criticisms by the critics), there is progression (e.g., the progress made by noise music for a new way to appreciate noise in music).

And in relation to the same-difference principle, if there is similarity in outcome (e.g., the contribution to the molding and regulation of beliefs and values, regardless of whether this be done by way of traditional melodic music or noise music), there is difference in outcome (e.g., the contribution to the molding and regulation of beliefs and values by way of traditional melodic music for a more conservative musical lifestyle—but the contribution to the molding and regulation of beliefs and values by way of noise music for a more radical musical lifestyle). And the reverse direction also holds true.

The Disturbance of Sound

This comprehensive analysis of the possibility and desirability of sound, from the perspectives of the mind, nature, society, and culture, is to show us the different ways in which sound is both possible and desirable, but not to the extent that the spokespersons from each side would like us to believe.

Sound thus has its disturbance, just as it has its orderliness too, as the other side of the same coin.

But this is only one part of a larger story, since there is the other side of the story, which concerns silence, the opposite of sound. Since this chapter already analyzes sound, the next chapter will therefore deal with the issue of silence.

The importance of these dual analyses should not be underestimated, because those on the side of sound often treat silence merely as the opposite of sound without really appreciating it from the standpoint of silence too.

With this in mind, let's now turn to Chapter Three for the study of silence and its duality—for which we now turn to chapter three.

Table 2.1. Sound and Its Doubleness

- **Sound and the Mind**
  —Ex: sound, perception limits, and the psychoacoustic model
  —Ex: sound, noise sensibility, and extroversion-introversion

- **Sound and Nature**
  —Ex: sound, animals, and survival
  —Ex: sound, earthquakes, and seismology

- **Sound and Society**
  —Ex: sound, underwater acoustics, and naval warfare
  —Ex: sound, small talk, and social skills

- **Sound and Culture**
  —Ex: sound, harmonic series, and musical culture
  —Ex: sound, and noise aesthetics

Notes: The examples in the categories are solely illustrative (not exhaustive), and the comparison is relative (not absolute), nor are they necessarily mutually exclusive. And some can be easily re-classified elsewhere. As generalities, they allow exceptions.
Source: A summary of Ch.2 of *FPHA*

Table 2.2. Sound and Human Perception Limits

- **Four Limits of Human Perception**
  —Ex: high frequency limit
  —Ex: absolute threshold of hearing
  —Ex: temporal masking
  —Ex: simultaneous masking

- **A Response from the Critics**
  —Firstly, one major criticism of the psychological analysis of the limits of human hearing is that psychoacoustics focuses primarily on the "psychological and physiological responses associated with sound"—but not on other aspects of sound which fall outside the domains of psychology and physiology, for example, in relation to the societal, cultural, and natural contexts. (WK 2010d)
  —Secondly, another major criticism of the psychological analysis of the limits of human hearing is that, even among psychoacousticians, there are disagreements on different issues.
  —And thirdly, still another major criticism of the psychological analysis of the limits of human hearing is that much work remains to be done to understand better the limits of human hearing.

Notes: The examples in the categories are solely illustrative (not exhaustive), and the comparison is relative (not absolute), nor are they necessarily mutually exclusive. And some can be easily re-classified elsewhere. As generalities, they allow exceptions.
Source: From *Sec. 2.2.1* of *FPHA*. See book for citations.

Table 2.3. Sound, Underwater Acoustics, and Naval Warfare

- **Sound in Water**
  —Ex: sound waves
  —Ex: sound absorption
  —Ex: sound reflection scattering
  —Ex: sound propagation

- **Sound for Naval Warfare**
  —Ex: surface and near-surface navigation
  —Ex: deep water navigation

- **Problems with Submarine Navigation**
  —Firstly, one major problem with submarine navigation is that it does not always work, because submarines collide underwater from time to time with other submarines, drilling rigs, undersea mountains, and so on.
  —Secondly, another major problem with submarine navigation is that submarines also become the target of "anti-submarine warfare detection systems such as radar and satellite surveillance." (WK 2010r)
  —And thirdly, still another major problem with submarine navigation is that submarines can be sunk by anti-submarine weapons and can also suffer substantial damage from internal explosion or from fire onboard sometimes.

Notes: The examples in the categories are solely illustrative (not exhaustive), and the comparison is relative (not absolute), nor are they necessarily mutually exclusive. And some can be easily re-classified elsewhere. As generalities, they allow exceptions.
Source: From *Sec. 2.4.1* of *FPHA*. See book for citations.

Table 2.4. Sound, Small Talk, and Social Skills (Part I)

- **Social Functions of Small Talk**
  —Ex: "it helps new acquaintances to explore and categorize each other's social position." (J. Laver 1975)
  —Ex: "small talk is closely related to the need for people to maintain positive face—to feel approved-of by those who are listening to them."
  —Ex: "it lubricates social interactions in a very flexible way, although the desired function is often dependent on the point in the conversation at which the small talk occurs." (J. Holmes 2000)

- **Topics of Small Talk**
  —Ex: the weather
  —Ex: television and films
  —Ex: sports

- **Patterns of Small Talk**
  —Ex: the first move
  —Ex: the second move
  —Ex: subsequent moves

- **Social Structure of Small Talk**
  —Ex: "speech patterns between women tend to be more collaborative than those of men, and tend to support each other's involvement in the conversation." (WK 2010s)
  —Ex: "men's small talk tends to be more competitive." (WK 2010s)

(*continued on next page*)

Table 2.4. Sound, Small Talk, and Social Skills (Part II)

- **Problems with the Discourse on Small Talk**
  - —Firstly, one major criticism of the discourse on small talk is that it can degenerate into a waste of time for much gossiping about nothing.
  - —Secondly, another major criticism of the discourse on small talk is that there can be alternative social functions not listed above.
  - —Thirdly, still another major criticism of the discourse on small talk is that it may well be much ado about nothing, especially when it ends up wasting much time in some cases.
  - —And fourthly, one more major criticism of the discourse on small talk is that whether it is appropriate or not to engage in small talk about a certain topic can be culturally relative.

Notes: The examples in the categories are solely illustrative (not exhaustive), and the comparison is relative (not absolute), nor are they necessarily mutually exclusive. And some can be easily re-classified elsewhere. As generalities, they allow exceptions.
Source: From *Sec. 2.4.2* of *FPHA*. See book for citations.

Table 2.5. Sound, Harmonic Series, and Musical Culture

- **The Physical Aspect**
  —Ex: "[w]henever two different pitches are played at the same time, their sound waves interact with each other—the highs and lows in the air pressure reinforce each other to produce a different sound wave. As a result, any given sound wave which is more complicated than a sine [basic] wave can be modelled by many different sine waves of the appropriate frequencies and amplitudes (a frequency spectrum)." (WK 2010u)

- **The Psychological Aspect**
  —Ex: what is perceived as "sound" has to do with "[v]ariations in air pressure against the ear drum, and the subsequent physical and neurological processing and interpretation," which then "give rise to the subjective experience called sound." (WK 2010u)

- **The Musical Aspect**
  —Ex: What is perceived as "harmony" in music is precisely the "combination of composite waves with short fundamental frequencies and shared or closely related partials." (WK 2010u)

- **Problems with the Sound of Music**
  —Firstly, one major criticism of this mainstream analysis of the sound of music is that what is perceived as "harmonic" also depends on "hearing sensitivity." (WK 2010q; W. Richardson 1995)
  —Secondly, another major criticism of this mainstream analysis of the sound of music is that whether or not harmonic music is beautiful or not also depends on differences in cultural tradition.

Notes: The examples in the categories are solely illustrative (not exhaustive), and the comparison is relative (not absolute), nor are they necessarily mutually exclusive. And some can be easily re-classified elsewhere. As generalities, they allow exceptions.
Source: From *Sec. 2.5.1* of *FPHA*. See book for more details and citations.

Table 2.6. Sound and Noise Aesthetics

- **Definitions**
  - —Ex: I use the term "noise aesthetics" here to relate it to the historical phenomenon of the exploitation of sound (especially noise) for music, also known as "noise music," which refers to the "varieties of avant-garde music and sound art that may use elements such as cacophony, dissonance, atonality, noise, indeterminacy, and repetition in their realization." (WK 2010ee)

- **Characteristics**
  - —Firstly, one major characteristic of noise music is that it gives noise a more positive evaluation in the context of aesthetics via music (than is the case in traditional melodic music).
  - —And secondly, another major characteristic of noise music is that it makes good use of different technologies for noise creation.

- **Historical Examples**
  - —Ex: the art of noises
  - —Ex: found sound
  - —Ex: experimental music
  - —Ex: postmodern developments
  - —Ex: industrial music
  - —Ex: noise rock and no wave music
  - —Ex: post-industrial and post-digital music

- **Problems with Noise Aesthetics**
  - —Ex: Firstly, one major criticism of noise aesthetics is that it celebrates a mixture of whatever techniques and tools available to make new kinds of noise, to the point that it can easily degenerate into the vice of what I originally called "irreverent eclecticism" in Chapter 4 of *The Future of Human Civilization*.
  - —Ex: Secondly, another major criticism of noise aesthetics is that it does not have universal appeal.

Notes: The examples are solely illustrative (not exhaustive), and the comparison is relative (not absolute), nor are they necessarily mutually exclusive. And some can be easily re-classified elsewhere. As generalities, they allow exceptions.

Source: From *Sec. 2.5.2* of *FPHA*. See book for citations.

• Part Three •

Silence

CHAPTER 3
SILENCE AND ITS DUALITY

> There is a silence where hath been no sound, There is a silence where no sound may be,—In the cold grave, under the deep, deep sea, Or in the wide desert where no life is found.
> —Thomas Hood (BT 2010c)

The Serenity of Silence

Acoustics deals with the study of both sound and silence (which is "the relative or total lack of audible sound," dependent on the sound perception of an organic species or an electrical equipment)—as already analyzed in *Sec. 1.2*.

This relationship between sound and silence is true, even though silence does not always have to be an absolute lack of audible sound, so the distinction between sound and silence can be one of degree.

That said—any study of sound requires an exploration of the nature of silence too, as one cannot be completely disconnected from the other.

With this in mind, this book offers a dialectic treatment of sound and silence in that, if sound has its orderliness, it has its disturbance too—just as, if silence has its serenity, it likewise has its turmoil.

The inquiry concernng silence is the subject matter of this chapter (Chapter Two), whereas that of sound was already addressed in Chapter Two.

Besides, the relationship between sound and silence constitutes a scholarly challenge, in that one cannot analyze silence (or sound, for that matter) completely separately.

Consequently, this then means that the issues concerning silence in this chapter (Chapter Three) and sound in Chapter Three can overlap, as each issue on one can be relevant to the other too.

With this clarification in mind—a good way to examine the duality of silence is by way of the evaluation of the extent to which it (silence) is in fact both possible and desirabe.

This chapter thus takes the challenging task to provide a comprehensive analysis of silence in relation to acoustics from the four perspectives of the mind, nature, society, and culture (in accordance to my sophisticated methodological holism, as explained in *Sec. 1.8*).

Thus, this chapter is organized in four main sections, in relation to (3.2) silence and the mind, (3.3) silence and nature, (3.4) silence and society, and (3.5) silence and culture—to be analyzed hereafter, respectively (and summarized in *Table 3.1*).

Silence and the Mind

Silence, when examined from the perspective of the mind, can reveal something fascinating about its possibility and desirability, and this can be shown by way of two case studies, namely, (3.2.1) silence, deafness, and the mask of benevolence and (3.2.2) silence, stuttering, and speech disorder—to be addressed below, in that order.

Silence, Deafness, and the Mask of Benevolence

The word "deafness" (also known as "hearing impairment," "hearing loss," or "hard of hearing") refers to "conditions in which individuals are fully or partially unable to detect or perceive at least some frequencies of sound which can typically be heard by members of their species." (WK 2010v)

However, some deaf people do not like the term "impaired" to be applied to them; instead, they prefer the terms "deaf" and "hard of hearing," for instance. (WK 2010v)

On Definition

The first issue to deal with hearing impairment is about its definition, in relation to the test of hearing sensitivity.

Hearing sensitivity is often correlated with the way that "animals use sound to communicate," and therefore "hearing in that type of animal is most sensitive for the frequencies produced by calls, or in the case of humans, speech." (WK 2010v)

So, to define "hearing impairment" is to identify "when an animal has diminished sensitivity to the sounds normally heard by its species. In humans, the term hearing impairment is usually reserved for people who have relative insensitivity to sound in the speech frequencies. The severity of a hearing impairment is categorized according to the increase in volume that must be made above the usual level before the listener can detect it. In profound deafness, even the loudest sounds that can be produced by an audiometer (an instrument used to measure hearing) may not be detected." (WK 2010v)

An alternative definitional concern has to do with to "the perceived clarity of a sound rather than its amplitude. In humans, that aspect is usually measured by tests of speech perception. These tests measure one's ability to understand speech, not to merely detect sound. There are very rare types of hearing impairments which affect speech understanding alone." (WK 2010v)

On Classification

The second issue to deal with hearing impairment is about its classification into four major categories, namely, (a) by type, (b) by severity, (c) by age of onset, and (d) by number of impaired ears, as described below (and summarized in *Table 3.2*).

(a) By Type

Hearing impairment can be classified by type, which can be "conductive, sensorineural, or both." (WK 2010v)

In regard to the first type, "a conductive hearing impairment is an impairment resulting from dysfunction in any of the mechanisms that normally conduct sound waves through the outer ear, the eardrum or the bones of the middle ear." (WK 2010v)

In regard to the second type, "[a] sensorineural hearing impairment is one resulting from dysfunction in the inner ear, especially the cochlea where sound vibrations are converted into neural signals, or in any part of the brain that subsequently processes these signals....Sensorineural loss...is by far the most common type of hearing loss." (WK 2010v)

And in regard to the third type, some individuals have the misfortune to suffer from both, not just from conductive or sensorineural hearing impairment.

(b) By Severity

Hearing impairment can be classified by severity, which "is ranked according to the loudness (measured in decibels (dB)) a sound must be before being detected by an individual. Hearing impairment may be ranked as mild, moderate, severe or profound as defined below": (WK 2010v)

- "Mild"
  - –"for adults: between 27 and 40 dB"
  - –"for children: between 20 and 40 dB"
- "Moderate"
  - –"between 41 and 55 dB"
- "Moderately Severe"
  - –"between 56 and 70 dB"
- "Severe"
  - –"between 71 and 90 dB"
- "Profound"
  - –"90 dB or greater"

However, "[f]or certain legal purposes such as insurance claims, hearing impairments are described in terms of percentages….[T]he idea of a percentage of hearing loss is somewhat arbitrary, but where decibels of loss are converted via a recognized legal formula, it is possible to calculate a standardized 'percentage of hearing loss' which is suitable for legal purposes only." (WK 2010v)

Another way to classify severity is "a Speech-in-Noise test. As the name implies, a Speech-in-Noise test will give you an indication of how well you can understand speech in a noisy environment. A person with a hearing loss will often be less able to understand speech, especially in noisy conditions." (WK 2010v)

(c) By Age of Onset

Hearing impairment can be classified by age of onset, in terms of two main types, namely, "pre-lingual deafness" and "post-lingual deafness." (WK 2010v)

In regard to "pre-lingual deafness"—the hearing impairment "is sustained prior to the acquisition of language, which can occur as a result of a congenital condition or through hearing loss in early infancy. Prelingual deafness impairs an individual's ability to acquire a spoken language, but children born into signing families rarely have delays in language development. Most pre-lingual hearing impairment is acquired via either disease or trauma rather than genetically inherited, so families with deaf children nearly always lack previous experience with sign language." (WK 2010v)

And in regard to "post-lingual deafness"—the hearing impairment "is sustained after the acquisition of language, which can occur as a result of disease, trauma, or as a side-effect of a medicine. Typically, hearing loss is gradual and often detected by family and friends of affected individuals long before the patients themselves will acknowledge the disability.

Common treatments include hearing aids and learning lip reading." (WK 2010v)

In general, "post-lingual deafness is far more common than pre-lingual deafness." (WK 2010v)

(d) By Number of Impaired Ears

Hearing impairment can be classified by number of impaired ears, namely, either "unilateral" or "bilateral." (WK 2010v)

In regard to "unilateral" hearing impairment—the hearing impairment is called "single sided deafness/SSD" and has "an impairment in only one ear. This can impair a person's ability to localize sounds (e.g., determining where traffic is coming from) and distinguish sounds from background noise in noisy environments." (WK 2010v)

And in regard to "bilateral" hearing impairment—the hearing impairment has an impairment in two ears.

On Causes

The third issue to deal with hearing impairment is about its classification into different causes, namely, (a) by age, (b) by long-term exposure to environmental noise, (c) by genetics, (d) by disease or illness, (e) by medications, (f) by exposure to ototoxic chemicals, and (g) by physical trauma, as described below (and summarized in *Table 3.2*).

(a) By Age

Hearing impairment can be classified in terms of its cause by age. For instance, "presbycusis, the progressive loss of ability to hear high frequencies with increasing age, begins in early adulthood, but does not usually interfere with ability to understand conversation until much later." (WK 2010v)

(b) By Long-Term Exposure to Environmental Noise

Hearing impairment can be classified in terms of its cause by long-term exposure to environmental noise. For instance, "[p]opulations of people living near airports or freeways are exposed to levels of noise typically in the 65 to 75 dB(A) range. If lifestyles include significant outdoor or open window conditions, these exposures over time can degrade hearing." (WK 2010v)

(b) By Genetics

Hearing impairment can be classified in terms of its cause by genetics. For instance, "[i]f a family has a dominant gene for deafness it will persist across generations because it will manifest itself in the offspring even if it is inherited from only one parent. If a family had genetic hearing impairment caused by a recessive gene it will not always be apparent as it will have to be passed onto offspring from both parents." (WK 2010v)

(d) By Disease or Illness

Hearing impairment can be classified in terms of its cause by disease or illness. For example, any of the following diseases or illnesses can have negative impact on hearing: Measles, Meningitis, Autoimmune disease, Mumps (Epidemic parotitis), Presbycusis, Adenoids, AIDS, HIV, Chlamydia, Fetal alcohol syndrome, Syphilis, Otosclerosis, etc. (WK 2010v)

(e) By Medications

Hearing impairment can be classified in terms of its cause by medications. For instance, "[s]ome medications cause irreversible damage to the ear, and are limited in their use for this reason. The most important group is the aminoglycosides (main member gentamicin) and platinum based chemotherapeutics such as cisplatin. Various other medications may reversibly affect hearing. This includes some diuretics, aspirin and NSAIDs, and macrolide antibiotics." (WK 2010v)

(f) By Exposure to Ototoxic Chemicals

Hearing impairment can be classified in terms of its cause by exposure to ototoxic chemicals. For instance, any of the following ototoxic chemicals can have negative impact on hearing: (WK 2010v)

- "Drugs"—"antimalarial, antibiotics, anti-inflammatory (non-steroidal), antineoplastic, diuretics"
- "Solvents"—"toluene, styrene, xylene, n-hexane, ethyl benzene, white spirits/Stoddard, carbon disulfide, fuels, perchloroethylene, trichloroethylene, p-xylene"
- "Asphyxiants"—"carbon monoxide, hydrogen cyanide"
- "Metals"—"lead, mercury, organotins (trimethyltin)"
- "Pesticides/Herbicides"—"paraquat, organophosphates"

Surely, the list is not complete but illustrative only.

(g) By Physical Trauma

Hearing impairment can be classified in terms of its cause by physical trauma, especially in relation to the "damage either to the ear itself or to the brain centers that process the aural information conveyed by the ears." (WK 2010v)

For instance, "people who sustain head injury are especially vulnerable to hearing loss or tinnitus, either temporary or permanent," and "exposure to very loud noise (90 dB or more, such as jet engines at close range) can cause progressive hearing loss. Exposure to a single event of extremely loud noise (such as explosions) can also cause temporary or permanent hearing loss. A typical source of acoustic trauma is an excessively loud music concert." (WK 2010v)

On Treatment

The fourth issue to deal with hearing impairment is about its classification into different types of treatment, namely, (a) by approaches, (b) by gene therapy, (c) by assistive techniques and devices, and (d) by resources and interventions, as described below (and summarized in *Table 3.2*).

(a) By Approaches

The treatment of hearing impairment can be classified in terms of different approaches. For instance, "[i]f the hearing loss occurs at a young age, interference with the acquisition of spoken language and social skills may occur. Hearing aids, which amplify the incoming sound, may alleviate some of the problems caused by hearing impairment, but are often insufficient. Cochlear implants artificially stimulate the VIIIth Nerve by providing an electric impulse substitution for the firing of hair cells." (WK 2010v)

(b) By Gene Therapy

The treatment of hearing impairment can be classified in terms of gene therapy. For instance, "a 2008 study has shown that gene therapy targeting Atoh1 can cause hair cell growth and attract neuronal processes in embryonic mice. It is hoped that a similar treatment will one day ameliorate hearing loss in humans." (WK 2010v)

(c) By Assistive Techniques and Devices

The treatment of hearing impairment can be classified in terms of different assistive techniques and devices. For instance, any of the following techniques and devices can assist those who suffer from hearing impairment: (WK 2010v)

- "Individuals can communicate by telephone using Telecommunications Device for the Deaf (TDD). These devices look like typewriters or word processors and transmit typed text over regular telephone lines. Other names in common use are textphone and minicom."

- "Mobile textphone devices came onto the market as of 2004, allowing simultaneous two way text communication."
- "Videophones and similar video technologies can be used for distance communication using sign language. Video conferencing technologies permit signed conversations as well as permitting a sign language-English interpreter to voice and sign conversations between a hearing impaired person and that person's hearing party, negating the use of a TTY device or computer keyboard."
- "Video Relay Service and Video Remote Interpreting services also use a third-party telecommunication service to allow a deaf or hard-of-hearing person to communicate quickly and conveniently with a hearing person, through a sign language interpreter."
- "Hearing dogs are a specific type of assistance dog specifically selected and trained to assist the deaf and hearing impaired by alerting their handler to important sounds, such as doorbells, smoke alarms, ringing telephones, or alarm clocks."
- "Other assistive devices include those that use flashing lights to signal events such as a ringing telephone, a doorbell, or a fire alarm."
- "The advent of the Internet's World Wide Web and closed captioning has given the hearing impaired unprecedented access to information. Electronic mail and online chat have reduced the need for deaf and hard-of-hearing people to use a third-party Telecommunications Relay Service in order to communicate with the hearing and other hearing impaired people."

Of course, there can be other techniques and devices, so the list above is illustrative only.

(d) By Resources and Interventions

The treatment of hearing impairment can be classified in terms of different resources and interventions. For instance, there are "wireless devices" like "the FM system, the audio induction loop and the infra red system. Each system has advantages and benefits for particular uses." (WK 2010v)

In addition, "[i]nside the classroom, children with hearing impairments may also benefit from interventions. These include providing favorable seating for the child. This can be done by having the student sit as close to the teacher as possible so that they will be able to hear the teacher, or read the teacher's lips more easily....When teaching students with hearing impairments, try to use overheads as much as possible. Overheads allow the teacher to write, as well as maintain visual focus on the hearing impaired student. For those students who are completely deaf, one of the most common interventions is having the child communicate with others through an interpreter using sign language." (WK 2010v; UIUC 2010)

Problems in Dealing with Hearing Impairment

But these different ways to deal with hearing impairment are not without their own problems. Consider, for illustration, some major problems below.

Firstly, one major problem in dealing with hearing impairment is that it can be costly.

For instance, "Cochlear implants are not only expensive, but require sophisticated programming in conjunction with patient training for effectiveness." (WK 2010v) It is thus not surprising that "[t]here has been considerable controversy within the culturally Deaf community over cochlear implants," especially when there are "the chances of a head injury, which carries a greater risk where the implant is involved." (WK 2010v)

Secondly, another major problem in dealing with hearing impairment is that there is no guarantee for success.

For instance, in regard to gene therapy, "it is important to note, however, that the regrowth of cochlear hair cells does not imply the restoration of hearing sensitivity as the sensory cells may or may not make connections with neurons that carry the signals from hair cells to the brain." (WK 2010v)

Thirdly, still another major problem in dealing with hearing impairment is that the aid to those who suffer from hearing impairment can degenerate into a new kind of patronizing and oppressive maltreatment of them, as pointed out by the critics, like Harlan Lane (1992:48-9) in his 1992 book titled *The Mask of Benevolence.*

And fourthly, one more major problem in dealing with hearing impairment is that, contrary to the popular belief in mainstream society, "[i]n Deaf culture, deafness is not considered a condition that needs to be fixed" and is not regarded as "a pathological condition, viewing it instead as a locus of pride." (WK 2010y; P. Ladd 2003; A. Mindess 2006)

A Didactic Exercise

These problems in dealing with hearing-impaired people do not mean to dismiss the usefulness of the different treatments but to show us the two opposing sides of the debate, such that the possibility and desirability of silence (from the perspective of the mind with silence, deafness, and the mask of benevolence as a case study here) are not to the extent that the respective defenders on both sides would like us to believe.

More importantly, the analysis of silence, deafness, and the mask of benevolence can reveal the relevance of the ontological principles in existential dialectics, and good examples include the absoluteness-relativeness principle, the predictability-unpredictability principle, the explicability-inexplicability principle, the preciseness-vagueness principle, the simpleness-complicatedness principle, the openness-hiddenness principle, the denseness-emptiness principle, the change-constancy principle, the slowness-quickness principle, the expansion-contraction principle, the theory-praxis principle, the convention-novelty principle, the evolution-transformation principle, the symmetry-asymmetry principle, the softness-hardness principle, the regression-progression principle, and the same-difference principle.

For instance, in relation to the absoluteness-relativeness principle, if there is absoluteness (e.g., the absolute concern, in a definition of hearing impairment, with hearing sensitivity), there is relativeness (e.g., what is the right concern with hearing sensitivity in a definition of hearing impairment is not necessarily so for those who focus on hearing clarity in an alternative definition). And the reverse direction also holds true.

In relation to the predictability-unpredictability principle, if there is predictability (e.g., the predictable tendency of those who accept gene therapy to seek the regrowth of cochlear hair cells), there is unpredictability (e.g., the more difficult task to predict exactly which particular gene therapy will be successful, under what conditions, since "the restoration of hearing sensitivity as the sensory cells may or may not make connections with neurons that carry the signals from hair cells to the brain"). And the reverse direction also holds true.

In relation to the explicability-inexplicability principle, if there is explicability (e.g,. the explanation by those who care for deaf people that hearing impairment is something to be corrected for the elimination or reduction of their suffering), there is inexplicability (e.g,. the lack of convincing explanation, within this assertive view, of why their care is necessarily needed, since, contrary to the popular view in mainstream society, "[i]n Deaf culture, deafness is not considered a condition that needs to be fixed" and is not regarded as "a pathological condition, viewing it instead as a locus of pride.") And the reverse direction also holds true.

In relation to the preciseness-vagueness principle, if there is preciseness (e.g., the precise classification of the five types of severity in hearing impairment measurement, namely, "mild," "moderate," "moderately severe," "severe," and "profound"), there is vagueness (e.g., the vagueness in the classification, since it is not clear why there are only five types, not four, six, seven, etc.). And the reverse direction also holds true.

In relation to the simpleness-complicatedness principle, if there is simpleness (e.g., the relatively simple analysis of deafness in terms of the major issues as described above), there is complicatedness (e.g., the relatively more complicated analysis of deafness by questioning the assumptions and claims in regard to the major issues as described above, as shown in the criticisms by the critics). And the reverse direction also holds true.

In relation to the openness-hiddenness principle, if there is openness (e.g., the open exploration of the need for society to aid hearing-impaired people), there is hiddenness (e.g., the hidden bias in this social aid to hearing-impaired people, since it often degenerate into a new kind of patronizing and oppressive maltreatment of them, as what Lane called "the mask of benevolence"). And the reverse direction also holds true.

In relation to the denseness-emptiness principle, if there is denseness (e.g., the relatively denser, or more pervasive, spread of people with normal hearing in society), there is emptiness (e.g., the relatively less dense, or more empty, spread of hearing-impaired people in society). And the reverse direction also holds true.

In relation to the change-constancy principle, if there is change (e.g. the ever new ways for hearing-impaired people to be treated), there is constancy (e.g., the ever constant existence of ways for hearing-impaired people to be treated). And the reverse direction also holds true.

In relation to the slowness-quickness principle, if there is slowness (e.g., the relatively slower speed of people to have hearing loss by long-term exposure to environmental noise like living near airports or freeways), there is quickness (e.g., the relatively faster speed of people to have hearing loss by head injury in an explosion). And the reverse direction also holds true.

In relation to the expansion-contraction principle, if there is expansion (e.g., the relatively more developed ability of those hearing–impaired people who solicit the help of others in society to seek different assistive techniques, resources, and intervention), there is contraction (e.g., the relatively less developed ability of those hearing–impaired people who solicit the help of others in society to counter the patronizing mask of benevolence from society, as pointed out by Lane). And the reverse direction also holds true.

In relation to the theory-praxis principle, if there is theory (e.g., the theoretical construction of genetics), there is praxis (e.g., the practical application of genetics to the field of hearing impairment for gene therapy). And the reverse direction also holds true.

In relation to the convention-novelty principle, if there is convention (e.g., the conventional wisdom that hearing-impaired people need help from others and society in general), there is novelty (e.g., the alternative novel challenge to this conventional wisdom about the benevolent help of society, by the new view that this help often degenerates into a new kind of patronizing and oppressive maltreatment of them, as what Lane called "the mask of benevolence"). And the reverse direction also holds true.

In relation to the evolution-transformation principle, if there is evolution (e.g., the natural evolution of humans to suffer from hearing loss in the state of nature), there is transformation (e.g., the technical transformation of human ability to counter hearing loss by the invention of gene therapy, among other methods). And the reverse direction also holds true.

In relation to the symmetry-asymmetry principle, if there is symmetry (e.g., the co-existence of different individuals in society), there is asymmetry (e.g., some individuals have severe hearing loss—but some other individuals do not have severe hearing loss). And the reverse direction also holds true.

In relation to the softness-hardness principle, if there is softness (e.g., the benevolence of those who care about hearing-impaired people), there is hardness (e.g., the mask of benevolence of those who care about hearing-impaired people, as pointed out by Lane). And the reverse direction also holds true.

In relation to the regression-progression principle, if there is regression (e.g., the regression made by the different treatments of hearing loss, as shown in the criticisms by the critics), there is progression (e.g., the progress made by the different treatments of hearing loss for the improvement of hearing). And the reverse direction also holds true.

And in relation to the same-difference principle, if there is similarity in outcome (e.g., the contribution to the molding and control of beliefs and behaviors, regardless of whether this be done by way of social help to deaf people or self-help instead), there is difference in outcome (e.g., the contribution to the molding and control of beliefs and behaviors by way of social help to deaf people for a relatively more cooperative society—but the contribution to the molding and control of beliefs and behaviors by way of self-help for a relatively more competitive society). And the reverse direction also holds true.

Silence, Stuttering, and Speech Disorder

Another way to evaluate the possibility and desirability of silence from the perspective of the mind has to do with silence, stuttering, and speech disorder.

The Issue of Definition

The term “stuttering” (also known as “stammering”) refers to “a speech disorder in which the flow of speech is disrupted by involuntary repetitions and prolongations of sounds, syllables, words or phrases, and involuntary silent pauses or blocks in which the stutterer is unable to produce sounds.” (WK 2010w)

In popular usage, “the term 'stuttering'...covers a wide spectrum of severity: it may encompass individuals with barely perceptible impediments, for whom the disorder is largely cosmetic, as well as others with extremely severe symptoms, for whom the problem can effectively prevent most oral communication.” (WK 2010w)

However, the tem “stuttering” should not be confused with the related term “developmental stuttering,” which refers to “stuttering that originates when a child is learning to speak and develops as the child matures into adulthood. Other speech disorders with symptoms resembling stuttering are cluttering, Parkinson's speech, essential tremor, palilalia, spasmodic dysphonia, selective mutism and social anxiety.” (WK 2010w)

The Issue of Characteristics

Stuttering can be described in terms of three major characteristics, namely, (a) primary behaviors, (b) variability, and (c) feelings and attitudes—to be addressed in what follows (and summarized in *Table 3.3*).

(a) Primary Behaviors

The first major characteristic of stuttering concerns its “primary behaviors,” which “are the overt, observable signs of speech fluency breakdown, including repeating sounds, syllables, words or phrases, silent blocks and prolongation of sounds. These differ in from the normal disfluencies found in all speakers in that stuttering disfluencies may last longer, occur more frequently, and are produced with more effort and strain.” (WK 2010w; D. Ward 2006)

In general, stuttering has three distinct symptoms, as shown below. (WK 2010w)

- "Repetition"—"occurs when a unit of speech, such as a sound, syllable, word, or phrase is repeated and are typical in children who are beginning to stutter. For example, 'to-to-to-tomorrow.'" (WK 2010w)
- "Prolongations"—"are the unnatural lengthening of continuant sounds, for example, 'mmmmmmmmmilk.' Prolongations are also common in children beginning to stutter." (WK 2010w)
- "Blocks of Silence"—"are inappropriate cessation of sound and air [silence], often associated with freezing of the movement of the tongue, lips and/or vocal folds. Blocks often develop later, and can be associated with muscle tension and effort." (WK 2010w; B. Guitar 2005)

Besides these three characteristics, stuttering also differs in terms of severity.

(b) Variability

The second major characteristic of stuttering concerns its variability (or severity), because "[t]he severity of a stutter is often not constant even for severe stutterers. Stutterers commonly report dramatically increased fluency when talking in unison with another speaker, copying another's speech, whispering, singing, and acting or when talking to pets, young children, or themselves. Other situations, such as public speaking and speaking on the telephone are often greatly feared by stutterers, and increased stuttering is reported." (WK 2010w; D. Ward 2006)

(c) Feelings and Attitudes

The third major characteristic of stuttering concerns feelings and attitudes, because "stuttering may have a significant negative cognitive and affective impact on the stutterer." (WK 2010w)

For isntance, "Joseph Sheehan, a prominent researcher in the field, has described stuttering in terms of the well-known analogy to an iceberg, with the immediately visible and audible symptoms of stuttering above the waterline and a broader set of symptoms such as negative emotions hidden below the surface. Feelings of embarrassment, shame, frustration, fear, anger, and guilt are frequent in stutterers, and may actually increase tension and effort, leading to increased stuttering." (WK 2010w; J. Kalinowski 2006; D. Ward 2006; B. Guitar 2005)

Over time, "continued exposure to difficult speaking experiences may crystallize into a negative self-concept and self-image. A stutterer may project his or her attitudes onto others, believing that they think he or she is nervous or stupid. Such negative feelings and attitudes may need to be a major focus of a treatment program. Many stutterers report about a high emotional cost, including jobs or promotions not received, as well as relationships broken or not pursued." (WK 2010w; A. Pollack 2006; B. Guitar 2005)

The Issue of Causes

Stuttering can be explained in terms of three major causes, namely, (a) metabolic, (b) developmental, and (c) acquired—to be addressed in what follows, respectively (and summarized in *Table 3.3*).

(a) Metabolic

The first major cause of stuttering is "metabolic" (or genetic), because "[s]ome inherited forms of stuttering have been traced to variations in genes governing lysosomal metabolism." (WK 2010w; NEJM 2010)

Gerald Maguire at the University of California, Irvine claimed that "[t]his builds upon the belief that has been growing over the last 10 years that stuttering is not a behavioral disorder, but a genetic disorder with manifestation in the brain," (WK 2010w; SM 2010)

(b) Developmental

The second major cause of stuttering is "development," not "genetic," in that it can be cured, in certain cases.

For instance, "[s]tuttering is...a developmental disorder beginning in early childhood and continuing into adulthood in at least 20% of affected children." (WK 2010w; N. Gordon 2002; A. Craig 2005)

Although in cases like this, "the rate of early recovery is very high"—"with time a young stutterer may transition from easy, relaxed repetition to more tense and effortful stuttering, including blocks and prolongations. Some propose that parental reaction may affect the development of chronic stutter. Recommendations to slow down, take a breath, say it again, etc. may increase the child's anxiety and fear, leading to more difficulties with speaking and, in the 'cycle of stuttering' to ever yet more fear, anxiety and expectation of stuttering." (WK 2010w; N. Gordon 2002; D. Ward 2006)

(c) Acquired

The third major cause of stuttering is "acquired," due to some unexpected events in life.

For instance, "[i]n rare cases, stuttering may be acquired in adulthood as the result of a neurological event such as a head injury, tumour, stroke or drug use. The stuttering has different characteristics from its developmental equivalent: it tends to be limited to part-word or sound repetitions, and is associated with a relative lack of anxiety and secondary stuttering behaviors. Techniques such as altered auditory feedback, which may promote fluency in stutterers with the developmental condition, are not effective with the acquired type." (WK 2010w; N. Gordon 2002; D. Ward 2006; A. Craig 2005)

The Issue of Treatment

Stuttering can be treated in six major ways, namely, (a) fluency shaping therapy, (b) stuttering modication therapy, (c) electronic fluency devices, (d) anti-stuttering medications, (e) support groups and the self-help movement, and (f) diaphragmatic breathing—to be addressed hereafter, in that order (and summarized in *Table 3.3*).

(a) Fluency Shaping Therapy

The first major treatment of stuttering makes use of "fluency shaping therapy" (also known as "speak more fluently," "prolonged speech" or "connected speech"), which "trains stutterers to speak fluently by controlling their breathing, phonation, and articulation (lips, jaw, and tongue). It is based on operant conditioning techniques." (WK 2010w; D. Ward 2006)

In this therapy, "[s]tutterers are trained to reduce their speaking rate by stretching vowels and consonants, and using other fluency techniques such as continuous airflow and soft speech contacts. The result is very slow, monotonic, but fluent speech used only in the speech clinic. After the stutterer masters these fluency skills, the speaking rate and intonation are increased gradually....Fluency shaping approaches are often taught in intensive group therapy programs, which may take two to three weeks to complete, but more recently the Camperdown program...us[es] a much shorter schedule...." (WK 2010w; D. Ward 2006)

(b) Stuttering Modification Therapy

The second major treatment of stuttering makes use of "stuttering modification therapy," which "is not to eliminate stuttering but to modify it so that stuttering is easier and less effortful. The rationale is that since fear and anxiety causes increased stuttering, using easier stuttering and with less fear and avoidance, stuttering will decrease. The most widely known approach was published by Charles Van Riper in 1973 and is also known as block modification therapy." (WK 2010w; D. Ward 2006)

(c) Electronic Fluency Devices

The third major treatment of stuttering makes use of "electronic fluency devices," which provide "altered auditory feedback, so that stutterers hear their voice differently" and "have been used for over 50 years in the treatment of stuttering." (WK 2010w; A. Bothe 2007)

However, "[s]tudies of these techniques have had mixed results, with some stutterers showing substantial reductions in stuttering, while others improved only slightly or not at all. In a 2006 review of the efficacy of stuttering treatments, none of the studies on altered auditory feedback met the criteria for experimental quality, such as the presence of control groups." (WK 2010w; A. Bothe 2006 &2007)

(d) Anti-Stuttering Medications

The fourth major treatment of stuttering makes use of "anti-stuttering medicaitons," like "benzodiazepines, anti-convulsants, anti-depressants, antipsychotic and antihypertensive medications, and dopamine antagonists." (WK 2010w)

However, "[a] comprehensive review of pharmacological treatments of stuttering in 2006 concluded that few of the drug trials were methodologically sound. Of those that were, only one, not unflawed study, showed a reduction in the frequency of stuttering to less than 5% of words spoken. In addition, potentially serious side effects of pharmacological treatments were noted, such as weight gain and the potential for blood pressure increases." (WK 2010w; G. Maguire 2000; A. Bothe 2006)

(e) Support Groups and the Self-Help Movement

The fifth major treatment of stuttering makes use of "support groups and the self-help movement," which "gain popularity and support by professionals and people who stutter. One of the basic tenets behind the self-help movement is that since a cure does not exist, quality of living can be improved by improved acceptance of self and stuttering." (WK 2010w)

(f) Diaphragmatic Breathing

And the sixth major treatment of stuttering makes use of "diaphragmatic breathing," which is "costal breathing…as a means by which stuttering can be controlled." (WK 2010w)

Problems in Dealing with Stuttering

However, these different treatments are not without their own problems. Consider, for illustration, a few of them below.

Firstly, one major criticism against these different treatments is that they do not completely restore or create speech naturalness.

For instance, in regard to "fluency shaping therapy"— the "lack of speech naturalness at the end of treatment remains a frequent criticism."

(WK 2010w) And in regard to "anti-stuttering medications"—"few of the drug trials were methodologically sound." (WK 2010w)

Secondly, another major criticism against these different treatments is that they have negative side-effects.

For instance, in regard to "anti-stuttering medications"—"serious side effects of pharmacological treatments were noted, such as weight gain and the potential for blood pressure increases." (WK 2010w; G. Maguire 2000; A. Bothe 2006)

And thirdly, still another major criticism against these different treatments is that they can be time-consuming.

For instance, in regard to the "fluency shaping therapy"—although the training programs can take only a few weeks in the speech clinic, the exercise can continue outside the speech clinic after the patient resumes his daily life in society.

Going Further

Of course, these criticisms against the different treatments of stuttering do not imply their utter uselesness but only show us the two opposing sides of the debate, such that the possibility and desirability of silence (from the perspective of the mind with silence, stuttering, and speech disorder as a case study here) are not to the extent that the respective defenders on both sides would like us to believe.

In addition, the analysis of silence, stuttering, and speech disorder can teach us something valuable about the ontological principles in existential dialectics, and good examples include the absoluteness-relativeness principle, the predictability-unpredictability principle, the explicability-inexplicability principle, the preciseness-vagueness principle, the simpleness-complicatedness principle, the openness-hiddenness principle, the denseness-emptiness principle, the change-constancy principle, the slowness-quickness principle, the expansion-contraction principle, the theory-praxis principle, the convention-novelty principle, the evolution-transformation principle, the symmetry-asymmetry principle, the softness-hardness principle, the regression-progression principle, and the same-difference principle.

For instance, in relation to the absoluteness-relativeness principle, if there is absoluteness (e.g., the absolute concern, of some parents, with stuttering in their children, such that they encourage the children "to slow down, take a breath, say it again, etc." as a way to resolve the problem), there is relativeness (e.g., what is the right thing to do for some parents who encourage their children "to slow down, take a breath, say it again, etc." as a way to resolve the problem is not necessarily so for the children who suffer from stuttering, because this encouragement can instead "increase the child's anxiety and fear, leading to more difficulties with speaking and, in the 'cycle of stuttering' to ever yet more fear, anxiety and expectation"). And the reverse direction also holds true.

In relation to the predictability-unpredictability principle, if there is predictability (e.g., the predictable tendency of a stutterer to receive any of the six major treatments), there is unpredictability (e.g, the more difficult task to predict exactly which treatment will yield success to a given stutterer under what conditions, since there are problems in the treatments, as shown in the criticisms by the critics). And the reverse direction also holds true.

In relation to the explicability-inexplicability principle, if there is explicability (e.g., the explanation, by those in support groups and the self-help movement, that, since "existing behavioral, prosthetic, and pharmaceutical treatments provid[e]...limited relief from the overt symptoms of stuttering" and "since a cure does not exist, quality of living can be improved by improved acceptance of self and stuttering"), there is inexplicability (e.g., the lack of convincing explanation, by those in support groups and the self-help movement, of why self-help is the way to go or is better, simply because existing treatments provide only limited relief and a cure does not really exist, since, for the critics, it is questionable if self-help can do much of anything under these circumstances). And the reverse direction also holds true.

In relation to the preciseness-vagueness principle, if there is preciseness (e.g., the precise classification of the three major characteristics of stuttering), there is vagueness (e.g., the vagueness in the classification of the three major characteristics of stuttering, since it is not clear why there are only three characteristics, not four, five, six, and so on). And the reverse direction also holds true.

In relation to the simpleness-complicatedness principle, if there is simpleness (e.g., the relatively simple recommendation, by the advocates, of the six different treatments for stuttering), there is complicatedness (e.g., the relatively more complicated analysis of the six different treatments for stuttering by questioning their claims and assumptions, as shown in the criticisms by the critics). And the reverse direction also holds true.

In relation to the openness-hiddenness principle, if there is openness (e.g., "the immediately visible and audible symptoms of stuttering" as the portion of an iceberg above the waterline, according to the analogy by Joseph Sheehan), there is hiddenness (e.g., "a broader set of symptoms such as negative emotions hidden below the surface" of the iceberg, as an analogy by Sheehan). And the reverse direction also holds true.

In relation to the denseness-emptiness principle, if there is denseness (e.g., the relatively denser concentration of individuals who prefer "anti-stuttering medications" as a treatment of stuttering to use "benzodiazepines, anti-convulsants, anti-depressants, antipsychotic and antihypertensive medications, and dopamine antagonists," for instance), there is emptiness (e.g., the relatively less dense, or more empty, concentraton of individuals who prefer support groups and the self-help movement to use anti-stuttering medications). And the reverse direction also holds true.

In relation to the change-constancy principle, if there is change (e.g., the ever new ways to treat stuttering, as shown in the six different treatments proposed over time), there is constancy (e.g., the ever constant existence of ways to treat stuttering). And the reverse direction also holds true.

In relation to the slowness-quickness principle, if there is slowness (e.g., the relatively slower speed of traditional group therapy programs for fluency shaping therapy to complete, since they normally take "two to three weeks to complete"), there is quickness (e.g., the relatively faster speed of the more recent Camperdown program for fluency shaping therapy to complete, since they "use a much shorter schedule" than the traditional group therapy programs). And the reverse direction also holds true.

In relation to the expansion-contraction principle, if there is expansion (e.g., the relative spread of the influence of "support groups and the self-help movement" nowadays), there is contraction (e.g., the relatively decline of the influence of traditional "behavioral, prosthetic, and pharmaceutical treatments" nowadays, because they "provid[e]...limited relief from the overt symptoms of stuttering"). And the reverse direction also holds true.

In relation to the theory-praxis principle, if there is theory (e.g., the theoretical construction of electronic devices for auditory purposes in electronical engineering), there is praxis (e.g, the practical application of electronic engineering to the field of stuttering for electronic fluency devices as a treatment of stuttering). And the reverse direction also holds true.

In relation to the convention-novelty principle, if there is convention (e.g., the conventional wisdom of treating stuttering "around eighteenth and nineteenth century Europe," like "cutting the tongue with scissors, removing a triangular wedge from the posterior tongue, and cutting nerves, or neck and lip muscles"), there is novelty (e.g., the alternative novel challenge to the conventional wisdom of treating stuttering, by the six new ways of treating stuttering nowadays). (WK 2010w) And the reverse direction also holds true.

In relation to the evolution-transformation principle, if there is evolution (e.g., the natural evolution of humans to have some members suffering from stuttering in primitive times), there is transformation (e.g., the technical transformation of human ability to treat stuttering by the invention of "anti-stuttering medications" and "electronic fluency devices" for the treatment). And the reverse direction also holds true.

In relation to the symmetry-asymmetry principle, if there is symmetry (e.g., the co-existence of different stutterers in society), there is asymmetry (e.g., some stutterers have only "barely perceptible impediments, for whom the disorder is largely cosmetic"—but other stutterers have "extremely severe symptoms, for whom the problem can effectively prevent most oral communication"). And the reverse direction also holds true.

In relation to the softness-hardness principle, if there is softness (e.g., the parental support of stutterers by encouragin them "to slow down, take a breath, say it again, etc."), there is hardness (e.g., the making fun of the stutterers by others). And the reverse direction also holds true.

In relation to the regression-progression principle, if there is regression (e.g., the regression made by different treatments of stuttering, as shown in the criticisms by the critics), there is progression (e.g., the progress made by different treatments of stuttering, as shown by the "relief from the overt symptoms of stuttering"). And the reverse direction also holds true.

And in relation to the same-difference principle, if there is similarity in outcome (e.g., the contribution to the molding and regulation of beliefs and behaviors, regardless of whether this be done by way of the treatment of stuttering or by way of the acceptance of stuttering), there is difference in outcome (e.g., the contribution to the molding and regulation of beliefs and behaviors by way of the treatment of stuttering for the cultivation of more polished civil manners—but the contribution to the molding and regulation of beliefs and behaviors by way of the acceptance of stuttering for the tolerance of more impulsive civil manners). And the reverse direction also holds true.

Silence and Nature

Silence, when examined from the perspective of nature, can reveal in an amazing way its possibility and desirability, and this can be done by way of a fascinating case study, namely, silence, infrasound, animals, and ghosts—to be addressed below.

Silence, Infrasound, Animals, and Ghosts

A fascinating way to evaluate the possibility and desirability of silence from the perspective of nature has to do with silence, infrasound, animals, and ghosts. (as summarized in *Table 3.4*).

Silence and Infrasound

The word "infrasound" here refers to a "sound that is lower in frequency than 20 Hz (Hertz) or cycles per second, the 'normal' limit of human hearing," which is why humans normally do not hear infrasound, as it becomes non-audible or silent to the human ears—except with the use of special equipments. (WK 2010x)

In other words, "[h]earing becomes gradually less sensitive as frequency decreases, so for humans to perceive infrasound, the sound pressure must be sufficiently high. The ear is the primary organ for sensing infrasound, but at higher levels it is possible to feel infrasound vibrations in various parts of the body." (WK 2010x)

And "[t]he study of such sound waves is sometimes referred to as infrasonics, covering sounds beneath 20 Hz down to 0.001 Hz. This frequency range is utilized for monitoring earthquakes, charting rock and petroleum formations below the earth, and also in ballistocardiography and seismocardiography to study the mechanics of the heart. Infrasound is characterized by an ability to cover long distances and get around obstacles with little dissipation." (WK 2010x)

Infrasound and Animals

In the state of nature, infrasound comes from different sources, be they in relation to natural events or to animals.

For instance, in relation to natural events, "[i]nfrasound sometimes results naturally from severe weather, surf, lee waves, avalanches, earthquakes, volcanoes, bolides, waterfalls, calving of icebergs, aurorae, lightning and upper-atmospheric lightning. Nonlinear ocean wave interactions in ocean storms produce pervasive infrasound vibrations around 0.2 Hz, known as microbaroms." (WK 2010x; M. Garcesc 2003 & 2006; H. Haak 2006; UA 2010)

And in relation to animals, "[w]hales, elephants, hippopotamuses, rhinoceros, giraffes, okapi, and alligators are known to use infrasound to communicate over distances—up to hundreds of miles in the case of whales. It has also been suggested that migrating birds use naturally generated infrasound, from sources such as turbulent airflow over mountain ranges, as a navigational aid." (WK 2010x)

The case of elephants is telling, because "[e]lephants, in particular, produce infrasound waves that travel through solid ground and are sensed by other herds using their feet, although they may be separated by hundreds of kilometres. These calls range from 15–35 Hz and can be as loud as 117 dB, allowing communication for many kilometres, with a possible maximum range of around 10 km (6 mi). These calls may be used to coordinate the movement of herds and allow male elephants to find mates." (WK 2010x; W. Langbauer 1991; D. Larom 1997)

In general, "[a]nimals have been known to perceive the infrasonic waves going through the earth by natural disasters and can use these as an early warning. A recent example of this is the 2004 Indian Ocean earthquake and tsunami. Animals were reported to flee the area long before the actual tsunami hit the shores of Asia. It is not known for sure if this is the exact reason, as some have suggested that it was the influence of electromagnetic waves, and not of infrasonic waves, that prompted these animals to flee." (WK 2010x; C. Kenneally 2004; PBS 2005)

Infrasound and Ghosts

Even more fascinatingly, the study of infrasound in nature can explain something about "ghosts" or "supernatural beings."

To start, it is important to explain the unusualness of infrasound, which is often inaudible to humans, because "20 Hz is considered the normal low frequency limit of human hearing"—unless some special equipments are used. (WK 2010x)

However, this can be qualified, because, under some ideal conditions, for instance, "[w]hen pure sine waves are reproduced under ideal conditions and at very high volume, a human listener will be able to identify tones as low as 12 Hz. Below 10 Hz it is possible to perceive the single cycles of the sound, along with a sensation of pressure at the eardrums." (WK 2010x; H. Olson 1967)

That said—"combined with the natural spread in thresholds within a population, it may have the effect that a very low frequency sound which is inaudible to some people may be loud to others. Infrasound has been known to cause feelings of awe or fear in humans. Since it is not consciously perceived, it can make people feel vaguely that supernatural events are taking place. Some film soundtracks make use of infrasound to produce unease or disorientation in the audience. *Irréversible* is one such movie, as is *Paranormal Activity*." (WK 2010x; MSNBC 2007)

Precisely here in this context, infrasound has been used to explain the phenomena of "ghosts."

For instance, "[r]esearch by Vic Tandy, a lecturer at Coventry University, suggested that an infrasonic signal of 19 Hz might be responsible for some ghost sightings. Tandy was working late one night alone in a supposedly haunted laboratory at Warwick, when he felt very anxious and could detect a grey blob out of the corner of his eye. When Tandy turned to face the grey blob, there was nothing." (WK 2010x)

Then, in the next day, "Tandy was working on his fencing foil, with the handle held in a vise. Although there was nothing touching it, the blade started to vibrate wildly. Further investigation led Tandy to discover that the extractor fan in the lab was emitting a frequency of 18.98 Hz, very close to the resonant frequency of the eye given as 18 Hz by NASA. This was why Tandy had seen a ghostly figure—it was an optical illusion caused by his eyeballs resonating. The room was exactly half a wavelength in length, and the desk was in the centre, thus causing a standing wave which was detected by the foil." (WK 2010x; SD 2010)

After this experience, "Tandy investigated this phenomenon further and wrote a paper entitled *The Ghost in the Machine*. Tandy carried out a number of investigations at various sites believed to be haunted, including the basement of the Tourist Information Bureau next to Coventry Cathedral and Edinburgh Castle." (WK 2010x; V. Tandy 1998 & 2000; C. Arnot 2000; S. Dudman 2003; IAWM 2007)

Problems with the Studies

However, there are problems with these studies. Consider, for illustration, a few of them below, as summarized in *Table 3.4*.

Firstly, one major criticism against the studies is that the explanations may not be true. For instance, in regard to animals, the allegation about the use of infrasound to predict disaster remains to be proved, because "[i]t is not known for sure if this is the exact reason, as some have suggested that it was the influence of electromagnetic waves, and not of infrasonic waves, that prompted these animals to flee." (WK 2010x; C. Kenneally 2004; PBS 2005)

Secondly, another major criticism against the studies is that there are alternative explanations instead. For instance, in regard to the phenomenon of ghosts (or supernatural events), Tandy's explanation is not satisfactory because it tries to explain away the existence of "ghosts" as "an optical illusion"; but, for the critics (or believers of anomalous expereince), there is something more than just "an optical illusion" in anomalous experience like "ghosts" (or "supernatural beings").

And thirdly, still another major criticism against the studies is that the explanation explains the wrong phenomenon. For instance, in regard to the phenomenon of ghosts (or supernatural events), Tandy's explanation, for the critics, is not really about "ghosts," but about something which sounds like "ghosts."

Transcending the Debate

Once again, these criticisms against the studies do not mean to dismiss their validity outright but only show us the two opposing sides of the debate, such that the possibility and desirability of silence (from the perspective of nature with silence, infrasound, animals, and ghosts as a case study here) are not to the extent that the respective defenders on both sides would like us to believe.

More importantly, the analysis of silence, infrasound, animals, and ghosts reveals something vital about the ontological principles in existential dialectics, and good examples include the formalness-informalness principle, the absoluteness-relativeness principle, the predictability-unpredictability principle, the explicability-inexplicability principle, the preciseness-vagueness principle, the simpleness-complicatedness principle, the openness-hiddenness principle, the denseness-emptiness principle, the slowness-quickness principle, the expansion-contraction principle, the theory-praxis principle, the convention-novelty principle, the evolution-transformation principle, the symmetry-asymmetry principle, the softness-hardness principle, the regression-progression principle, and the same-difference principle.

For instance, in relation to the formalness-informalness principle, if there is formalness (e.g., the formal logical requirement of completeness in scientific work, like the research on animals using infrasound to predict disaster), there is informalness (e.g., the non-formal existence of incompleteness in the research on animals using infrasound to predict disaster, because "[i]t is not known for sure if this is the exact reason, as some have suggested that it was the influence of electromagnetic waves, and not of infrasonic waves, that prompted these animals to flee"). And the reverse direction also holds true.

In relation to the absoluteness-relativeness principle, if there is absoluteness (e.g., the absolute advantage of elephants to sense infrasound waves produced at a distance of hundreds of kilometres), there is relativeness (e.g., what is possible for elephants to hear infrasound waves produced at a distance of hundreds of kilometres is not so for humans, because "20 Hz is considered the normal low frequency limit of human hearing," and also not at a distance of hundreds of kilometres away). And the reverse direction also holds true.

In relation to the predictability-unpredictability principle, if there is prediction (e.g., the predictable tendency of someone like Tandy to claim that ghost sightings are optical illusions), there is unpredictability (e.g., the more difficult task for Tandy to predict exactly what particular ghosts, not just illusive or apparent ghostly figures, will be sighted whenever infrasound is emitted). And the reverse direction also holds true.

In relation to the explicability-inexplicability principle, if there is explicability (e.g., the explanation, by the advocates, that some animals can use infrasound to predict disaster), there is inexplicability (e.g., the lack of sufficient explanation, within this view, of why this must be true, since "some have suggested that it was the influence of electromagnetic waves, and not of infrasonic waves, that prompted these animals to flee"). And the reverse direction also holds true.

In relation to the preciseness-vagueness principle, if there is preciseness (e.g., the precise identification of a distance of "several hundreds of kilometres" away that elephants can hear infrasound), there is vagueness (e.g., the vagueness in the identification, since it is not clear exactly how many kilometres away the elephants can hear infrasound). And the reverse direction also holds true.

In relation to the simpleness-complicatedness principle, if there is simpleness (e.g., the relatively simple analysis of ghost sightings as optical illusions by Tandy), there is complicatedness (e.g., the relatively more complicated analysis of ghost sightings as optical illusions by questioning the assumptions and claims used by Tandy, as shown in the criticisms by the critics). And the reverse direction also holds true.

In relation to the openness-hiddenness principle, if there is openness (e.g., the open exploration of the relationship between ghost sightings and optical illusions by Tandy), there is hiddenness (e.g., the hidden bias in Tandy's research, because of his obsession with apparent ghostly figures under the conditions that he encountered, which are not the same of ghosts in the literature of anomalous experience). And the reverse direction also holds true.

In relation to the denseness-emptiness principle, if there is denseness (e.g., the relatively denser concentration of animals to be able to naturally hear infrasound over distances), there is emptiness (e.g., the relatively less dense, or more empty, concentration of humans to be able to naturally hear infrasound over distances, without the help of special equipments). And the reverse direction also holds true.

In relation to the slowness-quickness principle, if there is slowness (e.g., the relatively slower speed of humans "to flee the area long before the actual tsunami hit the shores" during the 2004 Indian Ocean earthquake and tsunami), there is quickness (e.g., the relatively quicker speed of animals "to flee the area long before the actual tsunami hit the shores" in the 2004 Indian Ocean earthquake and tsunami). And the reverse direction also holds true.

In relation to the expansion-contraction principle, if there is expansion (e.g., the relative expansion of the influence of science in trying to explain ghost sightings nowadays), there is contraction (e.g., the relative decline of the influence of esoteric religion to explain ghost sightings nowadays). And the reverse direction also holds true.

In relation to the theory-praxis principle, if there is theory (e.g., the theoretical construction of infrasound in infrasonics), there is praxis (e.g., the practical application of infrosonics to the field of nuclear weapons, as "the Comprehensive Nuclear-Test-Ban Treaty Organization uses infrasound as one of its monitoring technologies"). (WK 2010x) And the reverse direction also holds true.

In relation to the convention-novelty principle, if there is convention (e.g., the conventional view about "ghosts" in esoteric religion), there is novelty (e.g., the alternative novel challenge to the conventional wisdom about "ghosts" in esoteric religion, by Tandy's new view that ghost sightings are optical illusions). And the reverse direction also holds true.

In relation to the evolution-transformation principle, if there is evolution (e.g., the natural evolution of humans to be unable to naturally hear infrasound at low frequences), there is transformation (e.g., the technical transformation of human ability to hear infrasound at low frequences by the invention of special equipments as in ballistocardiography and seismocardiography). And the reverse direction also holds true.

In relation to the symmetry-asymmetry principle, if there is symmetry (e.g., both humans and animals have hearing ability), there is asymmetry (e.g., "sound that is lower in frequency than 20 Hz (Hertz) or cycles per second" is "the 'normal' limit of human hearing"—but elephants can hear sounds at frequencies lower than 20 Hz). And the reverse direction also holds true.

In relation to the softness-hardness principle, if there is softness (e.g., some sounds are pleasant for humans to hear, as in hamonic music), there is hardness (e.g., some sounds cause "feelings of anxiety or fear in humans" to hear, as in infrasound). And the reverse direction also holds true.

In relation to the regression-progression principle, if there is regression (e.g., the regression made by Tandy's work on ghost sightings, as shown in the criticisms by the critics), there is progression (e.g., the progress made by Tandy's work for a new way to understand ghost sightings). And the reverse direction also holds true.

And in relation to the same-difference principle, if there is similarity in outcome (e.g., the contribution to the molding and regulation of beliefs and values, regardless of whether this be done by way of science or by way of esoteric religion, both in regard to the research on ghost sightings), there is difference in outcome (e.g., the contribution to the molding and regulation of beliefs and values by way of science for a more secular worldview—but the contribution to the molding and regulation of beliefs and values by way of esoteric religion for a more sacred worldview). And the reverse direction also holds true.

Silence and Society

Silence, when examined from the perspective of society, can cast an illuminating light on its possibility and desirability, and this can be shown by way of two case studies, namely, (3.4.1) silence, noise-control, and architectural acoustics, and (3.4.2) silence, the law, and the right to quiet enjoyment—to be addressed below, respectively.

Silence, Noise-Control, and Architectural Acoustics

The term "architectural acoustics," as already defined in *Sec. 1.4*, refers to "the science of noise control within buildings." (WK 2010f)

In fact, "he first application of architectural acoustics was in the design of opera houses and then concert halls. More widely, noise suppression is critical in the design of multi-unit dwellings and business premises that generate significant noise, including music venues like bars. The more mundane design of workplaces has implications for noise health effects. Architectural acoustics includes room acoustics, the design of recording and broadcast studios, home theaters, and listening rooms for media playback." (WK 2010f)

Consider, for illustration, four major ways that architectural acoustics are applied in society for noise-control, namely, (a) building skin envelope, (b) inter-space noise control, (c) interior space acoustics, and (d) mechanical equipment noise—to be addressed below, in that order (and summarized in *Table 3.5*).

Building Skin Envelope

The first illustration of the application of architectural acoustics to noise control in society has to do with "building skin envelope." (WK 2010f)

The term "building skin envelope" refers to a "science" which "analyzes noise transmission from building exterior envelope to interior and vice versa. The main noise paths are roofs, eaves, walls, windows, door and penetrations. Sufficient control ensures space functionality and is often required based on building use and local municipal codes." (WK 201f)

For instance, this science becomes necessary when working on "a suitable design for a home which is to be constructed close to a high volume roadway, or under the flight path of a major airport, or of the airport itself." (WK 2010f)

Inter-Space Noise Control

The second illustration of the application of architectural acoustics to noise control in society has to do with "inter-space noise control." (WK 2010f)

The term "inter-space noise control" refers to a "science" which analyzes "limiting and/or controlling noise transmission from one building space to another to ensure space functionality and speech privacy. The typical sound paths are room partitions, acoustic ceiling panels (such as wood dropped ceiling panels), doors, windows, flanking, ducting and other penetrations." (WK 2010f)

For instance, this science becomes necessary when working on a "suitable party wall design in an apartment complex to minimise the mutual disturbance due to noise by residents in adjacent apartments." (WK 2010f)

Interior-Space Acoustics

The third illustration of the application of architectural acoustics to noise control in society has to do with "interior-space acoustics." (WK 2010f)

The term "interior-space acoustics" refers to a "science" which analyzes "controlling a room's surfaces based on sound absorbing and reflecting properties. Excessive reverberation time, which can be calculated, can lead to poor speech intelligibility." (WK 2010f)

The properties of "sound reflection" and "sound absorption are important enough.

Firstly, in regard to "sound reflection"—it helps to "create standing waves that produce natural resonances that can be heard as a pleasant sensation or an annoying one. Reflective surfaces can be angled and coordinated to provide good coverage of sound for a listener in a concert hall or music recital space. To illustrate this concept consider the difference between a modern large office meeting room or lecture theater and a traditional classroom with all hard surfaces." (WK 2010f; G. Ballou 1998)

Secondly, in regard to "sound absorption"—it helps to absorb unwanted sound for noise control. For instance, "[i]nterior building surfaces can be constructed of many different materials and finishes. Ideal acoustical panels are those without a face or finish material that interferes with the acoustical infill or substrate. Fabric covered panels are one way to heighten acoustical absorption. Finish material is used to cover over the acoustical substrate. Mineral fiber board, or Micore, is a commonly used acoustical substrate. Finish materials often consist of fabric, wood or acoustical tile. Fabric can be wrapped around substrates to create what is referred to as a 'pre-fabricated panel' and often provides the good noise absorption if laid onto a wall. Prefabricated panels are limited to the size of the substrate ranging from 2'x 4' to 4' x 10'. Fabric retained in a wall-mounted perimeter track system, is referred to as 'on-site acoustical wall panels.' This is constructed by framing the perimeter track into shape, infilling the acoustical substrate and then stretching and tucking the fabric into the perimeter frame system. On-site wall panels can be constructed to accommodate door frames, baseboard, or any other intrusion. Large panels (generally, greater than 50 square feet) can be created on walls and ceilings with this method. Wood finishes can consist of punched or routed slots and provide a natural look to the interior space, although acoustical absorption may not be great." (WK 2010f)

In any event, when interior-space acoustics is applied for noise control in workplace, there are three major ways to achieve the goal, also known as "the ABCs," as shown below: (WK 2010f)

- "A = Absorb {via drapes, carpets, ceiling tiles, etc.)"
- "B = Block (via panels, walls, floors, ceilings and layout)"
- "C = Cover-up (via sound masking)"

Of course, "[w]hile all three of these are recommended to achieve optimal results, C = Cover-up by increasing background sound produces the most dramatic improvement in speech privacy—with the least disruption and typically the lowest cost." (WK 2010f)

Mechanical Equipment Noise

The fourth illustration of the application of architectural acoustics to noise control in society has to do with "mechanical equipment noise." (WK 2010f)

The term "mechanical equipment noise" refers to a "science" which analyzes "building services noise control" for "controlling noise produced by" mechanical equipments like the following: (WK 2010f)

- "ACMV (air conditioning and mechanical ventilation) systems in buildings, termed HVAC in North America"
- "Elevators"
- "Electrical generators positioned within or attached to a building"
- "Any other building service infrastructure component that emits sound"

This kind of noise control is important, because "[i]nadequate control may lead to elevated sound levels within the space which can be annoying and reduce speech intelligibility. Typical improvements are vibration isolation of mechanical equipment, and sound traps in ductwork. Sound masking can also be created by adjusting HVAC noise to a predetermined level." (WK 2010f)

Problems with the Application of Architectural Acoustics

But the application of architectural acoustics to noise control in society is not without its own problems. Consider, for illustration, a few of them below, as summarized in *Table 3.5*.

Firstly, one major criticism of the application of architectural acoustics to the domain of noise control in society is that health effects from noise can be exaggerated to the point that other real concerns are neglected.

For instance, "hearing loss could be attributed to a variety of factors including age, rather than solely due to excessive exposure to noise." (WK 2010z; C. Schmidt 2005; S. Staples 1996)

Secondly, another major criticism of the application of architectural acoustics to the domain of noise control in society is that government regulations are often not comprehensive enough and not fully enforceable.

For instance, in the U.S., in spite of "the National Environmental Policy Act (NEPA) in 1969 and the Noise Pollution and Abatement Act, more commonly called the Noise Control Act (NCA), in 1972—even "the United States Environmental Protection Agency (EPA) had testified before Congress that 30 million Americans are exposed to non-occupational noise high enough to cause hearing loss and 44 million Americans live in homes impacted by aircraft or highway noise." (WK 2010aa; USSPWC 1972)

In addition, "[m]any municipalities do not follow up on complaints. Even where a municipality has an enforcement office, it may only be willing to issue warnings, since taking offenders to court is expensive." (WK 2010aa)

Thirdly, still another major criticism of the application of architectural acoustics to the domain of noise control in society is that the different ways to conrol noise in society can be very costly.

For instance, it is not surprising that "undeveloped countries" are too poor to be on the frontline of applying architectural acoustics for noise control in a comprehensive way and therefore often "lag...behind" in its application. (WK 2010aa)

And fourthly, one more major criticism of the application of architectural acoustics to the domain of noise control in society is that not all cultures equally dislike noise to the same extent.

For instance, in many Latin-American cultures, the lifestyle is often noisier, be it about how people live together or how they have fun—relatively speaking of course, when compared with those in a Buddhist monastery.

Maneuvering between Two Opposing Sides

Once more, these criticisms against the application of architectural acoustics to the domain of noise control in society do not mean to totally reject its usefulness but only show us the two opposing sides of the debate, such that the possibility and desirability of silence (from the perspective of nature with silence, noise control, and architectural acoustics as a case study here) are not to the extent that the respective defenders on both sides would like us to believe.

Furthermore, the analysis of silence, noise control, and architectural acoustics can teach us something valuable about the ontological principles in existential dialectics, and good examples include the absoluteness-relativeness principle, the predictability-unpredictability principle, the explicability-inexplicability principle, the preciseness-vagueness principle, the simpleness-complicatedness principle, the openness-hiddenness principle, the denseness-emptiness principle, the slowness-quickness principle, the expansion-contraction principle, the theory-praxis principle, the convention-novelty principle, the evolution-transformation principle, the symmetry-asymmetry principle, the softness-hardness principle, the regression-progression principle, and the same-difference principle.

For instance, in relation to the absoluteness-relativeness principle, if there is absoluteness (e.g., the absolute concern of those in "building skin envelope" with "noise transmission from building exterior envelope to interior and vice versa," like working on "a suitable design for a home… close to a high volume roadway, or under the flight path of a major airport"), there is relativeness (e.g., what is the right focus for those in "building skin envelope" is not necessarily so for those in "mechanical equipment noise," who are more concerned with mechanical equipments like "air conditioning and mechanical ventilation systems in buildings," "elevators," etc.). And the reverse direction also holds true.

In relation to the predictability-unpredictability principle, if there is predictability (e.g., the predictable tendency of government regulaion in the U.S. nowadays for noise control), there is unpredictability (e.g., the more difficult task to predict exactly to what extent government regulations will be comprehensive enough and fully enforceable, because of the continued problems with government regulations, as shown in the criticisms by the critics). And the reverse direction also holds true.

In relation to the explicability-inexplicability principle, if there is explicability (e.g., the explanation by those in architectural acoustics that noise control is necessary to minimize negative health effects from noise), there is inexplicability (e.g., the lack of sufficient explanation, within this view, of why noise control is really so important, since, for the critics, other real concerns are neglected, because "hearing loss could be attributed to a variety of factors including age, rather than solely due to excessive exposure to noise"). And the reverse direction also holds true.

In relation to the preciseness-vagueness principle, if there is preciseness (e.g., the precise classification of three major ways to achieve noise control in workplace, namely, "the ABCs"), there is vagueness (e.g., the vagueness in the classification, since it is not clear why there are only three major ways, not four, five, six, and so on). And the reverse direction also holds true.

In relation to the simpleness-complicatedness principle, if there is simpleness (e.g., the relatively simple analysis of the four applications of architectural acoustics to the domain of noise control in society), there is complicatedness (e.g., the relatively more complicated analysis of the application of architectural acoustics to the domain of noise control in society by questioning its assumptions and claims, as shown in the criticisms by the critics). And the reverse direction also holds true.

In relation to the openness-hiddenness principle, if there is openness (e.g., the open exploration of the relationship between architectural acoustics and the minimization of negative health effects from noise), there is hiddenness (e.g., the hidden bias in architectural acoustics, because of its questionable assumptions, as shown in the criticisms by the critics). And the reverse direction also holds true.

In relation to the denseness-emptiness principle, if there is denseness (e.g., the relatively denser concentration of government regulations in the U.S. for higher noise-control standards), there is emptiness (e.g., the relatively less dense, or more empty, concentration of government regulations in "undeveloped countries" for higher noise standards). And the reverse direction also holds true.

In relation to the slowness-quickness principle, if there is slowness (e.g., the relatively slower speed of European countries to set up noise-control standards nowadays, which "generally lag by 10 to 20 years," when compared with those in the U.S.), there is quickness (e.g., the relatively faster speed of the U.S. to set up noise-control standards nowadays, which generally are ahead of European countries "by 10 to 20 years"). And the reverse direction also holds true.

In relation to the expansion-contraction principle, if there is expansion (e.g., the relative spread of government regulations for noise control in society nowadays, especially in the U.S.), there is contraction (e.g., the relative loss of the ability of citizens to engage in noisy activities nowadays without government regulations, especially in the U.S.). And the reverse direction also holds true.

In relation to the theory-praxis principle, if there is theory (e.g., the theoretical construction of acoustics, especially in relation to sound absorption, sound reflection, and sound masking), there is praxis (e.g., the practical application of acoustics to the field of architecture for noise control). And the reverse direction also holds true.

In relation to the convention-novelty principle, if there is convention (e.g., the conventional wisdom in the older days, for instance, "[i]n the 1960s and earlier," when "few people recognized that citizens might be entitled to be protected from adverse sound level exposure"), there is novelty (e.g., the alternative novel challenge to this conventional wisdom about sound level exposure, by "the National Environmental Policy Act in

1969 and the Noise Pollution and Abatement Act, more commonly called the Noise Control Act, in 1972," which set up government regulation for noise control). And the reverse direction also holds true.

In relation to the evolution-transformation principle, if there is evolution (e.g., the natural evolution of humans to live in the state of nature, at times with unwanted noise), there is transformation (e.g., the technical transformation of human living together in society, by the invention of government regulation for noise control). And the reverse direction also holds true.

In relation to the symmetry-asymmetry principle, if there is symmetry (e.g., the co-existence of different government regulations in different countries), there is asymmetry (e.g., the noise-control standards in American government regulation are higher than European government regulation, with a gap "by 10 to 20 years"). And the reverse direction also holds true.

In relation to the softness-hardness principle, if there is softness (e.g., the making of pleasant sound), there is hardness (e.g., the making of annoying sound). And the reverse direction also holds true.

In relation to the regression-progression principle, if there is regression (e.g., the regression made by architectural acoustics, as shown in the criticisms by the critics), there is progression (e.g., the progress made by architectural acoustics for noise control in society). And the reverse direction also holds true.

And in relation to the same-difference principle, if there is similarity in outcome (e.g., the contribution to the molding and regulation of beliefs and behaviors, regardless of whether this be done by way of noise control through government regulation or by way of noise management through free market), there is difference in outcome (e.g., the contribution to the molding and regulation of beliefs and behaviors by way of noise control through government regulation for a relatively more powerful political society—but the contribution to the molding and regulation of beliefs and behaviors by way of noise management through free market for a relativley more powerful civil society). And the reverse direction also holds true.

Silence, the Right to Quiet Enjoyment, and the Law

Another way to evaluate the possibility and desirability of silence from the perspective of society has to do with silence, the right to quiet enjoyment, and the law.

Definition

The term “right to quiet enjoyment” is covered under “nuisance” in tort law (especially under the common law), which, as already defined in *Sec. 1.4*, refers to “that which causes offence, annoyance, trouble or injury,” in that “persons in possession of real property (land owners, lease holders etc.) are entitled to the quiet enjoyment of their lands.” (WK 2010g)

But “this doesn't include visitors or those who aren't considered to have an interest in the land. If a neighbour interferes with that quiet enjoyment, either by creating smells, sounds, pollution or any other hazard that extends past the boundaries of the property, the affected party may make a claim in nuisance.” (WK 2010g)

Under the common law, the term “nuisance” is used “in three ways,” as shown below: (WK 2010g)

- “Activity or Condition”—“that is harmful or annoying to others (e.g., indecent conduct, a rubbish heap or a smoking chimney)”
- “Harm”—that is “caused by the before-mentioned activity or condition (e.g., loud noises or objectionable odors)”
- “Legal Liability”—“that arises from the combination of the two”

And “a nuisance can be either public (also 'common') or private,” as described below: (WK 2010g)

- “Public Nuisance”—refers to, as English scholar Sir J. F. Stephen defined it, “an act not warranted by law, or an omission to discharge a legal duty, which act or omission obstructs or causes inconvenience or damage to the public in the exercise of rights common to all His Majesty's subjects.”
- “Private Nuisance”—refers to “the interference with the right of specific people,” which means “a violation of one's use of quiet enjoyment of land. It doesn't include trespass.”

In other words, “[t]he law of nuisance was created to stop such bothersome activities or conduct when they unreasonably interfered either with the rights of other private landowners (i.e., private nuisance) or with the rights of the general public (i.e., public nuisance).” (WK 2010g)

However, to avoid misunderstanding, there are three distinctions to be made hereafter.

Firstly, one distinction to be clarified is that the “interference” in the definition “was not the result of a neighbor stealing land or trespassing on the land,"which falls under the jursdiction of property law, not tort law. (WK 2010g; W. Prosser 1966)

Secondly, another distinction to be clarified is that "the level of interference must rise above the merely aesthetic. For example: if your neighbour paints their house purple, it may offend you; however, it doesn't rise to the level of nuisance. In most cases, normal uses of a property that can constitute quiet enjoyment cannot be restrained in nuisance either. For example, the sound of a crying baby may be annoying, but it is an expected part of quiet enjoyment of property and does not constitute a nuisance." (WK 2010g)

And thirdly, still another distinction to be clarified is that "any affected property owner has standing to sue for a private nuisance. If a nuisance is widespread enough, but yet has a public purpose, it is often treated at law as a public nuisance. Owners of interests in real property (whether owners, lessors, or holders of an easement or other interest) have standing only to bring private nuisance suits." (WK 2010g)

With these distinctions in mind—"[n]uisance is one of the oldest causes of action known to the common law, with cases framed in nuisance going back almost to the beginning of recorded case law." (WK 2010g)

History of the Legal Development of Nuisance

At the beginning, especially "[i]n the late 19th and early 20th centuries, the law of nuisance became difficult to administer, as competing property uses often posed a nuisance to each other, and the cost of litigation to settle the issue grew prohibitive." (WK 2010g)

Because of this probem (among others), the law of nuisance becomes more complicated over time. Consider two examples below, for illustration.

Firstly, one example is to distinguish the law of nuisance from zoning laws, because "most jurisdictions now have a system of land use planning (e.g. zoning) that describes what activities are acceptable in a given location. Zoning generally overrules nuisance. For example: if a factory is operating in an industrial zone, neighbours in the neighbouring residential zone can't make a claim in nuisance. Jurisdictions without zoning laws, essentially leave land use to be determined by the laws concerning nuisance." (WK 2010g)

Secondly, another example is to distinguish the law of nuisance from environmental laws, because "modern environmental laws are an adaptation of the doctrine of nuisance to modern complex societies, in that a person's use of his property may harmfully affect another's property, or person, far from the nuisance activity, and from causes not easily integrated into historic understandings of nuisance law." (WK 2010g)

Remedies

These distinctions aside—once nuisance suits are brought to court, the issue of "remedies" is important in disputes about nuisance, and the remedies can be in terms of monetary compensation (as in "money damages") or of injunction (as in "equitable remedies"). (WK 2010g)

For example, "under the common law, the only remedy for a nuisance was the payment of damages. However, with the development of the courts of equity, the remedy of an injunction became available to prevent a defendant from repeating the activity that caused the nuisance, and specifying punishment for contempt if the defendant is in breach of such an injunction." (WK 2010g)

This then gave rise to "[t]he law and economics movement," which "has been involved in analyzing the most efficient choice of remedies given the circumstances of the nuisance. In Boomer v. Atlantic Cement Co. a cement plant interfered with a number of neighbors, yet the cost of complying with a full injunction would have been far more than a fair value of the cost to the plaintiffs of continuation. The New York court allowed the cement plant owner to 'purchase' the injunction for a specified amount—the permanent damages. In theory, the permanent damage amount should be the net present value of all future damages suffered by the plaintiff." (WK 2010g)

And the person who is in charge of the official duty of making investigations about nuisance used to be called "the Inspector of Nuisance," but this term is "now archaic," because "the position and/or term having been replaced by others. For example, in the United Kingdom, this office was generally associated with public health and sanitation. Both the 1847 Nuisances Removal and Diseases Prevention Act and the Metropolis Management Act 1855 defined such an office. Similar offices were established across the British Empire. The nearest modern equivalent of this position in the UK is an Environmental health officer. In the United States, a modern example is found in Section 3767 of the Ohio Revised Code which defines such a position to investigate nuisances, where this term broadly covers establishments in which lewdness and alcohol are found." (WK 2010g)

Problems with Public Nuisance

But there are problems with the enforcement of tort law in regard to nuisance (especially for "the right to quiet enjoyment"). Consider, for illustration, some of them below.

Firstly, one major problem with the enforcement of tort law in regard to nuisance (especially for "the right to quiet enjoyment") is that it can be very costly.

For instance, "in Boomer v. Atlantic Cement Co." (as mentioned above), "the cost of complying with a full injunction would have been far more than a fair value of the cost to the plaintiffs of continuation." (WK 2010g)

Secondly, another major problem with the enforcement of tort law in regard to nuisance (especially for "the right to quiet enjoyment") is that it becomes more and more complicated, to the point that it often falls under the jurisdiction of other laws.

For instance, the law of nuisance becomes overruled by other laws like "zoning laws" and "environmental laws" (as indicated above).

Thirdly, still another major problem with the enforcement of tort law in regard to nuisance (especially for "the right to quiet enjoyment") is that what counts as "damage" in tort law is subject to different contentious legal interpretations. (WK 2009)

For instance, "damage" can be interpreted in different ways, be they physical, economc, emotional, reputational, etc.; this disagreement is further compounded, when, in some cases, the "damages…cannot be quantified exactly in monetary terms." (WK 2009)

And the same can be said about the interpretation of "the duty of care" which is important in judging any case of nuisance, since there are different answers, depending on different cases; after all, the "guidelines for the courts in establishing a duty of care" are not universal, as "much of the principle is still at the discretion of judges." (WK 2009)

This issue of legal interpretation was already extensively analyzed in another book of mine titled *The Future of Post-Human Law* (2010g).

Looking Beyond

These criticisms against the enforcement of "the right to quiet enjoyment" are not meant to imply the uselessness of the law of nuisance but only show us the two opposing sides of the debate, such that the possibility and desirability of silence (from the perspective of society with silence, the right to quiet enjoyment, and the law as a case study here) are not to the extent that the respective defenders on both sides would like us to believe.

In addition, the analysis of silence, the right to quiet enjoyment, and the law can reveal something insightful about the ontological principles in existential dialectics, and good examples include the absoluteness-relativeness principle, the partiality-totality principle, the predictability-unpredictability principle, the explicability-inexplicability principle, the preciseness-vagueness principle, the simpleness-complicatedness principle, the openness-hiddenness principle, the denseness-emptiness principle, the slowness-quickness principle, the expansion-contraction principle, the theory-praxis principle, the convention-novelty principle, the evolution-transformation principle, the symmetry-asymmetry principle, the softness-hardness principle, the regression-progression principle, and the same-difference principle.

For instance, in relation to the absoluteness-relativeness principle, if there is absoluteness (e.g., the absolute tendency of a legal interpretation of what counts as "the duty of care" in tort law), there is relativeness (e.g., what counts as "the duty of care" by a judge in tort law is not necessarily so for another judge, because the "guidelines for the courts in establishing a duty of care" are not universal, as "much of the principle is still at the discretion of judges"). And the reverse direction also holds true.

In relation to the partiality-totality principle, if there is partiality (e.g., the relatively partial view from the individual interests of each landowner to purse activities which benefit him), there is totality (e.g., the relativley holistic vew from the social interests of society as a whole, such that the whole is not the sum of its parts, in that society also creates the law of nuisance "to stop…bothersome activities or conduct when they unreasonably interfered either with the rights of other private landowners…or with the rights of the general public"). And the reverse direction also holds true.

In relation to the predictability-unpredictability principle, if there is predictability (e.g. the predictable tendency of remedies to be settled on the basis of either monetary compensation, as in "money damages," or of injunction, as in "equitable remedies"), there is unpredictability (e.g., the more difficult task to predict exactly beforehand what the outcome of a particular nuisance case will turn out to be, because the remedies can be in different ways, just as a given case may not lead to the consideration of remedies as a solution at all). And the reverse direction also holds true.

In relation to the explicability-inexplicability principle, if there is explanation (e.g, the explanation that the principle of "the duty of care" is vital for judging a nuisance case in relation to the right to quiet enjoyment), there is inexplicability (e.g., the lack of sufficient explanation, within this view, of why the principle of "the duty of care" is the only one or the main one to be used in judging a nuisance case in the first place). And the reverse direction also holds true.

In relation to the preciseness-vagueness principle, if there is preciseness (e.g., the precise identification of the issue of "damage" as important in judging the resolution of a nuisance case in common law), there is vagueness (e.g., the vagueness in the identification of the issue of "damage" in tort law, because in some cases, the "damages…cannot be quantified exactly in monetary terms," and what counts as "damage" can be subject to different legal interpretations). And the reverse direction also holds true.

In relation to the simpleness-complicatedness principle, if there is simpleness (e.g., the relatively simple nature of the law of nuisance at the beginning of its historical development), there is complicatedness (e.g., the relatively more complicated nature of the law of nuisance, especially after "the late 19th and early 20th centuries," when "the law of nuisance became difficult to administer, as competing property uses often posed a nuisance to each other, and the cost of litigation to settle the issue grew prohibitive"; later, it was overruled or adapted by other laws like zoning laws and environmental laws). And the reverse direction also holds true.

In relation to the openness-hiddenness principle, if there is openness (e.g., the open exploration of the relationship between the understanding of "the duty of care" in tort law and the legal interpretation of such an understanding), there is hiddenness (e.g., the hidden bias in a given legal interpretation, since, in the case about "the duty of care," in general, "much of the principle is still at the discretion of judges"). And the reverse direction also holds true.

In relation to the denseness-emptiness principle, if there is denseness (e.g., the relatively denser concentration of nuisance cases nowadays which are overruled or adapted by other laws like zoning laws and environmental laws), there is emptiness (e.g., the relatively less dense, or more empty, concentration of nuisance cases nowadays which are totally independent of other laws like zoning laws and environmental laws). And the reverse direction also holds true.

In relation to the slowness-quickness principle, if there is slowness (e.g., the relatively slower speed of a nuisance case to be accepted in court, if "the level of interference" with the rights of others does not rise "above the merely aesthetic"), there is quickness (e.g., the relatively quicker speed of a nuisance case to be accepted in court, if "the level of interference" with the rights of others rises much "above the merely aesthetic"). And the reverse direction also holds true.

In relation to the expansion-contraction principle, if there is expansion (e.g., the relative spread, in our time, of the influence of "[t]he law and economics movement," which "has been involved in analyzing the most efficient choice of remedies given the circumstances of the nuisance"), there is contraction (e.g., the relative decline, in our time, of the influence of religion in judging nuisance cases). And the reverse direction also holds true.

In relation to the theory-praxis principle, if there is theory (e.g., the theoretical construction of efficiency in the law and economics movement), there is praxis (e.g., the practical application of efficiency in the law and economics movement to the field of tort law, like the case about the right to quiet enjoyment). And the reverse direction also holds true.

In relation to the convention-novelty principle, if there is convention (e.g., the conventional wisdom in the older days when the law of nuisance was not closely related to other laws like zoning laws and environmental laws), there is novelty (e.g., the alternative novel challenge to this conventional wisdom nowadays, when the law of nuisance can be overruled and adapted by other laws like zoning laws and environmental laws). And the reverse direction also holds true.

In relation to the evolution-transformation principle, if there is evolution (e.g., the natural evolution of humans to live together in the state of nature, with activities which sometimes annoy others), there is transformation (e.g., the technical transformation of human living together, by the invention of "the right to quiet enjoyment" in tort law which is to regulate activities which may annoy each other). And the reverse direction also holds true.

In relation to the symmetry-asymmetry principle, if there is symmetry (e.g., the existence of both private nuisance and public nuisance), there is asymmetry (e.g., "owners of interests in real property [whether owners, lessors, or holders of an easement or other interest] have standing only to bring private nuisance suits," even when "a nuisance is widespread enough" to cause a public nuisance too, in which case the government can come in and file a case for public nuisance). And the reverse direction also holds true.

In relation to the softness-hardness principle, if there is softness (e.g., the right to quiet enjoyment), there is hardness (e.g., the duty of care in the right to quiet enjoyment, which can be punished if violated, be it on the basis of money or injunction). And the reverse direction also holds true.

In relation to the regression-progression principle, if there is regression (e.g., the regression made by the enforcement of the law of nuisance, as shown in the problems pointed out by the critics), there is progression (e.g., the progress made by the enforcement of the law of nuisance, for the protection of the right to quiet enjoyment). And the reverse direction also holds true.

And in relation to the same-difference principle, if there is similarity in outcome (e.g., the contribution to the molding and regulation of beliefs and behaviors, regardless of whether this be done by way of tort law or by way of individual agreements, in regard to nuisance), there is difference in outcome (e.g., the contribution to the molding and regulation of beliefs and behaviors by way of tort law for the institutionalized order in political society—but the contribution to the molding and regulation of beliefs and behaviors by way of individual agreements for the non-institutioinalized order in the private sphere). And the reverse direction also holds true.

Silence and Culture

Silence, when examined from the perspective of culture, can cast an illuminating light on its possibility and desirability, and this can be shown by way of two case studies, namely, (3.5.1) silence, contemplation, and religion and (3.5.2) silence, and the rhetorical device in debate—to be addressed in what follows, respectively.

Silence, Contemplation, and Religion

A good case study here to understand silence from the perspective of culture has to do with silence, contemplation, and religion.

Silence Love in Religion

The word "contemplation" here has a different meaning when used in the context of religion to understand silence.

For instance, in Christian religion, especially in the version known as "Christian mysticism," the word "contemplation" refers to "a form of prayer distinct from vocal prayer (recitation of words) and from meditation in the strict sense (a form of mental prayer, also called methodical prayer, based on discursive reflection on various considerations)." (WK 2010bb; W. Johnston 1997; LEV 2010)

In accordance to the Catechism of the Catholic Church—"[t]he Christian tradition comprises three major expressions of the life of prayer:

vocal prayer, meditation, and contemplative prayer. They have in common the recollection of the heart." (WK 2010bb)

But contemplation differs from both vocal prayer (or recitation) and meditation, in that it is, in the Christian tradition, "a silent love." (WK 2010bb)

Meditation and Contemplation

The silent love in contemplation differs from "meditation" in different ways, according to different interpretations. Consider a few examples below, for illustration, as summarized in *Table 3.6*.

Firstly, in accordance to the Catechism of the Catholic Church, "[i]n discursive meditation, mind and imagination and other faculties are actively employed in an effort to understand our relationship with God." (WK 2010bb)

But "[i]n contemplative prayer, this activity is curtailed, so that contemplation has been described as 'a gaze of faith,' 'a silent love.'" (WK 2010bb)

Secondly, in accordance to Saint John of the Cross (1906), "the difference between discursive meditation and contemplation" is like this: "The difference between these two conditions of the soul is like the difference between working, and enjoyment of the fruit of our work; between receiving a gift, and profiting by it; between the toil of travelling and the rest of our journey's end." (WK 2010bb; B. Frost 1935)

Thirdly, in accordance to an Oriental Orthodox expert, Mattá al-Miskīn (2003), "[m]editation is an activity of one's spirit by reading or otherwise, while contemplation is a spontaneous activity of that spirit. In meditation, man's imaginative and thinking power exert some effort. Contemplation then follows to relieve man of all effort. Contemplation is the soul's inward vision and the heart's simple repose in God." (WK 2010bb)

But for al-Miskīn, "[t]here is no clear-cut boundary between Christian meditation and Christian contemplation, and they sometimes overlap. Meditation serves as a foundation on which the contemplative life stands, the practice by which someone begins the state of contemplation." (WK 2010bb)

Two Kinds of Contemplation

With these different interpretations in mind—there are two kinds of contemplation to be understood, namely, (a) natural and (b) supernatural—

as discussed below, in that order (and summarized in *Table 3.6*). (WK 2010bb; A. Poulain 1908)

(a) Natural Contemplation

The first kind of contemplation is called "natural" (also known as "acquired") and "[o]ver the centuries,...has been called by various names such as the Prayer of Faith, Prayer of the Heart, Prayer of Simplicity, Prayer of Simple Regard, Active Recollection, Active Quiet and Acquired Contemplation." (WK 2010bb; T. Keating 1991)

In any event, natural contemplation "has been comapred to the attitude of a mother watching over the cradle of her child: she thinks lovingly of the child without reflection and amid interruptions." (WK 2010bb)

As Saint Alphonsus Maria de Liguori argued, natural contemplation "consists in seeing at a simple glance the truths which could previously be discovered only through prolonged discourse"; so "reasoning is largely replaced by intuition," and "affections and resolutions, though not absent, are only slightly varied and expressed in a few words." (WK 2010bb)

But there is something central in natural contemplation, "known as the prayer of simplicity or of simple gaze," which "recurs constantly and easily (although with little or no development) amid many other thoughts, beneficial or otherwise. The prayer of simplicity often has a tendency to simplify itself even in respect to its object, leading one to think chiefly of God and of his presence, but in a confused manner." (WK 2010bb; A. Poulain 1908)

This simple devotion to God in natural contemplation is, for Adolphe Tanquerey, "a simple gaze on God and divine things proceeding from love and tending thereto" and, for Saint Francis de Sales, "a loving, simple and permanent attentiveness of the mind to divine things." (WK 2010bb; W. Johnston 2004)

Or, for St. Teresa, the prayer of simplicity "means taking time frequently to be alone with him who we know loves us" and seeking "him 'whom my soul loves.' It is Jesus, and in him, the Father. We seek him, because to desire him is always the beginning of love, and we seek him in that pure faith which causes us to be born of him and to live in him. In this inner prayer we can still meditate, but our attention is fixed on the Lord himself." (WK 2010bb)

And "[a] person is known to be called to natural contemplation because of succeeding in it with ease, and benefitting from it. This is especially so, if the person has a persistent attraction to this kind of prayer together with difficulty and distaste for discursive meditation." (WK 2010bb)

(b) Superatural Contemplation

The second kind of contemplation is called "supernatural" (and also known as "infused," "higher," "intuitive," "passive," and even "extraordinary") and refers to "a supernatural gift by which a person's mind and will become totally centered on God." (WK 2010bb)

Supernatural contemplation is therefore higher than natural contemplation, as it is "a form of mystical union with God, a union characterized by the fact that it is God, and God only, who manifests himself. Under this influence of God, which assumes the free cooperation of the human will, the intellect receives special insights into things of the spirit, and the affections are extraordinarily animated with divine love. This union that it entails may be linked with manifestations of a created object, as, for example, visions of the humanity of Christ or an angel or revelations of a future event, etc. They include miraculous bodily phenomena sometimes observed in ecstatics." (WK 2010bb; A. Poulain 1908)

Bernhard Häring once described supernatural contemplation in this beautiful way: "[I]n this mystical experience the 'divine partner in the dialogue' is in the foreground rather than the human response. In mystical prayer, the soul experiences the life in grace and from grace as a gift of divine love. It is the experience of faith in its most exalted realization." (WK 2010bb)

For Thomas Dubay (1989), supernatural contemplation is only among "those who try to live the whole Gospel wholeheartedly and who engage in an earnest prayer life"; and, for John Baptist Scaramelli, "asceticism and mysticism are two distinct paths to perfection, the former being the normal, ordinary end of the Christian life, and the latter something extraordinary and very rare." (WK 2010bb; J. Aumann 1985)

It is no wonder that, for some scholars like Jordan Aumann (1985), "this idea of the two paths was 'an innovation in spiritual theology and a departure from the traditional Catholic teaching.'" (WK 2010b)

And, for Saint Teresa of Avila, there are "four degrees or stages of mystical union" in supernatural contemplation, as shown below: (Wk 2010bb)

- "Incomplete Mystical Union"—is "the prayer of quiet or supernatural recollection, when the action of God is not strong enough to prevent distractions, and the imagination still retains a certain liberty"
- "Full or Semi-Ecstatic Union"—is the stage "when the strength of the divine action keeps the person fully occupied but the senses continue to act, so that by making an effort, the person can cease from prayer"

- "Ecstatic Union"—is the stateg "when communications with the external world are severed or nearly so, and one can no longer at will move from that state"
- "Transforming or Deifying Union"—is the "spiritual marriage (properly) of the soul with God"

Thus, "[t]he first three are weak, medium, and the energetic states of the same grace," but the last one is the final "spiritual marriage of the soul with God."

Problems with Silent Love

But this idea of the higher form of supernatural contemplation (for "silent love") is not with criticisms. Consider, for illustration, a few of them below, as summarized in *Table 3.6*.

Firstly, one major criticism of the idea of supernatural contemplation is that it is elitist, only for a few who can make it, while the rest of humanity are not qualified.

After all, Dubay (1989) himself acknowledged that supernatural contemplation is only among "those who try to live the whole Gospel wholeheartedly and who engage in an earnest prayer life"; and, for John Baptist Scaramelli, "asceticism and mysticism are two distinct paths to perfection, the former being the normal, ordinary end of the Christian life, and the latter something extraordinary and very rare." (WK 2010bb; J. Aumann 1985)

Secondly, another major criticism of the idea of supernatural contemplation is that it is not universally accepted in the Christian community.

After all, the idea of supernatural contemplation is popular within "Christian mysticism," which, as Jordan Aumann (1985) warned us, is "a departure from the traditional Catholic teaching.'" (WK 2010b)

And thirdly, still another major criticism of the idea of supernatural contemplation is that its distinction between meditation and contemplation is arbitrary from the perspective of Christian mysticism, with the inherent bias (or favoritism) for contemplation over meditation.

There are historical reasons for this bias for contemplation over meditation in Christian mysticism. Historically, one should remember that "[t]he non-discursive components or stages of meditative practice, which are often called 'contemplation' in Western traditions, were beneficiaries of 'a positive attitude toward contemplation [that] characterized the first fifteen centuries of the Christian era.' But 'a negative attitude has prevailed from the sixteenth century onward,' according to Thomas Keating [1997]." (WK 2010cc)

For Keating (1997), "the emergence of a negative attitude towards contemplation" had much to do with "the 17th century controversy over Quietism and an increasing tendency to distinguish between 'discursive meditation if thoughts predominated; affective prayer if the emphasis was on acts of the will; and contemplation if graces infused by God were predominant....This division of the development of prayer into compartmentalized units entirely separate from one another helped to further the....notion that contemplation was an extraordinary grace reserved to the few.'" (WK 2010cc)

Thus, the idea of supernatural contemplation is no longer popular nowadays as it used to be during "the first fifteen centuries of the Christian era."

Beyond Sacredness and Secularness

These criticisms against the idea of supernatural contemplation for "silent love" are not meant to totally discredit its persuasion but only show us the two opposing sides of the debate, such that the possibility and desirability of silence (from the perspective of culture with silence, contemplation, and religion as a case study here) are not to the extent that the respective defenders on both sides would like us to believe.

Furthermore, the analysis of silence, contemplation, and religion can teach us something valuable about the ontological principles in existential dialectics, and good examples include the formalness-informalness principle, the absoluteness-relativeness principle, the predictability-unpredictability principle, the explicability-inexplicability principle, the preciseness-vagueness principle, the simpleness-complicatedness principle, the openness-hiddenness principle, the denseness-emptiness principle, the slowness-quickness principle, the expansion-contraction principle, the theory-praxis principle, the convention-novelty principle, the evolution-transformation principle, the symmetry-asymmetry principle, the softness-hardness principle, the regression-progression principle, and the same-difference principle.

For instance, in relation to the formalness-informalness principle, if there is formalness (e.g., the formal logical requirement of completeness in a system of ideas, like the discourse on contemplation), there is informalness (e.g., the non-formal existence of incompleteness in the discourse on contemplation, to the point that it cannot prove itself, because it was based on an inherent bias for contemplation over meditation, due to some historical reasons as pointed out by Keating). And the reverse direction also holds true.

In relation to the absoluteness-relativeness principle, if there is absoluteness (e.g., the absolute positive view of contemplation during "the first fifteen centuries of the Christian era"), there is relativeness (e.g., what was true for contemplation to be preferred over meditation during "the first fifteen centuries of the Christian era" is not necessarily so in modern times, with "the emergence of a negative attitude towards contemplation"). And the reverse direction also holds true.

In relation to the predictability-unpredictability principle, if there is predictability (e.g., the predictable tendency of a believer in Christian mysticism to practice contemplation over meditation), there is unpredictability (e.g., the more difficult task to predict which particular believer will succeed in passing through which particular stages of "mystical union" with God, since only a few will ever make it, which is "something extraordinary and very rare"). And the reverse direction also holds true.

In relation to the explicability-inexplicability principle, if there is explicability (e.g., the explanation by John of the Cross that contemplation is higher than meditation in prayer because "the difference between discursive meditation and contemplation" is "like the difference between working, and enjoyment of the fruit of our work; between receiving a gift, and profiting by it; between the toil of travelling and the rest of our journey's end"), there is inexplicability (e.g., the lack of suffiicient explanation by John of the Cross of why this favoritism of contemplation over meditation is necessarily true, since, for the critics like Keating, the biased comparison could be explained by some contingent historical reasons). And the reverse direction also holds true.

In relation to the preciseness-vagueness principle, if there is preciseness (e.g., the precise classification, by Saint Teresa of Avila, of "mystical union" in supernatural contemplation in terms of "four degrees or stages"), there is vagueness (e.g., the vagueness in the classification, since it is not clear why there are only four degrees or stages, not five, six, seven, and so on). And the reverse direction also holds true.

In relation to the simpleness-complicatedness principle, if there is simpleness (e.g., the relatively simple analysis of contemplation in accordance to Christian mysticism), there is complicatedness (e.g., the relatively more complicated analysis of contemplation by questioning its assumptions and claims, as shown in the criticisms by the critics). And the reverse direction also holds true.

In relation to the openness-hiddenness principle, if there is openness (e.g., the open exploration of the relationships between meditation and contemplation in accordance to Christian mysticism), there is hiddenness (e.g., the hidden bias for contemplation in the analysis of the relationships, because of the contingent historical reasons as pointed out by Keating). And the reverse direction also holds true.

In relation to the denseness-emptiness principle, if there is denseness (e.g., the relatively denser concentration of Christian believers who preferred contemplation over meditation in "the first fifteen centuries of the Christian era"), there is emptiness (e.g., the relatively less dense, or more empty, concentration of Christian believers who prefer contemplation over meditation nowadays). And the reverse direction also holds true.

In relation to the slowness-quickness principle, if there is slowness (e.g., the relatively slower speed of someone in the final stage of "mystical union" in contemplative prayer to get distracted), there is quickness (e.g., the relatively faster speed of someone in the first stage of "mystical union" in contemplative prayer to get distracted). And the reverse direction also holds true.

In relation to the expansion-contraction principle, if there is expansion (e.g., the relative expansion of the influence of meditation from the 18th century until now, in the Christian tradition), there is contraction (e.g., the relative decline of the influence of contemplation from the 18th century until now, in the Christian tradition). And the reverse direction also holds true.

In relation to the theory-praxis principle, if there is theory (e.g., the theoretical construction of contemplation in philosophy, theology, and psychology, for instance), there is praxis (e.g,., the practical application of contemplation to the field of Christian mysticism for "mystical union" with God). And the reverse direction also holds true.

In relation to the convention-novelty principle, if there is convention (e.g., the conventional positive wisdom about contemplation over meditation in the older days), there is novelty (e.g., the alternative novel challenge to this conventional wisdom about contemplation, by the new idea that supernatural contemplation is higher than natural contemplation, as Aumann once pointed out that "this idea of the two paths was 'an innovation in spiritual theology and a departure from the traditional Catholic teaching'"). And the reverse direction also holds true.

In relation to the evolution-transformation principle, if there is evolution (e.g., the natural evolution of humans to use their mental faculties like imagination, reasoning, intuition, etc. in the state of nature), there is transformation (e.g., the technical transformation of human ability to use their mental faculties by the invention of systems of ideas like Christian mysticism for prefering contemplation over meditation). And the reverse direction also holds true.

In relation to the symmetry-asymmetry principle, if there is symmetry (e.g., the existence of different stages of mystical union in contemplative prayer), there is asymmetry (e.g., the first three stages are not fully in mystical union with God—but the final stage is fully in mystical union with God, in accordance to Christian mysticism). And the reverse direction also holds true.

In relation to the softness-hardness principle, if there is softness (e.g., the reward of mystical union with God, in accordance to Christian mysticism), there is hardness (e.g., only a few will make it to the final stage of mystical union with God, in accordance to Christian mysticism).

In relation to the regression-progression principle, if there is regression (e.g., the regression made by contemplative prayer, as shown in the criticisms by the critics), there is progression (e.g., the progress made by contemplative prayer for spiritual enlightenment). And the reverse direction also holds true.

And in relation to the same-difference principle, if there is similarity in outcome (e.g., the contribution to the molding and regulation of beliefs and behaviors, regardless of whether this be done by way of contemplative prayer or by way of discursive meditation), there is difference in outcome (e.g., the contribution to the molding and regulation of beliefs and behaviors by way of contemplative prayer for a sacred worldview based more on graces—but the contribution to the molding and regulation of beliefs and behaviors by way of discursive meditation for a sacred worldview based more on thoughts). And the reverse direction also holds true.

Silence, Abduction, and the Rhetorical Device in Debate

Another way to evaluate the possibility and desirability of silence from the perspective of culture concerns silence, abduction, and the rhetorical device in debate.

Argument from Silence, and Abduction

The term "argument from silence" (also known as *argumentum ex silentio* in Latin) refers to "a conclusion based on silence or lack of contrary evidence." (WK 2010dd; OED 1999)

For instance, in classical studies, "it often refers to the deduction from the lack of references to a subject in the available writings of an author to the conclusion that he was ignorant of it. When used as a logical proof in pure reasoning, the argument is classed among the fallacies, but an argument from silence can be a convincing form of abductive reasoning," which is neither deduction nor induction. (WK 2010dd; COD 2006)

A deductive argument may be valid or invalid, an inductive argument may be probable or improbable, but an abductive argument may be reasonable or unreasonable.

Some Textbook Examples of the Argument from Silence

For illustration, hereafter is a good example, where an argument from silence may not be reasonable: (WK 2010dd)

- Bobby: "I know where Mary lives."
- Billy: "Where?"
- Bobby: "I'm not telling you!"
- Billy: "You're just saying that because you don't know!"

In this example, "Billy's conclusion may not be justified: perhaps Bobby doesn't want to tell him." (WK 2010dd)

By contrast, consider the following example, where an argument from silence may be reasonable: (WK 2010dd)

- John: "Do you know any Spanish?"
- Jack: "Of course. I speak it like a native."
- John: "That's good, because I need to know the Spanish phrase for 'Happy Birthday.'"
- Jack: "Sorry, I don't have time for that right now. Maybe tomorrow. Bye."

In this example, "Jack continually refuses to give John the Spanish translation, either by ignoring John or by giving excuses. John then concludes, by argument from silence, that Jack does not in fact know Spanish or does not know it well. In other words, John believes that Jack's ignorance is the most plausible explanation for his silence. Use of argument from silence in this situation is reasonable given that the alternatives, that Jack either doesn't want or is afraid to translate, would be unreasonable without more information." (WK 2010dd)

Hereafter is still another example, where the argument from silence may not be reasonable: (WK 2010dd)

- John: "Do you know your wife's e-mail password?"
- Jack: "Yes, I do as a matter of fact."
- John: "What is it?"
- Jack: "Hey, that's none of your business."

In this example, "[w]hen John repeatedly asked for the password, Jack ignores him completely. Thus, using the argument from silence, John concludes that Jack does not actually know the password. Such an argument from silence, in contrast, may be considered unreasonable, since a password is a security feature not intended to be shared with a stranger simply because they asked. It may be reasonable, by contrast, to assume that Jack does indeed know the password but refuses to say it for legitimate security concerns." (WK 2010dd)

Scholarly Uses of the Argument from Silence

The argument from silence had been applied to scholarly works. Consider, for illustration, two historical examples, one where the argument from silence may be reasonable but the other where it may not.

In regard to a historical example where the argument from silence may be reasonable, it is known that "the editors of Yerushalmi and Bavli" did not mention each other, so "most scholars believe these documents were written independently. Louis Jacobs writes, 'If the editors of either had had access to an actual text of the other, it is inconceivable that they would not have mentioned this. Here the argument from silence is very convincing.'" (WK 2010dd; CC 1999)

And in regard to a historical example where the argument from silence may not be reasonable, it is known that "[i]n the Nova/Horizon episode 'The Case of the Bermuda Triangle,' David Kusche pointed out that the argument from silence has been used to create an improper shifting of the burden of proof in many Bermuda Triangle stories and theories." (WK 2010dd)

Here is how Kusche put it, "Say I claim that a parrot has been kidnapped to teach aliens human language and I challenge you to prove that is not true. You can even use Einstein's Theory of Relativity if you like. There is simply no way to prove such a claim untrue. The burden of proof should be on the people who make these statements, to show where they got their information from, to see if their conclusions and interpretations are valid, and if they have left anything out." (WK 2010dd)

Argument from Silence in Some Legal Traditions

In some legal traditions (like the German), "juries are explicitly instructed not to infer anything because of an accused person's silence; this is known as the right to silence. Thus, the jury may not infer anything from the accused's failure to testify. This in effect bars the use of argument from silence." (WK 2010dd)

However, "statements volunteered by the accused may normally be considered, and in such cases the argument from silence may apply in a limited form. If the accused chooses to testify, the right to silence is forfeited as regards that proceeding. Witnesses also normally have a right to silence as regards any question that is factually incriminating, but that right only bars the jury from making inferences about the witness's conduct. The range of inferences available about the defendant's conduct will vary." (WK 2010dd)

In addition, "[s]ome versions of the Hearsay Rule use a form of argument by silence in that if a party fails to respond to a statement that would normally provoke a denial, this statement is considered as having been adopted by that party and can generally be used against him or her at trial. The reasoning behind this is that if the party had disagreed with the statement, he or she would have objected." (WK 2010dd)

Problems with the Argument from Silence

But there are some problems with the argument from silence. Consider, for illustration, a few of them below.

Firstly, one major problem with the argument from silence is that it does not tell us whether or not a conclusion is true or false but only that it may be reasonable or unreasonable—as already illustrated in the textbook examples and the historical examples.

Secondly, another major problem with the argument from silence is that its application varies from one legal tradition to another and from one context to another—as already shown in the example about the German legal system, where the argument from silence is allowed only selectively.

And thirdly, still another major problem with the argument from silence is that, "in pure reasoning, the argument is classed among the fallacies." (WK 2010dd) So, what is regarded as "reasonable" in abduction is treated as "fallacious" in pure reasoning instead.

A Bigger Picture

Yet, these criticisms against the argument from silence do not imply to dismiss its usefulness but only show us the two opposing sides of the debate, such that the possibility and desirability of silence (from the perspective of culture with silence, abduction, and the argument from silence as a case study here) are not to the extent that the respective defenders on both sides would like us to believe.

Furthermore, the analysis of silence, abduction, and the argument from silence can reveal something important about the ontological principles in existential dialectics, and good examples include the absoluteness-relativeness principle, the predictability-unpredictability principle, the explicability-inexplicability principle, the preciseness-vagueness principle, the simpleness-complicatedness principle, the openness-hiddenness principle, the denseness-emptiness principle, the slowness-quickness principle, the expansion-contraction principle, the theory-praxis principle, the convention-novelty principle, the evolution-transformation principle, the symmetry-asymmetry principle, the softness-hardness principle, the regression-progression principle, and the same-difference principle.

For instance, in relation to the absoluteness-relativeness principle, if there is absoluteness (e.g., the absolute concern with reasonableness in abduction), there is relativeness (e.g., what is regarded as "reasonable" in abduction is treated insteead as "fallacious" in pure reasoning). And the reverse direction also holds true.

In relation to the predictability-unpredictability principle, if there is predictabiltiy (e.g., the predictable tendency of abduction to have a conclusion which is reasonable or unreasonable), there is unpredictability (.e.g, the more difficult task to predict beforehand which particular conversation will end up being abductive in nature, not deductive nor inductive, and which particular one will be reasonable or unreasonable). And the reverse direction also holds true.

In relation to the explicability-inexplicability principle, if there is explicability (e.g, the explanation by those using abductive reasoning that the conclusion may not reasonable under certain conditions, as shown in the textbook and historical examples), there is inexplicability (.e.g, the lack of convincing explanation, by those in abduction, of why being reasonable in abduction cannot be fallacious at all, at least for the critics in pure reasoning). And the reverse direction also holds true.

In relation to the preciseness-vagueness principle, if there is preciseness (e.g,. the precise classification of reasoning in terms of three kinds, namely, deduction, induction, and abduction), there is vagueness (e.g., the vagueness in the classification of reasoning in terms of three kinds, since it is not clear why the traditional classification of only two types, namely, deduction and induction, is not necessarily better, or why the kinds of reasoning cannot be four, five, and so on). And the reverse direction also holds true.

In relation to the simpleness-complicatedness principle, if there is simpleness (e.g., the relatively simple analysis of abduction by its advocates), there is complicatedness (e.g., the relatively more complicated analysis of abduction by questioning its assumptions and claims, as shown in the criticisms by the critics). And the reverse direction also holds true.

In relation to the openness-hiddenness principle, if there is openness (e.g., the open exploration of the three kinds of reasoning), there is hiddenness (e.g., the hidden bias in each kind of reasoning, because each is concerned with its own version of legitimacy in logic, be it about validity, probability, or reasonableness). And the reverse direction also holds true.

In relation to the denseness-emptiness principle, if there is denseness (e.g., the relatively denser concentration of those in pure reasoning to regard abductive reasoning as "fallacious"), there is emptiness (e.g., the relatively less dense, or more empty, concentration of those in abduction to regard abductive reasoning as "fallacious"). And the reverse direction also holds true.

In relation to the slowness-quickness principle, if there is slowness (e.g., the relatively slower readiness of a specialist in abduction to make a conclusion which is valid or invalid), there is quickness (e.g., the relatively quicker readiness of a specialist in abduction to make a conclusion which is reasonable or unreasonable). And the reverse direction also holds true.

In relation to the expansion-contraction principle, if there is expansion (e.g., the relatively more developed ability of an expert in abduction to analyze the reasonableness of an argument), there is contraction (e.g., the relatively less developed ability of an expert in abduction to analyze the validity or probability of an argument). And the reverse direction also holds true.

In relation to the theory-praxis principle, if there is theory (e.g., the theoretical construction of abduction in logic), there is praxis (e.g., the practical application of abduction to the field of legal systems in Germany, especially in relation to "statements volunteered by the accused"). And the reverse direction also holds true.

In relation to the convention-novelty principle, if there is convention (e.g., the conventional wisdom about reasoning in terms of deduction and induction), there is novelty (e.g., the alternative novel challenge to the conventional wisdom about reasoning, by the creation of abduction as the third kind of reasoning). And the reverse direction also holds true.

In relation to the evolution-transformation principle, if there is evolution (e.g., the natural evolution of humans to reason, at least in a primitive way, in the state of nature), there is transformation (e.g., the technical transformation of human ability to reason by the invention of some formal systems of reasoning, like deduction, induction, and abduction). And the reverse direction also holds true.

In relation to the symmetry-asymmetry principle, if there is symmetry (e.g., the existence of three different kinds of reasoning), there is asymmetry (e.g., deduction focuses on the validity of an argument, induction focuses on the probability of an argument, and abduction focuses on the reasonableness of an argument). And the reverse direction also holds true.

In relation to the softness-hardness principle, if there is softness (e.g., the praise of someone's argument as reasonable), there is hardness (e.g., the criticism of someone's argument as unreasonable). And the reverse direction also holds true.

In relation to the regression-progression principle, if there is regression (e.g., the regression made by abduction, as shown in the criticisms by the critics), there is progression (e.g., the progress made by abduction for a different kind of argument based on reasonableness). And the reverse direction also holds true.

And in relation to the same-difference principle, if there is similarity in outcome (e.g., the contribution to the molding and regulation of beliefs and values, regardless of whether this be done by way of pure reasoning or by way of abductive reasoning), there is difference in outcome (e.g., the contribution to the molding and regulation of beliefs and values by way of pure reasoning for a relatively strict form of rationality—but the contribution to the molding and regulation of beliefs and values by way of abductive reasoning for a relatively speculative form of rationality). And the reverse direction also holds true.

The Turmoil of Silence

This comprehensive analysis of the possibility and desirability of silence, from the perspectives of the mind, nature, society, and culture, has shown us the different ways in which silence is both possible and desirable, but not to the extent that the spokespersons from each side would like us to believe.

Silence has its turmoil as the dark side, just as it has its serenity as the bright side—so to speak.

But this is only one part of a larger story, since there is the other side of the story, which concerns sound, the opposite of silence—which, however, was already dealt with in Chapter Two.

These dual analyses are important enough, because those on the side of sound often downgrades silence merely as the opposite of sound without really appreciating it from the vantage point of silence too.

With this completion of the dual analyses here, what then can be concluded at the end of this intellectual journey?

This is the question that Chapter Four will answer—for which we now turn to the concluding chapter.

Table 3.1. Silence and Its Duality

- **Silence and the Mind**
  —Ex: silence, deafness, and the mask of benevolence
  —Ex: silence, stuttering, and speech disorder

- **Silence and Nature**
  —Ex: silence, infrasound, animals, and ghosts

- **Silence and Society**
  —Ex: silence, noise-control, and architectural acoustics
  —Ex: silence, the right to quiet enjoyment, and the law

- **Silence and Culture**
  —Ex: silence, contemplation, and religion
  —Ex: silence, abduction, and the rhetorical device in debate

Notes: The examples in the categories are solely illustrative (not exhaustive), and the comparison is relative (not absolute), nor are they necessarily mutually exclusive. And some can be easily re-classified elsewhere. As generalities, they allow exceptions.
Source: A summary of Ch.3 of *FPHA*

Table 3.2. Silence, Deafness, and the Mask of Benevolence (Part I)

- **On Classification**
  —Ex: by type
  —Ex: by severity
  —Ex: by age of onset
  —Ex: by number of impaired ears

- **On Causes**
  —Ex: by age
  —Ex: by long-term exposure to environmental noises
  —Ex: by genetics
  —Ex: by disease or illness
  —Ex: by medications
  —Ex: by exposure to ototoxic chemicals
  —Ex: by physical trauma

- **On Treatment**
  —Ex: by approaches
  —Ex: by gene therapy
  —Ex: by assistive techniques and devices
  —Ex: by resources and interventions

(*continued on next page*)

Table 3.2. Silence, Deafness, and the Mask of Benevolence (Part II)

- **Problems in Dealing with Hearing Impairment**
  - —Firstly, one major problem in dealing with hearing impairment is that it can be costly.
  - —Secondly, another major problem in dealing with hearing impairment is that there is no guarantee for success.
  - —Thirdly, still another major problem in dealing with hearing impairment is that the aid to those who suffer from hearing impairment can degenerate into a new kind of patronizing and oppressive maltreatment of them, as pointed out by the critics, like Harlan Lane (1992:48-9) in his 1992 book titled *The Mask of Benevolence*.
  - —And fourthly, one more major problem in dealing with hearing impairment is that, contrary to the popular belief in mainstream society, "[i]n Deaf culture, deafness is not considered a condition that needs to be fixed" and is not regarded as "a pathological condition, viewing it instead as a locus of pride." (WK 2010y; P. Ladd 2003; A. Mindess 2006)

Notes: The examples in the categories are solely illustrative (not exhaustive), and the comparison is relative (not absolute), nor are they necessarily mutually exclusive. And some can be easily re-classified elsewhere. As generalities, they allow exceptions.
Source: From *Sec. 3.2.1* of *FPHA*. See book for citations.

Table 3.3. Silence, Stuttering, and Speech Disorder

- **The Issue of Characteristics**
  —Ex: primary behaviors
  —Ex: variability
  —Ex: feelings and attitudes

- **The Issue of Causes**
  —Ex: metabolic
  —Ex: developmental
  —Ex: acquired

- **The Issue of Treatment**
  —Ex: fluency shaping therapy
  —Ex: stuttering modification therapy
  —Ex: electronic fluency devices
  —Ex: anti-stuttering medications
  —Ex: diaphragmatic breathing

- **Problems in Dealing with Stuttering**
  —Firstly, one major criticism against these different treatments is that they do not completely restore or create speech naturalness.
  —Secondly, another major criticism against these different treatments is that they have negative side-effects.
  —And thirdly, still another major criticism against these different treatments is that they can be time-consuming.

Notes: The examples in the categories are solely illustrative (not exhaustive), and the comparison is relative (not absolute), nor are they necessarily mutually exclusive. And some can be easily re-classified elsewhere. As generalities, they allow exceptions.
Source: From *Sec. 3.2.2* of *FPHA.* See book for citations.

Table 3.4. Silence, Infrasound, Animals, and Ghosts

- **Silence and Infrasound**
  —Ex: "[h]earing becomes gradually less sensitive as frequency decreases, so for humans to perceive infrasound, the sound pressure must be sufficiently high. The ear is the primary organ for sensing infrasound, but at higher levels it is possible to feel infrasound vibrations in various parts of the body." (WK 2010x) And "[t]he study of such sound waves is sometimes referred to as infrasonics, covering sounds beneath 20 Hz down to 0.001 Hz." (WK 2010x)

- **Silence and Animals**
  —Ex: in the state of nature, infrasound comes from different sources, be they in relation to natural events or to animals.

- **Silence and Ghosts**
  —Ex:even more fascinatingly, the study of infrasound in nature can explain something about "ghosts" or "supernatural beings." For instance, "[r]esearch by Vic Tandy, a lecturer at Coventry University, suggested that an infrasonic signal of 19 Hz might be responsible for some ghost sightings." (WK 2010x)

- **Problems with the Studies**
  —Firstly, one major criticism against the studies is that the explanations may not be true.
  —Secondly, another major criticism against the studies is that there are alternative explanations instead.
  —And thirdly, still another major criticism against the studies is that the explanation explains the wrong phenomenon.

Notes: The examples in the categories are solely illustrative (not exhaustive), and the comparison is relative (not absolute), nor are they necessarily mutually exclusive. And some can be easily re-classified elsewhere. As generalities, they allow exceptions.
Source: From *Sec. 3.3.1* of *FPHA*. See book for citations.

Table 3.5. Silence, Noise-Control, and Architectural Acoustics

- **The Applications of Architectural Acoustics**
  —Ex: building skin envelope
  —Ex: inter-space noise control
  —Ex: interior-space acoustics
  —Ex: mechanical equipment noise

- **Problems with the Applications of Architectural Acoustics**
  —Firstly, one major criticism of the application of architectural acoustics to the domain of noise control in society is that health effects from noise can be exaggerated to the point that other real concerns are neglected.
  —Secondly, another major criticism of the application of architectural acoustics to the domain of noise control in society is that government regulations are often not comprehensive enough and not fully enforceable.
  —Thirdly, still another major criticism of the application of architectural acoustics to the domain of noise control in society is that the different ways to conrol noise in society can be very costly.
  —And fourthly, one more major criticism of the application of architectural acoustics to the domain of noise control in society is that not all cultures equally dislike noise to the same extent.

Notes: The examples in the categories are solely illustrative (not exhaustive), and the comparison is relative (not absolute), nor are they necessarily mutually exclusive. And some can be easily re-classified elsewhere. As generalities, they allow exceptions.
Source: From *Sec. 3.4.1* of *FPHA*. See book for citations.

Table 3.6. Silence, Contemplation, and Religion (Part I)

- **Silent Love in Religion**
  —Ex: in accordance to the Catechism of the Catholic Church—"[t]he Christian tradition comprises three major expressions of the life of prayer: vocal prayer, meditation, and contemplative prayer. They have in common the recollection of the heart." (WK 2010bb) But contemplation differs from both vocal prayer (or recitation) and meditation, in that it is, in the Christian tradition, "a silent love." (WK 2010bb)

- **Meditation and Contemplation**
  —Ex: in accordance to an Oriental Orthodox expert, Mattá al-Miskīn (2003), "[m]editation is an activity of one's spirit by reading or otherwise, while contemplation is a spontaneous activity of that spirit. In meditation, man's imaginative and thinking power exert some effort. Contemplation then follows to relieve man of all effort. Contemplation is the soul's inward vision and the heart's simple repose in God." (WK 2010bb)

- **Two Kinds of Contemplation**
  —The first kind of contemplation is called "natural" (also known as "acquired") and, for Saint Alphonsus Maria de Liguori, consists in seeing at a simple glance the truths which could previously be discovered only through prolonged discourse"; so "reasoning is largely replaced by intuition," and "affections and resolutions, though not absent, are only slightly varied and expressed in a few words." (WK 2010bb) But there is something central in natural contemplation, "known as the prayer of simplicity or of simple gaze," which "recurs constantly and easily (although with little or no development) amid many other thoughts, beneficial or otherwise. The prayer of simplicity often has a tendency to simplify itself even in respect to its object, leading one to think chiefly of God and of his presence, but in a confused manner." (WK 2010bb; A. Poulain 1908)

(*continued on next page*)

Table 3.6. Silence, Contemplation, and Religion (Part II)

- **Two Kinds of Contemplation (*cont'd*)**
  —The second kind of contemplation is called "supernatural" (and also known as "infused") and refers to "a supernatural gift by which a person's mind and will become totally centered on God." (WK 2010bb) Supernatural contemplation is therefore higher than natural contemplation, and, for Saint Teresa of Avila, there are "four degrees or stages of mystical union" in supernatural contemplation," namely, (a) "incomplete mystical union," "full or semi-ecstatic union," "ecstatic union," and "transforming or deifying union." (Wk 2010bb)

- **Problems with Silent Love**
  —Firstly, one major criticism of the idea of supernatural contemplation is that it is elitist, only for a few who can make it, while the rest of humanity are not qualified.
  —Secondly, another major criticism of the idea of supernatural contemplation is that it is not universally accepted in the Christian community.
  —And thirdly, still another major criticism of the idea of supernatural contemplation is that its distinction between meditation and contemplation is arbitrary from the perspective of Christian mysticism, with the inherent bias (or favoritism) for contemplation over meditation.

Notes: The examples in the categories are solely illustrative (not exhaustive), and the comparison is relative (not absolute), nor are they necessarily mutually exclusive. And some can be easily re-classified elsewhere. As generalities, they allow exceptions.
Source: From *Sec. 3.5.1* of *FPHA*. See book for citations.

• Part Four •

Conclusion

CHAPTER 4
CONCLUSION—THE FUTURE OF ACOUSTICS

> I believe that the use of noise to make music will continue and increase until we reach a music produced through the aid of electrical instruments which will make available for musical purposes any and all sounds that can be heard.
> —John Cage (WK 2010ee)

Beyond Sound and Silence

The analyses of sound (in Chapter Two) and silence (in Chapter Three) are to reveal the extent of their possibility and desirability in acoustics.

These analyses should not be underestimated, in light of the opposing views on the nature of sound and silence (as already described in *Sec. 1.1* and the rest of the book). Contrary to the different opposing views (as introduced in this book), sound and silence are neither possible nor desirable to the extent that their respective ideologues would like us to believe.

Now, one should not misconstrue this critique of the conventional wisdom about the nature of sound and silence in acoustics as a suggestion that it is a worthless field of study, or that some fields of study (related to acoustics) like music, mathematical acoustics, psychoacoustics, architectural acoustics, and so on are to be ignored. Obviously, neither of these extreme views is reasonable either.

Instead, this book provides an alternative (better) way to understand the future of acoustics, especially in the dialectic context of sound and silence—while learning from different approaches in the literature but without favoring any one of them (nor integrating them, since they are not necessarily compatible with each other).

In other words, this book offers a new theory (that is, *the multilateral theory of acoustics*) to go beyond the existing approaches in the literature on acoustics in an original way.

My multilateral theory of acoustics has four major distinct features.

Firstly, my theory makes good use of all theoretical approaches in the literature on sound and silence, be they about the mental argument, the natural argument, the societal argument, or the cultural argument—especially from the most comprehensive combined perspectives of the mind, nature, society, and culture (as will be clear in Chapter Two and Chapter Three).

Secondly, just like many other theories of mine in my previous books, my theory here does not heavily favor any specific theory over others in the literature, nor trying to integrate them (as they are not necessarily compatible with each other). For this reason, the word "multilateral" in the title of the theory is suggestive.

Thirdly, my theory treats the issue of acoustics in the distinctive dialectic context of sound and silence, such that one is not to be reduced as part of the analysis of the other.

And fourthly, it contains seventeen major theses, namely, (a) the first thesis: the absoluteness-relativeness principle, (b) the second thesis: the predictability-unpredictability principle, (c) the third thesis: the explicability-inexplicability principle, (d) the fourth thesis: the preciseness-vagueness principle, (e) the fifth thesis: the simpleness-complicatedness principle, (f) the sixth thesis: the openness-hiddenness principle, (g) the seventh thesis: the denseness-emptiness principle, (h) the eighth thesis: the slowness-quickness principle, (i) the nineth thesis: the expansion-contraction principle, (j) the tenth thesis: the theory-praxis principle, (k) the eleventh thesis: the convention-novelty principle, (l) the twelveth thesis: the evolution-transformation principle, (m) the thirteenth thesis: the symmetry-asymmetry principle, (n) the fourteenth thesis: the softness-hardness principle, (o) the fifteenth thesis: the regression-progression principle, (p) the sixteenth thesis: the sameness-difference principle, and (q) the seventeenth thesis: the post-human creativeness—to be elaborated in the rest of the book, with a summary in the concluding chapter.

Of course, as this is something that I often emphasized in my previous books, other principles (besides the 16 as cited above) are also relevant, but these 16 are the most relevant in the current case study (in terms of the number of citations of each principle in the book).

Even then, in some cases, the difference between any two given principles, for instance, in terms of the number of citations in a book, is rather small, so the reason in those cases is more aesthetic (than otherwise), because it looks nicer to list only 16 theses for 16 principles (than 22 theses for all of the 22 principles) in the Table of Contents.

This is true, even if different studies of the same kind can yield different views about the degree of relevance for each principle, depending on the specific nature of a research in question, needless to say. So, if a different author analyzes the same subject matter in a different way, the relevance of the principles will be different.

With this clarification in mind—if successful, this seminal project is to fundamentally change the way that we think about acoustics, from the combined perspectives of the mind, nature, society, and culture, with enormous implications for the human future and what I originally called its "post-human" fate.

In the Category of Method

Firstly, in regard to the formalness-informalness principle (on the formal requirements of logical systems), if there is formalness (e.g., the formal logical requirement of completeness in scientific work, like the research on animals using infrasound to predict disaster, in *Sec. 3.3.1*; and the formal logical requirement of completeness in a system of ideas, like the discourse on contemplation, in *Sec. 3.5.1*), there is informalenss too (e.g., the non-formal existence of incompleteness in the research on animals using infrasound to predict disaster, because "[i]t is not known for sure if this is the exact reason, as some have suggested that it was the influence of electromagnetic waves, and not of infrasonic waves, that prompted these animals to flee," in *Sec. 3.3.1*; and the non-formal existence of incompleteness in the discourse on contemplation, to the point that it cannot prove itself, because it was based on an inherent bias for contemplation over meditation, due to some historical reasons as pointed out by Keating, in *Sec. 3.5.1*). And the reverse direction also holds true.

And secondly, in regard to the partiality-totality principle (on the relationships between parts and whole), if there is partial analysis (e.g., the more partial perspective of an introvert or, alternatively, an extrovert, in *Sec. 2.2.2*; and the relatively partial view from the individual interests of each landowner to purse activities which benefit him, in *Sec. 3.4.2*), there is also holistic analysis (e.g., the more holistic view of all the perspectives of both introverts and extroverts, such that the whole is not the sum of its parts, in that they do not all add up into a harmonious entirety, in *Sec. 2.2.2*; and the relativley holistic vew from the social interests of society as a whole, such that the whole is not the sum of its parts, in that society also creates the law of nuisance "to stop...bothersome activities or conduct when they unreasonably interfered either with the rights of other private landowners...or with the rights of the general public," in *Sec. 3.4.2*). And the reverse direction also holds true.

In the Category of Structure

Firstly, in regard to the finiteness-transfiniteness principle (on the nature of numbers), if there is finiteness (e.g., the finite number of categories on the single continuum, like "extroversion," "introversion," and "ambiversion," in *Sec. 2.2.2*; and the finite number of the kinds of animals like bats and others which eat moths, in *Sec. 2.3.1*), there is also transfiniteness (e.g., the transfinite number of points on the single continuum that any person can choose from, so as to identify with, in *Sec. 2.2.2*; and the transfinite number of individual moths which have been eaten by animals like bats and others, in the history of the earth, in *Sec. 2.3.1*). And the reverse direction also holds true.

In the Category of Process

Firstly, in regard to the change-constancy principle (on the alteration of things), if there is change (e.g., the ever new ways to measure absolute threshold of hearing, as shown by the use of different methods over time like "classical methods," "modified classical methods," "forced-choice methods," and "adaptive methods," in *Sec. 2.2.1*; the ever new methods used by submarines for navigation over time, in *Sec. 2.4.1*; the ever new topics of small talk over time, in *Sec. 2.4.2*; the ever new ways to do noise music, as shown in the seven historical examples, in *Sec. 2.5.2*; the ever new ways for hearing-impaired people to be treated, in *Sec. 3.2.1*; and the ever new ways to treat stuttering, as shown in the six different treatments proposed over time, in *Sec. 3.2.2*), there is constancy too (e.g., the ever constancy of the attempt to measure absolute threshold of hearing over time, in *Sec. 2.2.1*; the ever constancy of the availability of methods to be used by submarines for navigation, in *Sec. 2.4.1*; the ever constant existence of topics of small talk, in *Sec. 2.4.2*; the ever constancy of the appeal of noise as a basis for music for some individuals, in *Sec. 2.5.2*; the ever constant existence of ways for hearing-impaired people to be treated, in *Sec. 3.2.1*; and the ever constant existence of ways to treat stuttering, in *Sec. 3.2.2*). And the reverse direction also holds true.

And secondly, in relation to the order-chaos principle (on the pattern of things), if there is order (e.g., the relatively orderly arrangement of either "yes" or "no after each presentation of a sound stimuli in "the method of constant stimuli" to measure ATH, in *Sec. 2.2.1*; the relatively orderly patterns of harmonic series, like the "combination of composite waves with short fundamental frequencies and shared or closely related partials," in *Sec. 2.5.1*; and the relatively orderly nature of harmonic series in traditional melodic music, in *Sec. 2.5.2*), there is chaos too (e.g., the relatively disorderly or random presentation of the sound stimuli, since they are presented "at completely random order" by the tester in "the method of constant stimuli," in *Sec. 2.2.1*; the relatively disorderly nature of chaotic music by Cage, with the use of chance in *I Ching* and sometimes dice and coin too, in *Sec. 2.5.1*; and the relatively chaotic nature of dissonance in noise music, in *Sec. 2.5.2*). And the reverse direction also holds true.

In the Category of Agency

Firstly, in regard to the seriousness-playfulness principle (on the extent of seriousness), if there is seriousness (e.g., an extrovert can be serious in political public demonstrations, in *Sec. 2.2.2*; the serious business of a male whale to make sounds to a female whale for reproduction, in *Sec. 2.3.1*; the serious business of social functions in small talk, in *Sec. 2.4.2*; the serious nature of chaotic music for a different musical sensibility, in *Sec .2.5.1*; and the serious business by those in noise music to create a new kind of music, in *Sec. 2.5.2*), there is playfulness (e.g., an extrovert can enjoy himself by having fun in loud parties, like in political public demonstrations, in *Sec. 2.2.2*; the playful nature of a male whale to make sounds to a female whale for seduction, in *Sec. 2.3.1*; the playful nature of small talk, when men engage in "playful insults," in *Sec. 2.4.2*; the playful nature of chaotic music when it plays with chance in the composition, in *Sec. 2.5.1*; and the playful nature by those in noise music to play with different techniques and instruments in the process of creating new music, in *Sec. 2.5.2*). And the reverse direction also holds true.

1st Thesis: The Absoluteness-Relativeness Principle

The first thesis refers to the absoluteness-relativeness principle (on the multiplicity of things) in the category of "method" in existential dialectics, which was first proposed in *FPHK*, in that there is the multiplicity of things in reality, be they about entities, qualities (or properties), and relationships. If there is something absolute, there is likewise something relative. And there is no absoluteness without relativeness—and vice versa.

Both absoluteness and relativeness here are also relevant to different modalities often cited in the literature on ontology, such as possibility (e.g., something "can" happen) and its opposite (e.g., impossibility), probability (e.g., something "will" happen) and its opposite (e.g., improbability), and necessity (e.g., something "should" happen) and its opposite (e.g., contingency).

For instance, on the one hand, there are absolute viewpoints (e.g., the measurement of the absolute threshold of hearing by "the method of limits," with "several series of descending and ascending runs," in *Sec. 2.2.1*; the absolute view of an extrovert for more "social interaction" overall; in *Sec. 2.2.2*; the absolute orientation of bats to eat moths, in *Sec. 2.3.1*; the absolute tendency of "S-waves" which do not travel "in fluids with essentially no shear strength, such as air or water," in *Sec. 2.3.2*; the absolute tendency of submarines to not use active sonar systems and radar navigation when in non-friendly waters, in *Sec. 2.4.1*; the absolute norms of "male followers of the Islamic faith [to] avoid asking about another man's wife" in small talk, in *Sec. 2.4.2*; the absolute tendency of those who are fun with harmonic series in music to focus on the combination of sound waves which are "consonant," in *Sec. 2.5.1*; the absolute view of Russolo that "found traditional melodic music confining and envisioned noise music as its future replacement," in *Sec. 2.5.2*; the absolute concern, in a definition of hearing impairment, with hearing sensitivity, in *Sec. 3.2.1*; the absolute concern, of some parents, with stuttering in their children, such that they encourage the children "to slow down, take a breath, say it again, etc." as a way to resolve the problem, in *Sec. 3.2.2*; the absolute advantage of elephants to sense infrasound waves produced at a distance of hundreds of kilometres, in *Sec. 3.3.1*; the absolute concern of those in "building skin envelope" with "noise transmission from building exterior envelope to interior and vice versa," like working on "a suitable design for a home… close to a high volume roadway, or under the flight path of a major airport," in *Sec. 3.4.1*; the absolute tendency of a legal interpretation of what counts as "the duty of care" in tort law, in *Sec. 3.4.2*; the absolute positive view of contemplation during "the first fifteen centuries of the Christian era," in *Sec. 3.5.1*; and the absolute concern with reasonableness in abduction, in *Sec. 3.5.2*).

On the other hand, there are likewise relative counterparts (e.g., what is true for "the method of limits" in its use of "descending and ascending runs" is not so for "the method of constant stimuli" with "no ascending or descending trials," in *Sec. 2.2.1*; what is good for an extrovert for more "social interaction" overall is not necessarily so for an introvert for more "private space" overall, in *Sec. 2.2.2*; what is good for bats to eat moths is not necessarily so for moths to be eaten by bats, in *Sec. 2.3.1*; what is true for "S-waves" which travel only in solids is not so for "interface waves" which travel through solid-air, in *Sec. 2.3.2*; what is true in non-friendly waters when submarines do not use active sonar systems and radar navigation in order to avoid detection is not necessarily so in friendly waters when submarines can use active sonar systems and radar navigation to enter and exit ports, in *Sec. 2.4.1*; what is not appropriate for "male followers of the Islamic faith [to]...ask...about another man's wife" in small talk" is not necessarily so for members of other societies, say, in the U.S., who may not mind, in *Sec. 2.4.2*; what is beautiful for the fans of harmonic music to focus on consonant sounds is not necessarily so for the fans of John Cage, who prefer chaos and chance in music instead, in *Sec. 2.5.1*; what is appealing for Russolo in regard to noise music is not necessary so for the defenders of traditional melodic music who prefer harmonic series, in *Sec. 2.5.2*; what is the right concern with hearing sensitivity in a definition of hearing impairment is not necessarily so for those who focus on hearing clarity in an alternative definition, in *Sec. 3.2.1*; what is the right thing to do for some parents who encourage their children "to slow down, take a breath, say it again, etc." as a way to resolve the problem is not necessarily so for the children who suffer from stuttering, because this encouragement can instead "increase the child's anxiety and fear, leading to more difficulties with speaking and, in the 'cycle of stuttering' to ever yet more fear, anxiety and expectation," in *Sec. 3.2.2*; what is possible for elephants to hear infrasound waves produced at a distance of hundreds of kilometres is not so for humans, because "20 Hz is considered the normal low frequency limit of human hearing," and also not at a distance of hundreds of kilometres away, in *Sec. 3.3.1*; what is the right focus for those in "building skin envelope" is not necessarily so for those in "mechanical equipment noise," who are more concerned with mechanical equipments like "air conditioning and mechanical ventilation systems in buildings," "elevators," etc., in *Sec. 3.4.1*; what counts as "the duty of care" by a judge in tort law is not necessarily so for another judge, because the "guidelines for the courts in establishing a duty of care" are not universal, as "much of the principle is still at the discretion of judges," in *Sec. 3.4.2*; what was true for contemplation to be preferred over meditation during "the first fifteen centuries of the Christian era" is not necessarily so in modern times, with "the emergence of a negative attitude towards contemplation," in *Sec. 3.5.1*; and what is regarded as "reasonable" in abduction is treated insteead as "fallacious" in pure reasoning, in *Sec. 3.5.2*).

And the reverse direction also holds true.

However, it should be stressed (as this is something that I regularly did in my previous books, *almost verbatim*) that there are different shades of gray (or different degrees of truth) in the two opposites, and the classification is not necessarily mutually exclusive either. In fact, even when some combinations of the two occur, they only end up sharing the same dialectic relationship, but in a different degree.

Also, there are some other relationships (as a kind of family resemblance) which have something in common with the principle, but they are not exactly the same but only more or less comparable, which varies from case to case.

Good examples of family resemblance in relation to the principle are uniformity-diversity, internalness-externalness, immanence-transcendence, and so forth.

2nd Thesis: The Predictability-Unpredictability Principle

The second thesis is called the predictability-unpredictability principle (on the occurrence of events) in the category of "method" in existential dialectics (which was first proposed in *FC* and, later, other books of mine), in that both predictability and unpredictability have a major role to play in the occurrence of things, so that neither determinism nor indeterminism wins the centuries-old fight. There is no predictability without unpredictability—and vice versa.

There are events which are predictable, just as there are those which are not. Or what is regarded as unpredictable at one point in time may turn out to be predictable later, and, conversely, what is deemed as predictable may turn out to not be so predictable. Even in predictability, outcomes are subject to uncertainty, the degree of which varies from case to case.

For instance, on the one hand, there is predictability (e.g., the predictable tendency of a subject in "the method of constant stimuli" to respond either "yes" or "no" after each presentation of a sound stimuli, in *Sec. 2.2.1*; the predictable tendency of an extrovert to enjoy "partying," in *Sec. 2.2.2*; the predictable tendency of flying moths to take counter-measures like dropping a few inches in their flight to evade attack by bats, in *Sec. 2.3.1*; the predictable tendency of seismologists to engage in "[m]ore general forecasts" about "seismic hazard," in *Sec. 2.3.2*; the predictable tendency of submarines to operate in stealth mode when in non-friendly waters, in order to avoid detection, in *Sec. 2.4.1*; the predictable tendency of small talk to address "[w]eather…in regions where the climate has great variation and can be unpredictable," in *Sec. 2.4.2*; the predictable tendency of consonant sounds to be produced when they are "composite waves with short fundamental frequencies and shared or closely related partials," in *Sec. 2.5.1*; the predictability of those in post-industrial and post-digital music like "Richie Hawtin, Jan Jelinek, Ricardo Villalobos, Decomposed Subsonic, Trentemøller, and other Minimal techno and Microhouse DJs" to use "noise elements such as buzz, hum, and clicks as sonic flavor since the early 1990s," in *Sec. 2.5.2*; the predictable tendency of those who accept gene therapy to seek the regrowth of cochlear hair cells, in *Sec. 3.2.1*; the predictable tendency of a stutterer to receive any of the six major treatments, in *Sec. 3.2.2*; the predictable tendency of someone like Tandy to claim that ghost sightings are optical illusions, in *Sec. 3.3.1*; the predictable tendency of government regulaion in the U.S. nowadays for noise control, in *Sec. 3.4.1*; the predictable tendency of remedies to be settled on the basis of either monetary compensation, as in "money damages," or of injunction, as in "equitable remedies," in *Sec. 3.4.2*; the predictable tendency of a believer in Christian mysticism to practice contemplation over meditation, in *Sec. 3.5.1*; and the predictable tendency of abduction to have a conclusion which is reasonable or unreasonable, in *Sec. 3.5.2*).

On the other hand, there is unpredictability (e.g., the unpredictability of the presentation of the sound stimuli, which is given by the tester "at completely random order" in "the method of constant stimuli," in *Sec. 2.2.1*; the more difficult task to predict exactly which specific point on the single continuum of extroversion-introversion an extrovert is at a given time, in *Sec. 2.2.2*; the more difficult task to predict when a particular counter-measure by a flying moth will necessarily succeed in fighting off the attack by a bat, in *Sec. 2.3.1*; the more difficult task for seismologists to make forecasts about "seismic hazard" which are very accurate and very timely, in *Sec. 2.4.2*; the more difficult task to predict exactly which particular trip that a submarine takes in non-friendly waters will necessarily be successful, since it can collide underwater from time to time with other submarines, drilling rigs, undersea mountains, and so on, in *Sec. 2.4.1*; the more difficult task to predict exactly what particular aspects of weather a small talk at a given point will concern with, in *Sec. 2.4.2*;

the more difficult task to predict exactly which particular cultures in future eras will appreciate consonant sounds, under what occassions, etc., in *Sec. 2.5.1*; the more difficult task to predict exactly how long and to what extent this use of "buzz, hum, and clicks as sonic flavor" will last in which particular future eras, in *Sec. 2.5.2*; the more difficult task to predict exactly which particular gene therapy will be successful, under what conditions, since "the restoration of hearing sensitivity as the sensory cells may or may not make connections with neurons that carry the signals from hair cells to the brain," in *Sec. 3.2.1*; the more difficult task to predict exactly which treatment will yield success to a given stutterer under what conditions, since there are problems in the treatments, as shown in the criticisms by the critics, in *Sec. 3.2.2*; the more difficult task for Tandy to predict exactly what particular ghosts, not just illusive or apparent ghostly figures, will be sighted whenever infrasound is emitted, in *Sec. 3.3.1*; the more difficult task to predict exactly to what extent government regulations will be comprehensive enough and fully enforceable, because of the continued problems with government regulations, as shown in the criticisms by the critics, in *Sec. 3.4.1*; the more difficult task to predict exactly beforehand what the outcome of a particular nuisance case will turn out to be, because the remedies can be in different ways, just as a given case may not lead to the consideration of remedies as a solution at all, in *Sec. 3.4.2*; the more difficult task to predict which particular believer will succeed in passing through which particular stages of "mystical union" with God, since only a few will ever make it, which is "something extraordinary and very rare," in *Sec. 3.5.1*; and the more difficult task to predict beforehand which particular conversation will end up being abductive in nature, not deductive nor inductive, and which particular one will be reasonable or unreasonable, in *Sec. 3.5.2*).

And the reverse direction also holds true.

Again, it should be stressed, however, that there are different shades of gray (or different degrees of truth) in the two opposites, and the classification is not necessarily mutually exclusive either. In fact, even when some combinations of the two occur, they only end up sharing the same dialectic relationship, but in a different degree.

Also, there are some other relationships (as a kind of family resemblance) which have something in common with the principle, but they are not exactly the same but only more or less comparable, which varies from case to case.

Good examples of family resemblance in relation to the principle are sureness-arbitrariness, certainty-uncertainty, sureness-doubtfulness, likeliness-riskiness, and so forth.

3rd Thesis: The Explicability-Inexplicability Principle

The third thesis is called the explicability-inexplicability principle (on the underlying mechanisms of things) in the category of "method" in existential dialectics (which was proposed in *FPHU* and other books of mine), in that both explicability and inexplicability are part of the understanding of things. There is no explicability without inexplicability—and vice versa.

This principle tells us the dual nature of the research dilemma, in that, if reality can be explained in some ways, it also has its other ways which are not quite explainable, at a given point in time.

For instance, on the one hand, there is explicability (e.g., the explanation by psychoacousticians that humans have the audio range "between 20 Hz and 20,000 Hz," in *Sec. 2.2.1*; the explanation by Jung that the differences between extroversion and introversion have more to do with "a person's psychic energy," in *Sec. 2.2.2*; the explanation by the researchers in the two universities in Canada that "[t]he sounds are probably not advertising that the walnut sphinx caterpillars are distasteful" but that "[t]he birds simply appear to be startled," in *Sec. 2.3.1*; the explanation by seismologists that they understand the interior of the Earth better nowadays, because of "seismic tomography with earthquake waves," which allowed them, for instance, "to identify convection cells, mantle plumes and other large-scale features of the inner Earth," in *Sec. 2.3.2*; the explanation that sound reflection/scattering on the seabed "depends on the bottom material types and depth of the layers," as proposed by the theories developed by N. Chotiros and M. Buckingham, for instance, in *Sec. 2.4.1*; the explanation that small talk is not needed in situations involving "couples in an intimate relationship," which "can signal their level of closeness by a lack of small talk," in *Sec. 2.4.2*; the explanation that sound waves which are "consonant" are therefore "pleasant," as opposed to those dissonant ones which are "unpleasant" to hear, in *Sec. 2.5.1*; the explanation by those in noise music that dissonance is a desirable quality for music, in *Sec. 2.5.2*; the explanation by those who care for deaf people that hearing impairment is something to be corrected for the elimination or reduction of their suffering, in *Sec. 3.2.1*; the explanation, by those in support groups and the self-help movement, that, since "existing behavioral, prosthetic, and pharmaceutical treatments provid[e]...limited relief from the overt symptoms of stuttering" and "since a cure does not exist, quality of living can be improved by improved acceptance of self and stuttering," in *Sec. 3.2.2*; the explanation, by the advocates, that some animals can use infrasound to predict disaster, in *Sec. 3.3.1*; the explanation by those in architectural acoustics that noise control is necessary to minimize negative health effects from noise, in *Sec. 3.4.1*; the explanation that the principle of "the duty of care" is vital for judging a nuisance case in relation to the right to quiet enjoyment, in *Sec. 3.4.2*; the explanation by John of the Cross that contemplation is higher than meditation in prayer because "the difference between discursive meditation and contemplation" is "like the difference between working, and enjoyment of the fruit of our work; between receiving a gift, and profiting by it; between the toil of travelling and the rest of our journey's end," in *Sec. 3.5.1*; and the explanation by those using abductive reasoning that the conclusion may not reasonable under certain conditions, as shown in the textbook and historical examples, in *Sec. 3.5.2*).

On the other hand, there is inexplicability (e.g., the lack of sufficient explanation by psychoacousticians of why humans have this kind of audio range but other animals like "dogs can perceive vibrations higher than 20,000 Hz," in *Sec. 2.2.1*; the lack of sufficient explanation, if viewed within the framework of Jung, of why there cannot be other explanations on the basis of brain physiology as suggested by Eysenck or biological causes as proposed by others, for example, in *Sec. 2.2.2*; the lack of sufficient explanation by the researchers of why this explanation must be true, since there are alternative explanations, for example, in that the birds may get scared by the sounds, that the sounds may disorient their flight, and so on, in *Sec. 2.3.1*; the lack of sufficient explanation by seismologists of why they are still unable to predict earthquakes very accurately and very timely, if they think that they understand the interior of the Earth so well—since, for the critics, there can still be a lot of things that the sesimologists do not quite understand, in *Sec. 2.3.2*; the lack of sufficient explanation in the theories of why there are not other important factors in determining sound reflection/scattering on the seabed, except material types and depth of the layers, in *Sec. 2.4.1*; the lack of sufficient explanation, within this view, of why the absence of small talk among couples in an intimate relationship cannot mean something else instead, like a relationship in crisis, and so on—not that this must necessarily imply good intimate relatinship, in *Sec. 2.4.2*; the lack of sufficient explanation, within this view, of why "consonant" sounds in music are necessarily "pleasant" and "dissonant" ones are necessarily "unpleasant," as some people, for instance, enjoy chaotic music, in *Sec. 2.5.1*; the lack of sufficient explanation, by those in noise music, of why dissonance is necessarily desirable for music, since, for the critics, its positive evaluation is not universal, in *Sec. 2.5.2*; the lack of convincing explanation, within this assertive view, of why their care is necessarily needed, since, contrary to the popular view in mainstream society, "[i]n Deaf culture, deafness is not considered a condition that needs to be fixed" and is not regarded as "a pathological condition, viewing it instead as a locus of pride," in *Sec. 3.2.1*; the lack of convincing explanation, by those in support groups and the self-help movement, of why self-help is the way to go or is better, simply because existing treatments provide only limited relief and a cure does not really exist, since, for the critics, it is questionable if self-help can do much of anything under these circumstances, in *Sec. 3.2.2*; the lack of sufficient explanation, within this view, of why this must be true, since "some have suggested that it was the influence of electromagnetic waves, and not of infrasonic waves, that prompted these animals to flee," in *Sec. 3.3.1*; the lack of sufficient explanation, within this view, of why noise control is really so important, since, for the critics, other real concerns are neglected, because "hearing loss could be attributed to a variety of factors including age, rather than solely due to excessive exposure to noise," in *Sec. 3.4.1*; the lack of sufficient explanation, within this view, of why the principle of "the duty of care" is the only one or the main one to be used in judging a nuisance case in the first place, in *Sec. 3.4.2*; the lack of suffiicient explanation by John of the Cross of why this favoritism of contemplation over meditation is necessarily true, since, for the critics like Keating, the biased comparison could be explained by some contingent historical reasons, in *Sec. 3.5.1*; and the lack of convincing explanation, by those in abduction, of why being reasonable in abduction cannot be fallacious at all, at least for the critics in pure reasoning, in *Sec. 3.5.2*).

And the reverse direction also holds true.

Once more, it should be stressed, however, that there are different shades of gray (or different degrees of truth) in the two opposites, and the classification is not necessarily mutually exclusive either. In fact, even when some combinations of the two occur, they only end up sharing the same dialectic relationship, but in a different degree.

Also, there are some other relationships (as a kind of family resemblance) which have something in common with the principle, but they are not exactly the same but only more or less comparable, which varies from case to case.

Good examples of family resemblance in relation to the principle are underlyingness-regularness, causation-regularity, causation-correlation and so forth.

4th Thesis: The Preciseness-Vagueness Principle

The fourth thesis is called the preciseness-vagueness principle (on the refinement of things) in the category of "structure" in existential dialectics (which was first out worked out in *FIA*), in that both preciseness and vagueness are important, not that one is better than the other, but that both are used, in different degrees of preference, in accordance to the contextual application from the perspectives of nature, the mind, culture, and society. There is no preciseness without vagueness—and vice versa.

For instance, on the one hand, there is preciseness (e.g., the precise identification by psychoacousticians that "[t]he lower limit of audibility is [clearly] defined as 0 dB," in *Sec. 2.2.1*; the precise identification of "a genetic component of 39% to 58%" in some twin studies, in *Sec. 2.2.2*; the precise identification of "a few inches in their flight" that a moth may drop to evade attack by a bat, in *Sec. 2.3.1*; the precise identification by earthquake engineers of some forecasts which are to estimate "the probability of...a particular size affecting a particular location within a particular time span," in *Sec. 2.3.2*; the precise identification of the seven important methods used by submarines for surface and near-surface navigation, in *Sec. 2.4.1*; the precise identification of four major "safe" topics under "most circumstances," in *Sec. 2.4.2*; the identification of the "[v]ariations in air pressure against the ear drum, and the subsequent physical and neurological processing and interpretation," as the origination of "the subjective experience called sound," in *Sec. 2.5.1*; the precise classification of some techniques which are central for industrial music, like "assemblage, montage, bricolage, and appropriation," in *Sec. 2.5.2*; the precise classification of the five types of severity in hearing impairment measurement, namely, "mild," "moderate," "moderately severe," "severe," and "profound," in *Sec. 3.2.1*; the precise classification of the three major characteristics of stuttering, in *Sec. 3.2.2*; the precise identification of a distance of "several hundreds of kilometres" away that elephants can hear infrasound, in *Sec. 3.3.1*; the precise classification of three major ways to achieve noise control in workplace, namely, "the ABCs," in *Sec. 3.4.1*; the precise identification of the issue of "damage" as important in judging the resolution of a nuisance case in common law, in *Sec. 3.4.2*; the precise classification, by Saint Teresa of Avila, of "mystical union" in supernatural contemplation in terms of "four degrees or stages," in *Sec. 3.5.1*; and the precise classification of reasoning in terms of three kinds, namely, deduction, induction, and abduction, in *Sec. 3.5.2*).

On the other hand, there is vagueness (e.g., the vagueness in the identification, since psychoacousticians do not define clearly "the upper limit," which is treated instead as "a question of the limit where the ear will be physically harmed or with the potential to cause noise-induced hearing loss," which varies from person to person, in *Sec. 2.2.1*; the vagueness in the identification, since it is not clear what the exact degree of the genetic basis is, for a given twin, in *Sec. 2.2.2*; the vagueness in the identification, since it is not clear how many inches exactly are involved, or why only a few inches are necessary, in *Sec. 2.3.1*; the vagueness in this identification, since "[s]uch methods have yet to be generally accepted in the seismology community," as they are not very accurate, in *Sec. 2.3.2*; the vagueness in the identification of the seven important methods, since it is not clear why there are seven of them, not nine, ten, and so on—or which one is the best overall, etc., in *Sec. 2.4.1*; the vagueness in the identification, since it is not clear why there must be these four major "safe" topics, not others, whether more or less, in *Sec. 2.4.2*; the vagueness in the identification, since it is not clear how and why exactly a given sound is to be "interpreted" in different cultures and "sensitive" to different individuals, in *Sec. 2.5.1*; the vagueness in the classification, since it is not clear why only techniques like these aforecited, not some others, are included, in *Sec. 2.5.2*; the vagueness in the classification, since it is not clear why there are only five types, not four, six, seven, etc., in *Sec. 3.2.1*; the vagueness in the classification of the three major characteristics of stuttering, since it is not clear why there are only three characteristics, not four, five, six, and so on, in *Sec. 3.2.2*; the vagueness in the identification, since it is not clear exactly how many kilometres away the elephants can hear infrasound, in *Sec. 3.3.1*; the vagueness in the classification, since it is not clear why there are only three major ways, not four, five, six, and so on, in *Sec. 3.4.1*; the vagueness in the identification of the issue of "damage" in tort law, because in some cases, the "damages...cannot be quantified exactly in monetary terms," and what counts as "damage" can be subject to different legal interpretations, in *Sec. 3.4.2*; the vagueness in the classification, since it is not clear why there are only four degrees or stages, not five, six, seven, and so on, in *Sec. 3.5.1*; and the vagueness in the classification of reasoning in terms of three kinds, since it is not clear why the traditional classification of only two types, namely, deduction and induction, is not necessarily better, or why the kinds of reasoning cannot be four, five, and so on, in *Sec. 3.5.2*).

And the reverse direction also holds true.

Once more, it should be stressed, however, that there are different shades of gray (or different degrees of truth) in the two opposites, and the classification is not necessarily mutually exclusive either. In fact, even when some combinations of the two occur, they only end up sharing the same dialectic relationship, but in a different degree.

Also, there are some other relationships (as a kind of family resemblance) which have something in common with the principle, but they are not exactly the same but only more or less comparable, which varies from case to case.

Good examples of family resemblance in relation to the principle are clarity-ambiguity, directness-indirectness, quantitativeness-qualitativeness, specificity-obscurity, thickness-thinness, describability-nondescribability, specificity-generality, concretness-abstractness, and the like.

5th Thesis: The Simpleness-Complicatedness Principle

The fifth thesis refers to the simpleness-complicatedness principle (on the interconnection among things) in the category of "structure" in existential dialectics (which was first out worked out in *FIA*), in that both simpleness and complicatedness are vital, without favoring one over the other, and each is utilized, depending on the basis of the perspectives of nature, the mind, culture, and society. There is no simpleness without complicatedness—and vice versa.

For instance, on the one hand, there is simple analysis (e.g., the relatively simple analysis of the limits of human hearing by psychoacousticians, in *Sec. 2.2.1*; the relatively simple analysis of introversion and extroversion, in *Sec. 2.2.2*; the relatively simple analysis of the sounds of animals for different purposes, in *Sec. 2.3.1*; the relatively simple analysis of earthquake prediction by earthquake engineers, in *Sec. 2.3.2*; the relatively simple formulae which "have been developed for the target strength of various simple shapes as a function of angle of sound incidence," in *Sec. 2.4.1*; the relatively simple analysis of small talk in terms of its social functions, in *Sec. 2.4.2*; the relative simple nature of sine waves, in *Sec. 2.5.1*; at the beginning in the older days, the art of noises tended to be simple, pure, and sweet, in *Sec. 2.5.2*; the relatively simple analysis of deafness in terms of the major issues as described above, in *Sec. 3.2.1*; the relatively simple recommendation, by the advocates, of the six different treatments for stuttering, in *Sec. 3.2.2*; the relatively simple analysis of ghost sightings as optical illusions by Tandy, in *Sec. 3.3.1*; the relatively simple analysis of the four applications of architectural acoustics to the domain of noise control in society, in *Sec. 3.4.1*; the relatively simple nature of the law of nuisance at the beginning of its historical development, in *Sec. 3.4.2*; the relatively simple analysis of contemplation in accordance to Christian mysticism, in *Sec. 3.5.1*; and the relatively simple analysis of abduction by its advocates, in *Sec. 3.5.2*).

On the other hand, there is also complicated counterpart (e.g., the relatively more complicated analysis of the limits of human hearing by questioning the claims and assumptions used in psychoacoustics, as shown in the criticisms, in *Sec. 2.2.1*; the relatively more complicated analysis of introversion and extroversion by questioning the claims about different causes, which remain in contention, in *Sec. 2.2.2*; the relatively more complicated analysis of the sounds of animals by asking some critical questions about the claims by researchers, as shown in the criticisms, in *Sec. 2.3.1*; the relatively more complicated analysis of earthquake prediction by the critics who question the assumptions and claims by earthquake engineers, as shown in the criticisms, in *Sec. 2.3.2*; the relatively more complex formulae for more complex shapes which "may be approximated by combining these simple ones," in *Sec. 2.4.1*; the relatively more complicated analysis of small talk by questioning its assumptions and claims, as shown in the criticisms, in *Sec. 2.4.2*; the relatively more complicated nature of other waves which are "modelled by many different sine waves" of different "appropriate frequencies and amplitudes," in *Sec. 2.5.1*; nowadays, the art of noises "becomes continually more complicated, strives to amalgamate the most dissonant, strange and harsh sounds," in *Sec. 2.5.2*; the relatively more complicated analysis of deafness by questioning the assumptions and claims in regard to the major issues as described above, as shown in the criticisms by the critics, in *Sec. 3.2.1*; the relatively more complicated analysis of the six different treatments for stuttering by questioning their claims and assumptions, as shown in the criticisms by the critics, in *Sec. 3.2.2*; the relatively more complicated analysis of ghost sightings as optical illusions by questioning the assumptions and claims used by Tandy, as shown in the criticisms by the critics, in *Sec. 3.3.1*; the relatively more complicated analysis of the application of architectural acoustics to the domain of noise control in society by questioning its assumptions and claims, as shown in the criticisms by the critics, in *Sec. 3.4.1*; the relatively more complicated nature of the law of nuisance, especially after "the late 19th and early 20th centuries," when "the law of nuisance became difficult to administer, as competing property uses often posed a nuisance to each other, and the cost of litigation to settle the issue grew prohibitive," so, later, it was overruled or adapted by other laws like zoning laws and environmental laws, in *Sec. 3.4.2*; the relatively more complicated analysis of contemplation by questioning its assumptions and claims, as shown in the criticisms by the critics, in *Sec. 3.5.1*; and the relatively more complicated analysis of abduction by questioning its assumptions and claims, as shown in the criticisms by the critics, in *Sec. 3.5.2*).

And the reverse direction also holds true.

Again, it should be stressed, however, that there are different shades of gray (or different degrees of truth) in the two opposites, and the classification is not necessarily mutually exclusive either. In fact, even when some combinations of the two occur, they only end up sharing the same dialectic relationship, but in a different degree.

Also, there are some other relationships (as a kind of family resemblance) which have something in common with the principle, but they are not exactly the same but only more or less comparable, which varies from case to case.

Good examples of family resemblance in relation to the principle are inflexibility-flexibility, standardization-specialization, imperfectness-perfectness, superficiality-depth, shallowness-deepness, economicalness-elaboratedness, plainness-circumspection, onesidedness-multisidedness, and the like.

6th Thesis: The Openness-Hiddenness Principle

The sixth thesis refers to the openness-hiddenenss principle (on the detection of things) in the category of "structure" in existential dialectics (which was already worked out in my previous books, especially in *FPHU*), in that reality has its hidden face, just as it is open to outside view in some other ways. There is no openness without hiddenness—and vice versa.

For instance, on the one hand, there is openness (e.g., the open exploration of the relationship between the measurement of the absolute hearing threshold and the understanding of human "auditory system," in *Sec. 2.2.1*; the open exploration, by Jung, of the correlation between psychic energy and the differences between extroverts and introverts, in *Sec. 2.2.2*; the open, or detectable, nature of moths which can be detected by the ultrasound in the range emitted by bats during their attempt to attack moths, in *Sec. 2.3.1*; the open exploration of the correlation between earthquake prediction and the understanding of the interior of the Earth, in *Sec. 2.3.2*; the open operation by submarines in friendly waters, by using active sonar systems and radar navigation, which can be easily detected, in *Sec. 2.4.1*; the open exploration of the relationship between small talk and social skills, in *Sec. 2.4.2*; the open exploration of the correlation between consonant sound waves and pleasant sounds, in *Sec. 2.5.1*; the open exploration of sound for music, in *Sec. 2.5.2*; the open exploration of the need for society to aid hearing-impaired people, in *Sec. 3.2.1*; "the immediately visible and audible symptoms of stuttering" as the portion of an iceberg above the waterline, according to the analogy by Joseph Sheehan, in *Sec. 3.2.2*; the open exploration of the relationship between ghost sightings and optical illusions by Tandy, in *Sec. 3.3.1*; the open exploration of the relationship between architectural acoustics and the minimization of negative health effects from noise, in *Sec. 3.4.1*; the open exploration of the relationship between the understanding of "the duty of care" in tort law and the legal interpretation of such an understanding, in *Sec. 3.4.2*; the open exploration of the relationships between meditation and contemplation in accordance to Christian mysticism, in *Sec. 3.5.1*; and the open exploration of the three kinds of reasoning, in *Sec. 3.5.2*).

On the other hand, there is also hiddenness (e.g., the hidden bias in the measurement by each method, since each has its own critieria, assumptions, and disadvantages, in *Sec. 2.2.1*; the hidden bias in Jung's view, because of its focus on psychic energy, but not on other causes, in *Sec. 2.2.2;* the hidden, or evasive, nature of moths, like dropping a few inches in their flight to evade attack, in *Sec. 2.3.1*; the hidden bias in this supposed correlation, because of the assumptions and claims which have been questioned by the critics, such that earthquake prediction is not what the proponents would like us to believe, in *Sec. 2.3.2*; the stealth mode in which submarines travel when in non-friendly waters, in order to avoid detection, in *Sec. 2.4.1*; the hidden bias in the discourse on small talk and social skills, because it dogmatically assumes that small talk is necessarily functional for better social skills, in *Sec. 2.4.2*; the hidden bias in the exploration of this correlation between consonant sound waves and pleasant sounds, because it downgrades "dissonant" sound waves as "unpleasant," in *Sec. 2.5.1*; the hidden bias in sound music, because of its obsession with noise and dissonance, for instance, in *Sec. 2.5.2*; the hidden bias in this social aid to hearing-impaired people, since it often degenerate into a new kind of patronizing and oppressive maltreatment of them, as what Lane called "the mask of benevolence," in *Sec. 3.2.1*; "a broader set of symptoms such as negative emotions hidden below the surface" of the iceberg, as an analogy by Sheehan, in *Sec. 3.2.2*; the hidden bias in Tandy's research, because of his obsession with apparent ghostly figures under the conditions that he encountered, which are not the same of ghosts in the literature of anomalous experience, in *Sec. 3.3.1*; the hidden bias in architectural acoustics, because of its questionable assumptions, as shown in the criticisms by the critics, in *Sec. 3.4.1*; the hidden bias in a given legal interpretation, since, in the case about "the duty of care," in general, "much of the principle is still at the discretion of judges," in *Sec. 3.4.2*; the hidden bias for contemplation in the analysis of the relationships, because of the contingent historical reasons as pointed out by Keating, in *Sec. 3.5.1*; and the hidden bias in each kind of reasoning, because each is concerned with its own version of legitimacy in logic, be it about validity, probability, or reasonableness, in *Sec. 3.5.2*).

And the reverse direction also holds true.

Again, it should be stressed, however, that there are different shades of gray (or different degrees of truth) in the two opposites, and the classification is not necessarily mutually exclusive either. In fact, even when some combinations of the two occur, they only end up sharing the same dialectic relationship, but in a different degree.

Also, there are some other relationships (as a kind of family resemblance) which have something in common with the principle, but they are not exactly the same but only more or less comparable, which varies from case to case.

Good examples of family resemblance in relation to the principle are overtness-covertness, publicness-privateness, openness-closedness, transparency-secrecy, openness-biasedness, and so on.

7th Thesis: The Denseness-Emptiness Principle

The seventh thesis concerns the denseness-emptiness principle (on the distribution of entities in space) in the category of "structure" in existential dialectics, which is first proposed in *FPHUP*, in that both density and void are needed, in relation to the mind, nature, culture, and society, albeit in different ways. There is no denseness without emptiness—and vice versa.

Lest any misunderstanding occurs, the term "void" is used here only as an approximation of emptiness (depending on the degree of the lack of density), since, in physics, it is well known that "empty" space is not really empty all the way, because it can be full of energy (e.g., random quantum fluctuations at the sub-atomic level, and, for that matter, dark energy in the universe) and matter (e.g., different versions of sub-atomic particles, and, for that matter, dark matter in the universe). (F. Wilczek 2008)

For instance, on the one hand, there is denseness (e.g., the relatively denser concentration of sound waves at higher frequenccies, in *Sec. 2.2.1*; the relatively denser concentration of extroverts in America, which is regarded as "an extroverted society," in *Sec. 2.2.2*; the relatively denser concentration of moths to be the targets of being eaten by bats, in *Sec.*

2.3.1; the relatively denser concentration of S-waves in solids, in *Sec. 2.3.2*; the relatively denser concentration of submarines using active sonar systems and radar navigation in friendly waters, in *Sec. 2.4.1*; the relatively denser concentration of people who engage in small talk if they are not in an intimate relationship, in *Sec. 2.4.2*; the relatively denser concentration of people who like harmonic series to find consonant sounds pleasant, in *Sec. 2.5.1*; the relatively denser concentration of noise musicians in the second half of the 20th century in the West, in *Sec. 2.5.2*; the relatively denser, or more pervasive, spread of people with normal hearing in society, in *Sec. 3.2.1*; the relatively denser concentration of individuals who prefer "anti-stuttering medications" as a treatment of stuttering to use "benzodiazepines, anti-convulsants, anti-depressants, antipsychotic and antihypertensive medications, and dopamine antagonists," for instance, in *Sec. 3.2.2*; the relatively denser concentration of animals to be able to naturally hear infrasound over distances, in *Sec. 3.3.1*; the relatively denser concentration of government regulations in the U.S. for higher noise-control standards, in *Sec. 3.4.1*; the relatively denser concentration of nuisance cases nowadays which are overruled or adapted by other laws like zoning laws and environmental laws, in *Sec. 3.4.2*; the relatively denser concentration of Christian believers who preferred contemplation over meditation in "the first fifteen centuries of the Christian era," in *Sec. 3.5.1*; and the relatively denser concentration of those in pure reasoning to regard abductive reasoning as "fallacious," in *Sec. 3.5.2*).

On the other hand, there is emptiness (e.g., the relatively less dense, or more empty, concentration of sound waves at lower frequencies, in *Sec. 2.2.1*; the relatively less dense, or more empty, concentration of extroverts in "Japan or regions where Buddhism, Sufism etc. prevail" and "prize introversion," in *Sec. 2.2.2*; the relatively less dense, or more empty, concentration of lions to be the targets of being eaten by bats, in *Sec. 2.3.1*; the relatively less dense, or more emtpy, concentration of "S-waves" in liquids, in *Sec. 2.3.2*; the relatively less dense, or more empty, concentration of submarines using active sonar systems and radar navigation in non-friendly waters, in *Sec. 2.4.1*; the relatively less dense, or more empty, concentration of people who engage in small talk if they are in an intimate relationship, in *Sec. 2.4.2*; the relatively less dense, or more empty, concentration of people who, like Cage, prefers chaotic music to find consonant sounds pleasant, in *Sec. 2.5.1*; the relatively less dense, or more empty, concentration of noise musicians in the second half of the 20th century in Communist China and the Soviet Union, in *Sec. 2.5.2*; the relatively less dense, or more empty, spread of hearing-impaired people in society, in *Sec. 3.2.1*; the relatively less dense, or more empty, concentraton of individuals who prefer support groups and the self-help movement to use anti-stuttering medications, in *Sec. 3.2.2*; the relatively less dense, or more empty, concentration of humans to be able to naturally hear infrasound over distances, without the help of special equipments, in *Sec. 3.3.1*; the relatively less dense, or more empty, concentration of government regulations in "undeveloped countries" for higher noise standards, in *Sec. 3.4.1*; the relatively less dense, or more empty, concentration of nuisance cases nowadays which are totally independent of other laws like zoning laws and environmental laws, in *Sec. 3.4.2*; the relatively less dense, or more empty, concentration of Christian believers who prefer contemplation over meditation nowadays, in *Sec. 3.5.1*; and the relatively less dense, or more empty, concentration of those in abduction to regard abductive reasoning as "fallacious," in *Sec. 3.5.2*).

And the reverse direction also holds true.

Again, it should be stressed, however, that there are different shades of gray (or different degrees of truth) in the two opposites, and the classification is not necessarily mutually exclusive either. In fact, even when some combinations of the two occur, they only end up sharing the same dialectic relationship, but in a different degree.

Also, there are some other relationships (as a kind of family resemblance) which have something in common with the principle, but they are not exactly the same but only more or less comparable, which varies from case to case.

Good examples of family resemblance in relation to the principle are fullness-voidness, nearness-farness, concentration-dispersion, denseness-nothingness, somethingness-nothingness, and the like.

8th Thesis: The Slowness-Quickness Principle

The eighth thesis concerns the slowness-quickness principle (on the speed of change) in the category of "process" in existential dialectics (which was first worked out in *FIA*), in that both slowness and quickness co-exist, with their own internal tension, to the extent that each fights for its own relevance with the other, in accordance to the perspectives of nature, the mind, culture, and society, without one being the victor and the other being the vanquished in the long haul. There is no slowness without quickness—and vice versa.

For instance, on the one hand, there is slowness (e.g., the relatively slower speed of someone to hear a sound when there is a masking by other sounds to obscure it, in *Sec. 2.2.1*; the relatively slower speed for an extrovert to get turned off by loud parties, in *Sec. 2.2.2*; the relatively slower readiness of a flying noctuid moth to emit sounds to evade attack by a bat, in *Sec. 2.3.1*; the relatively slower speed of S-waves to travel through solids, when compared with P-waves, in *Sec. 2.3.2*; the relatively slower speed of old submarines in the early days when traveling in deep water, in *Sec. 2.4.1*; the relatively slower speed of someone like H. Grice to give more info than was explicitly asked for in a small talk, in *Sec. 2.4.2*; the relatively slower speed of people to have hearing loss by long-term exposure to environmental noise like living near airports or freeways, in *Sec. 3.2.1*; the relatively slower speed of traditional group therapy programs for fluency shaping therapy to complete, since they normally take "two to three weeks to complete," in *Sec. 3.2.2*; the relatively slower speed of humans "to flee the area long before the actual tsunami hit the shores" during the 2004 Indian Ocean earthquake and tsunami, in *Sec. 3.3.1*; the relatively slower speed of European countries to set up noise-control standards nowadays, which "generally lag by 10 to 20 years," when compared with those in the U.S., in *Sec. 3.4.1*; the relatively slower speed of a nuisance case to be accepted in court, if "the level of interference" with the rights of others does not rise "above the merely aesthetic," in *Sec. 3.4.2*; the relatively slower speed of someone in the final stage of "mystical union" in contemplative prayer to get distracted, in *Sec. 3.5.1*; and the relatively slower readiness of a specialist in abduction to make a conclusion which is valid or invalid, in *Sec. 3.5.2*).

On the other hand, there is quickness (e.g., the relatively quicker speed of someone to hear a sound if there is no masking by other sounds to obscure it, in *Sec. 2.2.1*; the relatively quicker speed for an introvert to get turned off by loud parties, in *Sec. 2.2.2*; the relatively quicker readiness of a flying noctuid moth to exercise "a reflex action…that cause[s] it to drop a few inches in its flight to evade attack," in *Sec. 2.3.1*; the relatively faster speed of P-waves to travel through solids, when compared with S-waves, in *Sec. 2.3.2*; the relatively quicker speed of new submarines nowadays when traveling in deep water, in *Sec. 2.4.1*; the relatively quicker speed of someone like Schneider to give more info than was explicitly asked, because "politeness in small talk is maximised by responding with a more substantial answer," in *Sec. 2.4.2*; the relatively faster speed of people to have hearing loss by head injury in an explosion, in *Sec. 3.2.1*; the relatively faster speed of the more recent Camperdown program for fluency shaping therapy to complete, since they "use a much shorter schedule" than the traditional group therapy programs, in *Sec. 3.2.2*; the relatively quicker speed of animals "to flee the area long before the actual tsunami hit the shores" in the 2004 Indian Ocean earthquake and tsunami, in *Sec. 3.3.1*; the relatively faster speed of the U.S. to set up noise-control standards nowadays, which generally are ahead of European countries "by 10 to 20 years," in *Sec. 3.4.1*; the relatively quicker speed of a nuisance case to be accepted in court, if "the level of interference" with the rights of others rises much "above the merely aesthetic," in *Sec. 3.4.2*; the relatively faster speed of someone in the first stage of "mystical union" in contemplative prayer to get distracted, in *Sec. 3.5.1*; and the relatively quicker readiness of a specialist in abduction to make a conclusion which is reasonable or unreasonable, in *Sec. 3.5.2*).

And the reverse direction also holds true.

Again, it should be stressed, however, that there are different shades of gray (or different degrees of truth) in the two opposites, and the classification is not necessarily mutually exclusive either. In fact, even when some combinations of the two occur, they only end up sharing the same dialectic relationship, but in a different degree.

Also, there are some other relationships (as a kind of family resemblance) which have something in common with the principle, but they are not exactly the same but only more or less comparable, which varies from case to case.

Good examples of family resemblance in relation to the principle are inconvenience-convenience, passiveness-activenessness, gradualness-abruptness, deceleration-acceleration, and the like.

9th Thesis: The Expansion-Contraction Principle

The nineth thesis is called the expansion-contraction principle in the category of "process" in existential dialectics, in that entities in the world can both expand in some ways and contract in other ones, as part of their nature. There is no expansion without contraction—and vice versa.

This principle, although not so explicitly called, was already used in my previous works on different topics (e.g., the theory of floating consciousness in *FCD* and *FPHC*, the union of the unions in *BWT*, and the cyclical progression of system fragmentation and integration in *FCD*).

For instance, on the one hand, there is expansion (e.g., the relatively more developed ability of humans to hear the audio range "between 20 Hz and 20,000 Hz," in *Sec. 2.2.1*; the relatively more developed ability of an extrovert to excel in social interaction, in *Sec. 2.2.2*; the relatively more developed ability of a flying noctuid moth to exercise "a reflex action…that cause[s] it to drop a few inches in its flight to evade attack," in *Sec. 2.3.1*; the relatively more developed ability of S-waves to travel through solids, in *Sec. 2.3.2*; the relative expansion of the influence of "global positioning system" or GPS for submarine navigation nowadays, in *Sec. 2.4.1*; the relatively more developed ability of those who indulge in small talk over time to improve their social skills, in *Sec. 2.4.2*; the relatively more developed ability of a musical composer of harmonic series to make use of consonant sounds, in *Sec. 2.5.1*; the relative rise of the influence of noise music in the West, especially though not exclusively during the 20th century, in *Sec. 2.5.2*; the relatively more developed ability of those hearing–impaired people who solicit the help of others in society to seek different assistive techniques, resources, and intervention, in *Sec. 3.2.1*; the relative spread of the influence of "support groups and the self-help movement" nowadays, in *Sec. 3.2.2*; the relative expansion of the influence of science in trying to explain ghost sightings nowadays, in *Sec. 3.3.1*; the relative spread of government regulations for noise control in society nowadays, especially in the U.S., in *Sec. 3.4.1*; the relative spread, in our time, of the influence of "[t]he law and economics movement," which "has been involved in analyzing the most efficient choice of remedies given the circumstances of the nuisance," in *Sec. 3.4.2*; the relative expansion of the influence of meditation from the 18th century until now, in the Christian tradition, in *Sec. 3.5.1*; and the relatively more developed ability of an expert in abduction to analyze the reasonableness of an argument, in *Sec. 3.5.2*).

On the other hand, there is contraction (e.g., the relatively less developed ability of humans to hear sounds outside the audio range "between 20 Hz and 20,000 Hz," in *Sec. 2.2.1*; the relatively less developed ability of an extrovert to excel in contemplation or meditation, in *Sec. 2.2.2*; the relatively less developed ability of a flying noctuid moth to emit sounds to evade attack by a bat, in *Sec. 2.3.1*; the relatively less developed ability of S-waves to travel through liquids, in *Sec. 2.3.2*; the relative contraction or decline of the influence of "terrestrial radio-based navigation systems" for submarine navigation nowadays, in *Sec. 2.4.1*; the relatively less developed ability of those who indulge in small talk over time to focus on serious conversation, in *Sec. 2.4.2*; the relatively less developed ability of a musical composer of harmonic series to make use of dissonant sounds, in *Sec. 2.5.1*; the relative decline of the influence of traditional melodic music in the West especially in the 20th century, in *Sec. 2.5.2*; the relatively less developed ability of those hearing–impaired people who solicit the help of others in society to counter the patronizing mask of benevolence from society, as pointed out by Lane, in *Sec. 3.2.1*; the relatively decline of the influence of traditional "behavioral, prosthetic, and pharmaceutical treatments" nowadays, because they "provid[e]…limited relief from the overt symptoms of stuttering," in *Sec.*

3.2.2; the relative decline of the influence of esoteric religion to explain ghost sightings nowadays, in *Sec. 3.3.1*; the relative loss of the ability of citizens to engage in noisy activities nowadays without government regulations, especially in the U.S., in *Sec. 3.4.1*; the relative decline, in our time, of the influence of religion in judging nuisance cases, in *Sec. 3.4.2*; the relative decline of the influence of contemplation from the 18th century until now, in the Christian tradition, in *Sec. 3.5.1*; and the relatively less developed ability of an expert in abduction to analyze the validity or probability of an argument, in *Sec. 3.5.2*).

And the reverse direction also holds true.

Again, it should be stressed, however, that there are different shades of gray (or different degrees of truth) in the two opposites, and the classification is not necessarily mutually exclusive either. In fact, even when some combinations of the two occur, they only end up sharing the same dialectic relationship, but in a different degree.

Also, there are some other relationships (as a kind of family resemblance) which have something in common with the principle, but they are not exactly the same but only more or less comparable, which varies from case to case.

Good examples of family resemblance in relation to the principle are conquest-autarky, rise-fall (or up-down), spread-shrink, extendingness-shorteningness, and so forth.

10th Thesis: The Theory-Praxis Principle

The tenth thesis is the theory-praxis principle (on the duality of knowledge) in the category of "agency" in existential dialectics (which was first worked out in *FPHE*), in that, if there is theoretical construction, there is likewise its practical application, both technical and normative. There is no theory without praxis—and vice versa.

For instance, on the one hand, there is theory (e.g., the theoretical construction of the limits of human hearing in psychoacoustics, in *Sec. 2.2.1*; the theoretical construction of the extroversion-introversion single continuum by psychologists like Jung and others, in *Sec. 2.2.2*; the theoretical construction of the nature of ultrasound by acousticians, in *Sec. 2.3.1*; the theoretical construction of body waves and surface waves in seismology, in *Sec. 2.3.2*; the theoretical construction of global positioning system or GPS, in *Sec. 2.4.1*; the theoretical construction of harmonic series, in *Sec. 2.5.1*; the theoretical construction of "airplane propellers" in aeronautics, in *Sec. 2.5.2*; the theoretical construction of genetics, in *Sec. 3.2.1*; the theoretical construction of electronic devices for auditory purposes in electronical engineering, in *Sec. 3.2.2*; the theoretical construction of infrasound in infrasonics, in *Sec. 3.3.1*; the theoretical construction of acoustics, especially in relation to sound absorption, sound reflection, and sound masking, in *Sec. 3.4.1*; the theoretical construction of efficiency in the law and economics movement, in *Sec. 3.4.2*; the theoretical construction of contemplation in philosophy, theology, and psychology, for instance, in *Sec. 3.5.1*; and the theoretical construction of abduction in logic, in *Sec. 3.5.2*).

On the other hand, there is praxis (e.g., the practical application of the theoretical construction of the limits of human hearing by acoustic engineers, who, "when compressing a digital signal,...can safely assume that any frequency beyond approximately 20 kHz will not have any effect on the perceived sound of the finished product, and thus use a high-shelving filter to cut everything outside this range," in *Sec. 2.2.1*; the practical application of the extroversion-introversion theoretical construct to the work on the geographical variation of the traits in the United States by Peter Rentfrow [2008] and others, in *Sec. 2.2.2*; the practical application of the understanding of ultrasound to the beef cattle industry to improve...the yield of cattle operations," in *Sec. 2.3.1*; the practical application of the knowledge of body waves and surface waves to the field of earthquake engineering for earthquake prediction, in *Sec. 2.3.2*; the practical application of GPS to submarine navigation for naval warfare, in *Sec. 2.4.1*; the practical application of harmonic series to the clinical use of music for musical therapy, in *Sec. 2.5.1*; the practical application of the sound of "airplane propellers" to the field of "found sound" in noise music, in *Sec. 2.5.2*; the practical application of genetics to the field of hearing impairment for gene therapy, in *Sec. 3.2.1*; the practical application of electronic engineering to the field of stuttering for electronic fluency devices as a treatment of stuttering, in *Sec. 3.2.2*; the practical application of infrosonics to the field of nuclear weapons, as "the Comprehensive Nuclear-Test-Ban Treaty Organization uses infrasound as one of its monitoring technologies," in *Sec. 3.3.1*; the practical application of acoustics to the field of architecture for noise control, in *Sec. 3.4.1*; the practical application of efficiency in the law and economics movement to the field of tort law, like the case about the right to quiet enjoyment, in *Sec. 3.4.2*; the practical application of contemplation to the field of Christian mysticism for "mystical union" with God, in *Sec. 3.5.1*; and the practical application of abduction to the field of legal systems in Germany, especially in relation to "statements volunteered by the accused," in *Sec. 3.5.2*).

And the reverse direction also holds true.

Again, it should be stressed, however, that there are different shades of gray (or different degrees of truth) in the two opposites, and the classification is not necessarily mutually exclusive either. In fact, even when some combinations of the two occur, they only end up sharing the same dialectic relationship, but in a different degree.

Also, there are some other relationships (as a kind of family resemblance) which have something in common with the principle, but they are not exactly the same but only more or less comparable, which varies from case to case.

Good examples of family resemblance in relation to the principle are discovery-application, knowledge-action, invention-innovation, and so forth.

11th Thesis: The Convention-Novelty Principle

The eleventh thesis is the convention-novelty principle (on the nature of creative thinking) in the category of "agency" in existential dialectics (which was first worked out in *FPHCT*), in that, if there is conventional wisdom, there is likewise novel challenge, to the extent that both convergent and divergent thinking are part of life. There is no convention without novelty—and vice versa.

As srummarized from *FPHCT* (almost verbatim here), there are (a) "creative techniques" and (b) "creative traits," which, when satisfied—in relation to the larger context of the mind, nature, society, and culture—can be used to enhance creative works.

In addition, creative thinking has its own possibilities and limits (in relation to invention), just as it has its own promises and pitfalls (in relation to innovation)— as already analyzed in *FPHCT*).

In the end, creative thinking has its own desirability and dark sides (as also already analyzed in *FPHCT*).

With these clarifications in mind—there are good empirical examples for the convention-novelty princple.

For instance, on the one hand, there is convention (e.g., the conventional wisdom about measuring ATH by the "classical methods" as first proposed by Gustav Theodor Fechner in his work *Elements of Psychophysics* in the 19th century, in *Sec. 2.2.1*; the conventional wisdom about the differences between extroverts and introverts, in *Sec. 2.2.2*; the conventional wisdom that many caterpillars make sounds, in *Sec. 2.3.1*; the conventional wisdom about the core of the Earth as solids, in *Sec. 2.3.2*; the conventional wisdom for ships in coastal and internal waters to rely on "the standard system of navigational aids…[like] buoys, navigational markers, lighthouses, etc.," in *Sec. 2.4.1*; the conventional wisdom about Grice's Maxim of Quantity, which suggests that responses should contain no more info than was explicitly asked for in small talk, in *Sec. 2.4.2*; the conventional wisdom about consonant sounds as pleasant for beautiful music, in *Sec. 2.5.1*; the conventional wisdom in traditional music, about the distinction between "musical tones" and "noise," in *Sec. 2.5.2*; the conventional wisdom that hearing-impaired people need help from others and society in general, in *Sec. 3.2.1*; the conventional wisdom of treating stuttering "around eighteenth and nineteenth century Europe," like "cutting the tongue with scissors, removing a triangular wedge from the posterior tongue, and cutting nerves, or neck and lip muscles," in *Sec. 3.2.2*; the conventional view about "ghosts" in esoteric religion, in *Sec. 3.3.1*; the conventional wisdom in the older days, for instance, "[i]n the 1960s and earlier," when "few people recognized that citizens might be entitled to be protected from adverse sound level exposure," in *Sec. 3.4.1*; the conventional wisdom in the older days when the law of nuisance was not closely related to other laws like zoning laws and environmental laws, in *Sec. 3.4.2*; the conventional positive wisdom about contemplation over meditation in the older days, in *Sec. 3.5.1*; and the conventional wisdom about reasoning in terms of deduction and induction, in *Sec. 3.5.2*).

On the other hand, there is novelty (e.g., the alternative novel challenge to the conventional wisdom about the "classical methods" by new ways of measuring ATH nowadays, like "forced-choice methods," and "adaptive methods," for instance, in *Sec. 2.2.1*; the alternative novel view by some scholars nowadays, who argued that "[a]cknowledging that introversion and extroversion are normal variants of behavior can help in self-acceptance and understanding of others," as "an extrovert can accept his or her introverted partner's need for space, while an introvert can acknowledge his or her extroverted partner's need for social interaction," in *Sec. 2.2.2*; the alternative novel challenge to the conventional wisdom by the researchers in two Canadian universities who "experimentally investigate[d] how these noises are made and what roles they might play," in *Sec. 2.3.1*; the alternative novel challenge to this conventional wisdom when "Richard Dixon Oldham in 1906 and...Harold Jeffreys in 1926" discovered that "the outer core of the Earth is liquid," in *Sec. 2.3.2*; the alternative novel challenge to this standard system by the new way of satellite navigation, in *Sec. 2.4.1*; the alternative novel challenge to this conventional wisdom about the Maxim of Quantity, by Schneider for a new view which suggested instead that it is polite to give more info than was explicitly asked for, in *Sec. 2.4.2*; the alternative novel challenge to this conventional wisdom about pleasant consonant sounds, by Cage with his new view that chaotic music can be beautiful too, in *Sec. 2.5.1*; the alternative novel challenge to this conventional wisdom about music by Varese in found music, who proposed a new way of doing music, that is, by incorporating noise into music itself, in *Sec. 2.5.2*; the alternative novel challenge to this conventional wisdom about the benevolent help of society, by the new view that this help often degenerates into a new kind of patronizing and oppressive maltreatment of them, as what Lane called "the mask of benevolence," in *Sec. 3.2.1*; the alternative novel challenge to the conventional wisdom of treating stuttering, by the six new ways of treating stuttering nowadays, in *Sec. 3.2.2*; the alternative novel challenge to the conventional wisdom about "ghosts" in esoteric religion, by Tandy's new view that ghost sightings are optical illusions, in *Sec. 3.3.1*; the alternative novel challenge to this conventional wisdom about sound level exposure, by "the National Environmental Policy Act in 1969 and the Noise Pollution and Abatement Act, more commonly called the Noise Control Act, in 1972," which set up government regulation for noise control, in *Sec. 3.4.1*; the alternative novel challenge to this conventional wisdom nowadays, when the law of nuisance can be overruled and adapted by other laws like zoning laws and environmental laws, in *Sec. 3.4.2*; the alternative novel challenge to this conventional wisdom about contemplation, by the new idea that supernatural contemplation is higher than natural contemplation, as Aumann once pointed out that "this idea of the two paths was 'an innovation in spiritual theology and a departure from the traditional Catholic teaching,'" in *Sec. 3.5.1*; and the alternative novel challenge to the conventional wisdom about reasoning, by the creation of abduction as the third kind of reasoning, in *Sec. 3.5.2*).

And the reverse direction also holds true.

Again, it should be stressed, however, that there are different shades of gray (or different degrees of truth) in the two opposites, and the classification is not necessarily mutually exclusive either. In fact, even when some combinations of the two occur, they only end up sharing the same dialectic relationship, but in a different degree.

Also, there are some other relationships (as a kind of family resemblance) which have something in common with the principle, but they are not exactly the same but only more or less comparable, which varies from case to case.

Good examples of family resemblance in relation to the principle are convergence-divergence (or convergent thinking vs. divergent thinking), normalness-nonnormalness, conformity-nonconformity, and so on.

12th Thesis: The Evolution-Transformation Principle

The twelveth thesis refers to the evolution-transformation principle (on the multiple kinds of agency). in the category of "agency" in existential dialectics (which was first worked out in *FAE* and then in other books of mine).

This principle—and the symmetry-asymmetry principle, for instance—are both about agency. More precisely, the word "agency," in a formal definition, refers to "a person or thing through which power is exerted or an end is achieved." (MWD 2007) It therefore does not have to necessarily involve an intelligent lifeform.

Because of this dual meaning in agency, the evolution-transformation principle is more concerned with *the kind of agency*, that is, both about the *evolution* in the state of nature (e.g., an object of natural beauty) and the *transformation* in the world of intelligent lifeforms (e.g., a work of art). There is no evolution without transformation—and vice versa.

In classical Darwinian evolutionary theory (as more detailedly analyzed in *BNN*), evolution is "blind." But in the human world, change often takes place because of the conscious intervention of humans in transforming society and culture, just to cite two instances.

And the transformative part of the principle precisely refers to the other dimension in the dual meaning of agency, in giving technology (as invented by intelligent lifeforms like humans and, soon, post-humans) a major role to play in the change of the world, which is something that I extensively analyzed in *FHC* in the context of the technophilic lifeworld, espeically though not exclusively since modern times.

For instance, on the one hand, there is evolution (e.g., the natural evolution of humans to hear the audio range "between 20 Hz and 20,000 Hz," in *Sec. 2.2.1*; the natural evolution of humans to understand each other in terms of their different personalities in the state of nature, in Sec. *2.2.2*; the natural evolution of cattles to live in the state of nature, with some in good health and others in bad one, in *Sec. 2.3.1*; the natural evolution of humans in the state of nature to sense earthquakes when they occurred, in *Sec. 2.3.2*; the natural evolution of humans to live on land, unlike fish living in water, in *Sec. 2.4.1*; the natural evolution of humans to engage in small talk over time in the state of nature, in *Sec. 2.4.2*; the natural evolution of humans to be sensitive to different sounds in the state of nature, in *Sec. 2.5.1*; the natural evolution of humans to hear sounds in the state of nature, in *Sec. 2.5.2*; the natural evolution of humans to suffer from hearing loss in the state of nature, in *Sec. 3.2.1*; the natural evolution of humans to have some members suffering from stuttering in primitive times, in *Sec. 3.2.2*; the natural evolution of humans to be unable to naturally hear infrasound at low frequencies, in *Sec. 3.3.1*; the natural evolution of humans to live in the state of nature, at times with unwanted noise, in *Sec. 3.4.1*; the natural evolution of humans to live together in the state of nature, with activities which sometimes annoy others, in *Sec. 3.4.2*; the natural evolution of humans to use their mental faculties like imagination, reasoning, intuition, etc. in the state of nature, in *Sec. 3.5.1*; and the natural evolution of humans to reason, at least in a primitive way, in the state of nature, in *Sec. 3.5.2*).

On the other hand, there is transformation (e.g., the technical transformation of human ability to hear sounds outside the audio range with the invention of instruments for ultrasonography and elastography to study "the ultrasonic frequency range," for instance, in *Sec. 2.2.1*; the technical transformation of human understanding of each other by the technical invention of such personality models as "Eysenck's three-factor model, Cattell's 16 personality factors, the Big Five personality traits, the four temperaments, the Minnesota Multiphasic Personality Inventory, the Myers Briggs Type Indicator, and Socionics," in *Sec. 2.2.2*; the technical transformation of cattles to survive, by the human invention of ultrasound technology "to improve…the yield of cattle operations," in *Sec. 2.3.1*; the technical transformation of human understanding of earthquakes by the invention of seismographs for earthquake prediction before they even occurred, in *Sec. 2.3.2*; the technical transformation of human ability to live in water by the invention of submarines, in *Sec. 2.4.1*; the technical transformation of human tendency to engage in small talk by the invention of some etiquettes like Grice's Maxim of Quantity and Schneider's counter-proposal, in *Sec. 2.4.2*; the technical transformation of human ability to listen to sounds by the invention of chaotic music, like the one by Cage, in *Sec. 2.5.1*; the technical transformation of human ability to hear sounds by the invention of different musical instruments, like those used by noise music as a modern addition, in *Sec. 2.5.2*; the technical transformation of human ability to counter hearing loss by the invention of gene therapy, among other methods, in *Sec. 3.2.1*; the technical transformation of human ability to treat stuttering by the invention of "anti-stuttering medications" and "electronic fluency devices" for the treatment, in *Sec. 3.2.2*; the technical transformation of human ability to hear infrasound at low frequences by the invention of special equipments as in ballistocardiography and seismocardiography, in *Sec. 3.3.1*; the technical transformation of human living together in society, by the invention of government regulation for noise control, in *Sec. 3.4.1*; the technical transformation of human living together, by the invention of "the right to quiet enjoyment" in tort law which is to regulate activities which may annoy each other, in *Sec. 3.4.2*; the technical transformation of human ability to use their mental faculties by the invention of systems of ideas like Christian mysticism for prefering contemplation over meditation, in *Sec. 3.5.1*; and the technical transformation of human ability to reason by the invention of some formal systems of reasoning, like deduction, induction, and abduction, in *Sec. 3.5.2*).

And the reverse direction also holds true.

Again, it should be stressed, however, that there are different shades of gray (or different degrees of truth) in the two opposites, and the classification is not necessarily mutually exclusive either. In fact, even when some combinations of the two occur, they only end up sharing the same dialectic relationship, but in a different degree.

Also, there are some other relationships (as a kind of family resemblance) which have something in common with the principle, but they are not exactly the same but only more or less comparable, which varies from case to case.

Good examples of family resemblance in relation to the principle are inorganicness-volition, naturalness-technologicalness, naturalness-nonnaturalness, nonwillingness-willingness, inorganicness-motivation, and so on.

13th Thesis: The Symmetry-Asymmetry Principle

The thirteenth thesis is labeled as the symmetry-asymmetry principle (on the relationships among existents) in the category of "agency" in existential dialectics (which was also already worked out in my previous books), in that there is no asymmetry without symmetry—and vice versa.

For instance, the Same can be symmetric and asymmetric towards the Others. But in case of asymmetry (as analyzed in *BDPD*), oppression and self-oppression can occur. So, when the Same is asymmetric towards the Others, the Same can also be relatively asymmetric towards itself in self-oppression, just as the Others can be likewise towards themselves. The subsequent oppressiveness is dualistic, as much by the Same against the Others and itself, as by the Others against the Same and themselves. Both oppression and self-oppression can be achieved by way of downgrading differences between the Same and the Others and of accentuating them.

This is true, even though not all forms of asymmetry have to be about oppression and self-oppression.

In addition, from Chapter Three of *FPHG*, symmetry is not perfect, to be understood in an approximate sense under many life circumstances. With this clarification in mind, hereafter are some empirical examples.

For instance, on the one hand, there is symmetry (e.g., the co-existence of different ways to measure ATH, like "the method of constant stimuli" and "the method of limits," in *Sec. 2.2.1*; the co-existence of different personality traits like extroversion and introversion, in *Sec. 2.2.2*; the co-existence of different counter-measures by moths to avoid being eaten by bats, in *Sec. 2.3.1*; the existence of body waves and surface waves, in *Sec. 2.3.2*; the co-existence of different navigation systems, like OMEGA and Alpha, in *Sec. 2.4.1*; the co-existence of different etiquettes about small talk, in *Sec. 2.4.2*; the co-existence of different overtones, in *Sec. 2.5.1*; the existence of different genres of music, in *Sec. 2.5.2*; the co-existence of different individuals in society, in *Sec. 3.2.1*; the co-existence of different stutterers in society, in *Sec. 3.2.2*; both humans and animals have hearing ability, in *Sec. 3.3.1*; the co-existence of different government regulations in different countries, in *Sec. 3.4.1*; the existence of both private nuisance and public nuisance, in *Sec. 3.4.2*; the existence of different stages of mystical union in contemplative prayer, in *Sec. 3.5.1*; and the existence of three different kinds of reasoning, in *Sec. 3.5.2*).

On the other hand, there is also asymmetry (e.g., "the method of constant stimuli" is better in one way, because its "catch trials" can "help to detect the amount of a listener's guessing"—but "the method of limits" is better in a different way, because it requires only a smaller number of trials, which can save time "required to complete the testing," in *Sec. 2.2.1*; "an extroverted person is likely to enjoy time spent with people and find less reward in time spent alone"—but "[i]ntroverts are people whose energy tends to expand through reflection and dwindle during interaction, in *Sec. 2.2.2*; the counter-measure by a flying noctuid moth to exercise "a reflex action…that cause[s] it to drop a few inches in its flight to evade attack"—but the counter-measure by a Tiger moth to "emit clicks which jam bats' echolocation," in *Sec. 2.3.1*; body waves travel through solids—but surface waves travel through solid-air, in *Sec. 2.3.2*; the preferred usage of OMEGA in the West—but the preferred usage of Alpha in Russia, in *Sec. 2.4.1*; Grice's Maxim of Quantity in the 1970's—but Schneider's counter-proposal in the 1980's, in *Sec. 2.4.2*; some overtones are harmonic—whereas others are inharmonic, because they are "created that are not anywhere near a harmonic," in *Sec. 2.5.1*; traditional melodic music focuses on harmonic series—but noise music focuses on dssonance, in *Sec. 2.5.2*; some individuals have severe hearing loss—but some other individuals do not have severe hearing loss, in *Sec. 3.2.1*; some stutterers have only "barely perceptible impediments, for whom the disorder is largely cosmetic"—but other stutterers have "extremely severe symptoms, for whom the problem can effectively prevent most oral communication," in *Sec. 3.2.2*; "sound that is lower in frequency than 20 Hz (Hertz) or cycles per second" is "the 'normal' limit of human hearing"—but elephants can hear sounds at frequencies lower than 20 Hz, in *Sec. 3.3.1*; the noise-control standards in American government regulation are higher than European government regulation, with a gap "by 10 to 20 years," in *Sec. 3.4.1*; "owners of interests in real property [whether owners, lessors, or holders of an easement or other interest] have standing only to bring private nuisance suits," even when "a nuisance is widespread enough" to cause a public nuisance too, in which case the government can come in and file a case for public nuisance, in *Sec. 3.4.2*; the first three stages are not fully in mystical union with God—but the final stage is fully in mystical union with God, in accordance to Christian mysticism, in *Sec. 3.5.1*; and deduction focuses on the validity of an argument, induction focuses on the probability of an argument, and abduction focuses on the reasonableness of an argument, in *Sec. 3.5.2*).

And the reverse direction also holds true.

Again, for the last time, it should be stressed, however, that there are different shades of gray (or different degrees of truth) in the two opposites, and the classification is not necessarily mutually exclusive either. In fact, even when some combinations of the two occur, they only end up sharing the same dialectic relationship, but in a different degree.

Also, there are some other relationships (as a kind of family resemblance) which have something in common with the principle, but they are not exactly the same but only more or less comparable, which varies from case to case.

Good examples of family resemblance in relation to the principle are strength-weakness, potence-impotence, balance-extremity, reflexiveness-unreflexiveness, equality-inequality, harmoniousness-unharrmoniousness, and something like that.

14th Thesis: The Softness-Hardness Principle

The fourteenth thesis refers to the softness-hardness principle (on the force of change) in the category of "agency" in existential dialectics (which was first worked out in *ALD*), in that any change by an agent, be it organic (like humans) or non-organic (like natural objects), can occur in a forceful (aggressive) or gentle (pacific) way, which can come in all shapes and sizes, of course. There is no softness without hardness—and vice versa.

For instance, on the one hand, there is softness (e.g., the use of ultrasound by whales to communicate with each other for mating, in *Sec. 2.3.1*; the use of GPS for civilian communication, in *Sec. 2.4.1*; the use of "compliments about some aspect of personal appearance" among women in small talk, in *Sec. 2.4.2*; consonant sounds, in *Sec. 2.5.1*; the pleasant sounds in harmonic series, in *Sec. 2.5.2*; the benevolence of those who care about hearing-impaired people, in *Sec. 3.2.1*; the parental support of stutterers by encouragin them "to slow down, take a breath, say it again, etc.," in *Sec. 3.2.2*; some sounds are pleasant for humans to hear, as in hamonic music, in *Sec. 3.3.1*; the making of pleasant sound, in *Sec. 3.4.1*; the right to quiet enjoyment, in *Sec. 3.4.2*; the reward of mystical union with God, in accordance to Christian mysticism, in *Sec. 3.5.1*; and the praise of someone's argument as reasonable, in *Sec. 3.5.2*).

On the other hand, there is hardness (e.g., the use of ultrasound by whales to hunt and kill, in *Sec. 2.3.1*; the use of GPS for navel warfare, in *Sec. 2.4.1*; the use of "verbal sparring matches, playful insults, and putdowns" among men in small talk, in *Sec. 2.4.2*; dissonant sounds, in *Sec. 2.5.1*; "the most dissonant, strange and harsh sounds" in the contemporary art of noises, in *Sec. 2.5.2*; the mask of benevolence of those who care about hearing-impaired people, as pointed out by Lane, in *Sec. 3.2.1*; the making fun of the stutterers by others, in *Sec. 3.2.2*; some sounds cause "feelings of anxiety or fear in humans" to hear, as in infrasound, in *Sec. 3.3.1*; the making of annoying sound, in *Sec. 3.4.1*; the duty of care in the right to quiet enjoyment, which can be punished if violated, be it on the basis of money or injunction, in *Sec. 3.4.2*; only a few will make it to the final stage of mystical union with God, in accordance to Christian mysticism, in *Sec. 3.5.1*; and the criticism of someone's argument as unreasonable, in *Sec. 3.5.2*).

And the reverse direction also holds true.

Again, it should be stressed, however, that there are different shades of gray (or different degrees of truth) in the two opposites, and the classification is not necessarily mutually exclusive either. In fact, even when some combinations of the two occur, they only end up sharing the same dialectic relationship, but in a different degree.

Also, there are some other relationships (as a kind of family resemblance) which have something in common with the principle, but they are not exactly the same but only more or less comparable, which varies from case to case.

Good examples of family resemblance in relation to the principle are peacefulness-violence, cooperation-competition, pacificity-aggression, dovishness-hawkishness, reward-punishment, peace-war, and the like.

15th Thesis: The Regression-Progression Principle

The fifteenth thesis is called the regression-progression princple (on the direction of history) in the category of "outcome" in existential dialectics (which was also already worked out in my previous books), in that neither the cyclical nor the linear views are adequate for explaining many phenomena at all levels. There is no regression without progression—and vice versa.

History progresses to more advanced forms, but with a regressive touch. Examples include no freedom without unfreedom, no equality without inequality, and no civilization without barbarity. This is not an inevitable law, but merely a highly likely empirical trend.

For instance, on the one hand, there is regression (e.g., the regression made by the psychoacoustic model to understand sounds, as shown in the criticisms by the critics, in *Sec. 2.2.1*; the regression made by each personality trait in relation to things which it may not excel in, like engaging in solitary activities by an extrovert, in *Sec. 2.2.2*; the regression made by the study of animal sounds, as shown by the criticisms about some major problems in the findings, in *Sec. 2.3.1*; the regression made by earthquake engineers, as shown by the problems pointed out by the critics, in *Sec. 2.3.2*; the regression made by submarine navigation, as shown in the problems pointed out by the critics, in *Sec. 2.4.1*; the regression made by small talk, as shown in the criticisms by the critics, in *Sec. 2.4.2*; the regression made by harmonic music, as shown in the criticisms by the critics, in *Sec. 2.5.1*; the progression made by noise music, as shown in the criticisms by the critics, in *Sec. 2.5.2*; the regression made by the different treatments of hearing loss, as shown in the criticisms by the critics, in *Sec. 3.2.1*; the regression made by different treatments of stuttering, as shown in the criticisms by the critics, in *Sec. 3.2.2*; the regression made by Tandy's work on ghost sightings, as shown in the criticisms by the critics, in *Sec. 3.3.1*; the regression made by architectural acoustics, as shown in the criticisms by the critics, in *Sec. 3.4.1*; the regression made by the enforcement of the law of nuisance, as shown in the problems pointed out by the critics, in *Sec. 3.4.2*; the regression made by contemplative prayer, as shown in the criticisms by the critics, in *Sec. 3.5.1*; and the regression made by abduction, as shown in the criticisms by the critics, in *Sec. 3.5.2*).

On the other hand, there is progress (e.g., the progress made by the psychoacoustic model to help us understand the limits of human hearing, in *Sec. 2.2.1*; the progress made by each personality trait in relation to things which it may excel in instead, like involving in large social gatherings by an extrovert, in *Sec. 2.2.2*; the regression made by the study of animal sounds, for a better understanding of animal communication and survival, in *Sec. 2.3.1*; the progress made by earthquake engineers for general earthquake forecasts, in *Sec. 2.3.2*; the progress made by submarine navigation for naval warfare, in *Sec. 2.4.1*; the progress made by small talk for the enhancement of social skills, in *Sec. 2.4.2*; the progress made by harmonic music for the cultivation of pleasant consonant sounds, in *Sec. 2.5.1*; the progress made by noise music for a new way to appreciate noise in music, in *Sec. 2.5.2*; the progress made by the different treatments of hearing loss for the improvement of hearing, in *Sec. 3.2.1*; the progress made by different treatments of stuttering, as shown by the "relief from the overt symptoms of stuttering," in *Sec. 3.2.2*; the progress made by Tandy's work for a new way to understand ghost sightings, in *Sec. 3.3.1*; the progress made by architectural acoustics for noise control in society, in *Sec. 3.4.1*; the progress made by the enforcement of the law of nuisance, for the protection of the right to quiet enjoyment, in *Sec. 3.4.2*; the progress made by contemplative prayer for spiritual enlightenment, in *Sec. 3.5.1*; and the progress made by abduction for a different kind of argument based on reasonableness, in *Sec. 3.5.2*).

And the reverse direction also holds true.

Again, it should be stressed, however, that there are different shades of gray (or different degrees of truth) in the two opposites, and the classification is not necessarily mutually exclusive either. In fact, even when some combinations of the two occur, they only end up sharing the same dialectic relationship, but in a different degree.

Also, there are some other relationships (as a kind of family resemblance) which have something in common with the principle, but they are not exactly the same but only more or less comparable, which varies from case to case.

Good examples of family resemblance in relation to the principle are undesirability-desirability, risk-opportunity, badness-goodness, cost-benefit, and so on.

16th Thesis: The Same-Difference Principle

The sixteenth thesis refers to the same-difference principle (on the metamorphosis of change) in the category of "outcome" in existential dialectics (which was worked out in *ALD* and other books of mine), in that an entity, as it evolves over time, can be both different from and similar to its opposing alternatives and does not have to be solely more different from them over time. There is no similarity without difference—and vice versa.

Opposites are not absolute in a black-or-white fashion; so, an entity can become relatively more similar to (or more different from) its opposite over time.

This is further constrained by another principle, that is, the symmetry-asymmetry principle about the relationships among existents under the category about agency in existential dialectics, in that if there is symmetry (equality) between two entities, there is likewise asymmetry (inequality) emerging in a different way.

For instance, on the one hand, there is similarity in outcome (e.g., the contribution to the molding and regulation of beliefs and values, regardless of whether this be done by way of understanding sounds from the scientific perspective of psychoacoustics or from the pre-scientific perspective of magic, shamanism, etc., in *Sec. 2.2.1*; the contribution to the molding and regulation of beliefs, values and behaviors, regardless of whether this be done by way of extroversion or introversion, in *Sec. 2.2.2*; the contribution to the molding and regulation of behaviors, regardless of whether this be done by way of evasive manoeuvres or by way of charming manoeuvres, in *Sec. 2.3.1*; the contribution to the molding and regulation of beliefs and values, regardless of whether this be done by way of earthquake engineering or by way of shamanism, in *Sec. 2.3.2*; the contribution to the molding and regulation of beliefs and behaviors, regardless of whether this be done by way of stealthy submarine navigation or by way of open commercial ship navigation, in *Sec. 2.4.1*; the contribution to the molding and regulation of beliefs and behaviors, regardless of whether this be done by way of small talk or substantial talk, in *Sec. 2.4.2*; the contribution to the molding and regulation of beliefs and values, regardless of whether this be done by way of harmonic music or chaotic music, in *Sec. 2.5.1*; the contribution to the molding and regulation of beliefs and values, regardless of whether this be done by way of traditional melodic music or noise music, in *Sec. 2.5.2*; the contribution to the molding and control of beliefs and behaviors, regardless of whether this be done by way of social help to deaf people or self-help instead, in *Sec. 3.2.1*; the contribution to the molding and regulation of beliefs and behaviors, regardless of whether this be done by way of the treatment of stuttering or by way of the acceptance of stuttering, in *Sec. 3.2.2*; the contribution to the molding and regulation of beliefs and values, regardless of whether this be done by way of science or by way of esoteric religion, both in regard to the research on ghost sightings, in *Sec. 3.3.1*; the contribution to the molding and regulation of beliefs and behaviors, regardless of whether this be done by way of noise control through government regulation or by way of noise management through free market, in *Sec. 3.4.1*; the contribution to the molding and regulation of beliefs and behaviors, regardless of whether this be done by way of tort law or by way of individual agreements, in regard to nuisance, in *Sec. 3.4.2*; the contribution to the molding and regulation of beliefs and behaviors, regardless of whether this be done by way of contemplative prayer or by way of discursive meditation, in *Sec. 3.5.1*; and the contribution to the molding and regulation of beliefs and values, regardless of whether this be done by way of pure reasoning or by way of abductive reasoning, in *Sec. 3.5.2*).

On the other hand, there is difference in outcome (e.g., the contribution to the molding and regulation of beliefs and values by way of understanding sounds from the scientific perspective of psychoacoustics for a more logical-rational worldview—but the contribution to the molding and regulation of beliefs and values by way of understanding sounds from the pre-scientific perspective of magic, shamanism, etc for a more religious worldview, in *Sec. 2.2.1*; the contribution to the molding and regulation of beliefs, values and behaviors by way of extroversion for a more social lifeworld—but the contribution to the molding and regulation of beliefs, values and behaviors by way of introversion for a more solitary lifeworld, in *Sec. 2.2.2*; the contribution to the molding and regulation of behaviors by way of evasive manoeuvres by a flying moth vis-à-vis a bat in the violent battle for survival—but the contribution to the molding and regulation of behaviors by way of charming manoeuvres by a male whale to a female whale in the seductive battle for reproduction, in *Sec. 2.3.1*; the contribution to the molding and regulation of beliefs and values by way of earthquake engineering for a more secular lifeworld—but the contribution to the molding and regulation of beliefs and values by way of shamanism for a more sacred lifeworld, in *Sec. 2.3.2*; the contribution to the molding and regulation of beliefs and behaviors by way of stealthy submarine navigation for a more secretive lifeworld—but the contribution to the molding and regulation of beliefs and behaviors by way of open commercial ship navigation for a more transparent lifeworld, in *Sec. 2.4.1*; the contribution to the molding and regulation of beliefs and behaviors by way of small talk for a relatively smoother social world—but the contribution to the molding and regulation of beliefs and behaviors by way of substantial talk for a relatively blunter social world, in *Sec. 2.4.2*; the contribution to the molding and regulation of beliefs and values by way of harmonic music for a more conservative aesthetic experience—but the contribution to the molding and regulation of beliefs and values by way of chaotic music for a more radical aesthetic experience, in *Sec. 2.5.1*; the contribution to the molding and regulation of beliefs and values by way of traditional melodic music for a more conservative musical lifestyle—but the contribution to the molding and regulation of beliefs and values by way of noise music for a more radical musical lifestyle, in *Sec. 2.5.2*; the contribution to the molding and control of beliefs and behaviors by way of social help to deaf people for a relatively more cooperative society—but the contribution to the molding and control of beliefs and behaviors by way of self-help for a relatively more competitive society, in *Sec. 3.2.1*; the contribution to the molding and regulation of beliefs and behaviors by way of the treatment of stuttering for the cultivation of more polished civil manners—but the contribution to the molding and regulation of beliefs and behaviors by way of the acceptance of stuttering for the tolerance of more impulsive civil manners, in *Sec. 3.2.2*; the contribution to the molding and regulation of beliefs and values by way of science for a more secular worldview—but the contribution to the molding and regulation of beliefs and values by way of esoteric religion for a more sacred worldview, in *Sec. 3.3.1*; the contribution to the molding and regulation of beliefs and behaviors by way of noise control through government regulation for a relatively more powerful political society—but the contribution to the molding and regulation of beliefs and behaviors by way of noise management through free market for a relativley more powerful civil society, in *Sec. 3.4.1*; the contribution to the molding and regulation of beliefs and behaviors by way of tort law for the institutionalized order in political society—but the contribution to the molding and regulation of beliefs and behaviors by way of individual agreements for the non-institutioinalized order in the private sphere, in *Sec. 3.4.2*; the contribution to the molding and regulation of beliefs and behaviors by way of contemplative prayer for a sacred worldview based more on graces—but the contribution to the molding and regulation of beliefs and behaviors by way of discursive meditation for a sacred worldview based more on thoughts, in *Sec. 3.5.1*; and the contribution to the molding and regulation of beliefs and values by way of pure reasoning for a relatively strict form of rationality—but the contribution to the molding and regulation of beliefs and values by way of abductive reasoning for a relatively speculative form of rationality, in *Sec. 3.5.2*).

And the reverse direction also holds true.

Again, it should be stressed, however, that there are different shades of gray (or different degrees of truth) in the two opposites, and the classification is not necessarily mutually exclusive either. In fact, even when some combinations of the two occur, they only end up sharing the same dialectic relationship, but in a different degree.

Also, there are some other relationships (as a kind of family resemblance) which have something in common with the principle, but they are not exactly the same but only more or less comparable, which varies from case to case.

Good examples of family resemblance in relation to the principle are homogeneity-heterogeneity, we-they, and so forth.

17th Thesis: The Post-Human Creativeness

And the seventeenth thesis is about the role of "post-humans," which I originally proposed in my previous books, starting with the first book titled *The Future of Human Civilization* in 2000 and all others afterwards.

As already pointed out in *Sec. 1.7*, I need to emphasize, as this is something that I used to repeat (*almost verbatim*) from my previous books, two clarifications here about the term "post-human" as a neologism in my works.

Firstly, the word "post-human" here should *not* be confused with another term which looks similar but has a totally different meaning in the literature of postmodernism, namely, "post-humanism"—which constitutes a critique of "humanism" as tradionally understood (especially, though not exclusively, in relation to the idea of progress in science and reason in the Enlightenment project). (WK 2008)

My works reject the project of "postmodernism" and propose the future world of what I orignally called "after-postmodernity" in *FHC* and *FCD*, for instance.

And secondly, the word "post-human" here should also *not* be confused with a similar term which is used to champion the ideology of technology for the future co-existence between humans and cyborgs in "trans-humanism." (WK 2008a)

Instead, my term "post-human" in relation to "posthuman-ism" also rejects "transhumanism" (especially, though not exclusively, in relation to the promises of technology) and refers to something else altogether, that is, the future extinction of humans and its post-human successors in deep space and beyond unto multiverses.

My critique of "transhumanism" was more extensively elabroated in *Sec. 2.4.1* of *BEPE*.

With these two clarifications in mind (as summarized in *Table 1.31*)—the post-human creativeness in acoustics can be addressed in relation to five main directions, based on the multilateral theory of acoustics as analysed in this book (together with my visions as already worked out in my previous books).

Firstly, the post-human creativeness in acoustics will learn from the multilateral theory of acoustics in relation to the dialectic context of sound and silence—especially in the larger dialectic context of no sound without silence (and vice versa), in the context of acoustics.

Secondly, the post-human creativeness in acoustics will learn from the multilateral theory of acoustics in relation to the four perspectives of the mind, nature, society, and culture.

Thirdly, the post-human creativeness in acoustics will learn from the multilateral theory of acoustics in relation to the non-privilege of any specific theory over others in the literature and the non-integration of them all (as they are not necessarily compatible with each other).

Fourthly, the post-human creativeness in acoustics will have to confront the new challenges as posed by the long-term civilizational development of intelligent life in the distant future, both here on this planet Earth and elsewhere in deep space until multiverses.

As this is something that I regularly pointed out in my previous books for background information (and summarize here, *almost verbatim*, with some updates), I already worked out, in my numerous books, what these new challenges will be and provide different original visions to meet them—especially, in regard to the future evolutions of the mind (e.g., "metamorphic humor," "the contrarian personality," "the hyper-martial body," "the hyper-sexual body," "floating consciousness," "hyper-spatial consciousness," "thinking machines," "unfolding unconsciousness," "thinking robots," "genetically altered superior beings," "cyborgs," and others), of nature (e.g., "resettlement geology," "post-cosmology," "the alteration of space-time," "the creation of new matter-energy," "selective geometry," and the like), of society (e.g., "heterodox education," "multifaceted war and peace," "virtual organizations," different versions of "post-capitalism" and "post-democracy," the movement of "cyclical progression" at both structural and systemic levels, "ambivalent technology," and so forth), and of culture (e.g., "post-human mind games," "comparative-substitutive religion," "post-ethics," "post-civilization," "transformative aesthetic experience," "contrastive mathematical-logic," and the whatnot) in history (e.g., the age of "after-postmodernity"), in the context of my approaches in relation to methodology (e.g., "sophisticated methodologial holism," "critical-dialectic formal science," etc,) and ontology (e.g., "existential dialectics," "contrastive rationality").

Of course, the examples (as listed above) are not exhaustive but illustrative, since my numerous books have worked out many other visions (as already roughly summarized in *Sec.1.7* and *Sec. 1.8*). Many, though not all, of my visions on the mind, nature, society, culture, history, methodology, and ontology (as cited above) are summarized in the tables as shown in Chapter One and Chapter Four (especially in *Table 4.43* about my original theories on numerous topics).

And finally, or fifthly, the post-human creativeness in acoustics will learn from the multilateral theory of acoustics in regard to the need to go beyond sound and silence (as further elaborated in the next section).

Towards the Post-Human Creativeness

This "post-human" creativeness in acoustics to transcend both sound and silence can be understood in terms of five great future transformations of acoustics, as explained below (and summarized in *Table 1.2*).

Five Great Future Transformations of Acoustics

The five great future transformations of acoustics can be explained hereafter, with the caveat, however, that the classification in terms of five is solely aesthetic and utilitarian (illustrative), as there is no objective basis that it must be classified in term of five.

The First Great Future Transformation of Acoustics

The first great future transformation of acoustics in the post-human era concerns what I originally call *new sound detection.*

This new sound detection allows future advanced humans and later their successors, namely, the post-humans, the ability to hear and detect sound in a way that current humans cannot.

There are already some indications that the future of acoustics is moving toward this direciton. Consider, for illustration, a few examples below.

Firstly, submarine undercover navigation systems and anti-submarine warfare detection systems nowadays (as already discussed in *Sec. 2.4.1*) point to the need for more sophisticated sound detection ability.

Secondly, the use of seismographs to detect seismic waves for earthquake prediction nowadays (as already discussed in *Sec. 2.4.2*) points to another need for more sophisticated sound detection ability.

Thirdly, the search for what I originally called "the potential of unfolding unconsciousness" (as already discussed in my book titled *The Future of Post-Human U nconsciousess* in 2008), especially in relation to "clair-audience" (in hearing), "clair-sentience" (in feeling), and "clair-cognizance" (in knowing), for instance, points to one additional need for more sophisticated sound detection ability.

All these are just some illustrative (relatively primitive) examples, of course, but the important point to remember here is that the quest for more sophisticated sound detection will further transform acoustics at the more advanced level in the post-human future that the human world has never known.

The Second Great Future Transformation of Acoustics

The second great future transformation of acoustics in the post-human era concerns what I originally call *new sound control.*

There are already some indications that the future of acoustics is moving toward this direciton.

Consider, for illustration, the advances in architectural acoustics nowadays (as already discussed in *Sec. 3.4.1*), especially in relation to the techniques and devices for "absorption," "blocking," and "masking" (also known as the "ABC's" in interior-space acoustics), together with other applications (as discussed in that section), point to the need for more sophisticated noise-control.

All these applications are just some illustrative (relatively primitive) examples, of course, but the important point to remember here is that the quest for more sophisticated sound control will further transform acoustics at the more advanced level in the post-human future that the human world has never known.

The Third Great Future Transformation of Acoustics

The third great future transformation of acoustics in the post-human era concerns what I originally call *new sound creation.*

There are already some indications that the future of acoustics is moving toward this direciton.

Consider, for illustration, the different historical examples of noise aesthetics especially in the 20th century until now (as already discussed in

Sec. 2.5.2), in relation to techniques (e.g., "cacophony," "atonality," "indeterminacy," "repetition," "randomness," "assemblage," "improvisation," etc.) and devices (e.g., "sound recorders," "live machines," "musical instruments," "noise software," "furnitures," "mechanical parts,"etc.), point to the need for more sophisticated noise-creation.

All these historical examples are just some illustrative (relatively primitive) examples, of course, but the important point to remember here is that the quest for more sophisticated sound creation will further transform acoustics at the more advanced level in the post-human future that the human world has never known.

The Fourth Great Future Transformation of Acoustics

The fourth great future transformation of acoustics in the post-human era concerns what I originally call *new sound horizon*.

There are already some indications that the future of acoustics is moving toward this direciton.

Consider, for illustration, the broader horizon of understanding sound (e.g., the use of "small talk" for social skills, as already discussed in *Sec. 2.4.2*—and the use of sounds by animals for survival, as already discussed in *Sec. 2.3.1*, just to cite two examples) and silence (e.g., the right to quiet enjoyment as already discussed in *Sec. 3.4.2*—and the appeal to abduction in some legal sytems as already discussed iin *Sec. 3.5.2*, just to cite two examples) point to the need for a broader understanding of sound horizon.

This need for a broader understanding of sound horizon is related to different versions of my vision on holism in my previous books, like "holistic humor" (as already discussed in *FPHH*), "holistic education" (as already discussed in *FPHEDU*), "holistic knowledge" (as already discussed in *The Future of Post-Human Knowledge* or *FPHK*, and other books of mine) and "holistic methodology" (as already discussed in many books of mine like *FCD, FPHC, BCPC, FC, FPHK*, etc.).

My vision of holism is related to the future evolutions of the mind in relation to perception, conception, imagination, intuition, emotions, and behaviors.

For example, my visions of the future evolutions of the mind (including the brain and the body) are especially relevant, especially in relation to "unfolding unconsciousness" (as already discussed in *FPHU*), "floating consciousness" (as already discussed in *FCD* and *FPHC*), "hyper-spatial consciousness" (as already discussed in *FPHG* and *FPHC*), "multilogical learning" (as already discussed in *FPHL*), "comprehensive creating thinking" (as already discussed in *FPHCT*), "hyper-martial body" (as already discussed in *FPHMA*), "hyper-sexual body" (as already discussed in *FPHS*), "transformative aesthetic experience" (as already discussed in *FAE*), "beyond normality and abnormality" (as already discussed in *FPHP*), "beyond morality and immorality" (as already discussed in *BEPE*), and the like.

In any event, all these uses for a broader understanding of sound horizon (in the context of acoustics) are just some illustrative (relatively primitive) examples, of course—but the important point to remember here is that the need for a broader understanding of sound horizon will further transform acoustics at the more advanced level in the post-human future that the human world has never known.

The Fifth Great Future Transformation of Acoustics

And the fifth great future transformation of acoustics in the post-human era concerns what I originally call *new sound spirituality*.

There are already some indications that the future of acoustics is moving toward this direciton.

Consider, for illustration, the search for sound spirituality (e.g., the exploration of meditation and contemplation for higher spirituality, as already discussed in *Sec. 3.5.1*, just to cite two different ways) points to the need for a more enlightened sound spirituality.

In addition, I already went to great lengths to show, in *The Future of Capitalism and Democracy* (2002), why and how there will be more spiritual transformations to come, especially in relation to the two value ideals of (i) transcending freedom in floating existence and (ii) transcending equality in the rivalry of cosmic hegemony.

Even then, as this is something I often stressed in many of my previous books, almost verbatim, that these two value ideals constitute only a form of transcendent state of higher spiritual concerns, with other more ideals to come in the distant future, especially when considered in conjunction with other future transformations of the mind, nature, society, and culture (as already addressed in my other books on numerous topics).

Some good examples of my other visions of the spiritual transformations to come include "spiritual education" (as already discussed in *FPHEDU*), "contrarian personality" (as already discussed in *FPHP*), "post-civlization" (as already discussed in *BCIV*), "post-ethics" (as already discussed in *BEPE*), "beyond aggression and pacificity" (as already discussed in *FPHWP*), "post-human religion" (as already discussed in *FPHR*), "post-human law" (as already discussed in *FPHLAW*), and so on.

All these explorations are just some illustrative (relatively primitive) examples, of course, but the important point to remember here is that the need for a more enlightened sound spirituality will further transform acoustics at the more advanced level in the post-human future that the human world has never known.

Enormous Implications

These future great transformations of sound and silence in acoustics have enormous implications for some of the profound questions which have intrigued some of the finest minds in human history—Is there "clair-audience" (in hearing) for intelligent life to hear in an unusual way never possible before? Can meditation and contemplation in a state of silence bring the mind of intelligent life to be in union with the cosmos (or something of that nature)? And….?

Questions like this will yield new answers in post-human acoustics in the current dialectic context of (a) treating both sound and silence and (b) going beyond sound and silence—such that their possibility and desirability are not to the extent that the respective defenders would like us to believe.

In fact, some of my other books titled *The Future of Post-Human Religion* (2010), *Beyond Cosmology to Post-Cosmology* (2010), and *The Future of Post-Human Unconsciousness* (2008) had already dealt with issues of this nature, though from different vantage points.

This clarification aside—an important point to remember here is that the five great future transformations of sound and silence in acoustics will bring neither utopia nor dystopia—as this is something that I have warned time and again in all of my previous books.

After all, it is not up to humans to judge whether or not, or to what extent, the post-human future of acoustics will be for better and for worse, because it will be decided by the values and beliefs of future post-humans in a way that the entire history of our human world hitherto existing has never encountered.

But it is already a giant step for us to take if we can acknowledge this bold vision of the five future transformation of acoustics, so as to contribute to the better understanding of the future of acoustics and its post-human fate.

Table 4.1. Sophisticated Methodological Holism (Part I)

• "My methodological holism implies the partiality-totality principle in the ontology of existential dialectics (see the table on the partiality-totality principle for summary), which is against the varieties of (a) reductionism and (b) reverse-reductionism, in relation to (i) concept, (ii) theory, (iii) methodology, and (iv) ontology." (*FC*)

• "[M]y methodological holism here is not opposed to methodological individualism but includes it (and, for that matter, other methodologies too)…." (*FPHC*) For this reason (and others too, as summarized hereafter), my version of methodological holism is sophisticated—not vulgar as sometimes used by inapt scholars using the same term. (*FC*)

• "[M]y methodological holism does not democratically presume that all levels are equally valid, as all levels are not created equal. In other words, in relation to issue X, level A may be more relevant than level B, but in relation to Y, level B can be more relevant than level A instead." (*FPHC*) One excellent example of this vulgar democratic presumption is what I called in *BNN* "the compromise fallacy." (*FC*)

• My methodological holism does not presume that a lower level of analysis is more important than a higher level, solely because the former serves as the foundation for the latter—and vice versa, for that matter. One excellent example of this reductionistic presumption is what I called in *FPHST* "the foundation fallacy." (*FPHST*, *FC*)

(continued on next page)

Table 4.1. Sophisticated Methodological Holism (Part II)

• "[M]y methodological holism does not make any a-priori postulation that there must be a definite (and, for that matter, indefinite) number of levels" in any analysis. (*FPHC*) Nor does it dogmatically require that there must be a certain combination of levels of analysis in a given inquiry. (*FC*)

• "[M]ethodological holism, in my usage, does not assume that all levels…can necessarily be integrated, since methodological holism is not aimed to search for the holy grail of 'an integral theory…' (as is the case for Wilber). In other words, it allows that sometimes some levels may experience irreducible gaps between them, to be understood, at best, as empirical correlations, not as causal relations…." (*FPHC*)

• "[D]ifferent levels may overlap and even interact with each other in a given context (but sometimes may not), and the fact that I even proposed different ways of re-classifying the levels (whenever needed) in *FDC* reinforces this point….The dual danger here is either forcefully making different levels interact when they are just different (or, metaphorically speaking, apples and oranges) or inappropriately ignoring their interactions when some situations instead require them." (*FPHC*)

• "[T]o understand different levels from their own (unique) perspectives (as required by my methodological holism) is not the same as trying to reduce them to a preferred level in the process of learning from other levels. This second kind of multidisciplinary work is not genuine and does no justice to the unique complexities and merits inherent at each level." (*FPHC*)

(continued on next page)

Table 4.1. Sophisticated Methodological Holism (Part III)

• "[My] methodological holism walks a fine line between the artificial classification (separation) of levels and the simultaneous incorporation of them, if only for the sake of human scholarly endeavor. It should be reminded that nature does not impose upon itself the academic classification of the levels of analysis as humans have. The enterprise of classification is therefore anthropocentric." (*FPHC*)

• "[M]y methodological holism advocates neither epistemic subjectivism nor epistemic non-subjectivism (e.g., realism, idealism, and historicism), neither epistemic relativism (e.g., subjectivism, historicism) nor epistemic absolutism (e.g., realism, positivism), neither epistemic reductionism nor epistemic emergencism, and neither epistemic objectivism (e.g., realism, idealism) nor epistemic historicism….Neither does methodological holism, in my usage, accept the false meta-conceptual dichotomy between nominalism and realism….These false dichotomies…are to be transcended. In other words, methodological holism does not fully accept epistemic realism, positivism (a form of epistemic idealism), historicism, subjectivism, and reductionism in epistemology and philosophy of science but learns from the strengths and weaknesses of all of the opposing approaches without siding with any of them…." (*FPHC*)

• "Sophisticated methodological holism is subject to the constraints as imposed by the syntax of existential dialectics (e.g., the partiality-totality principle and the predictability-unpredictability principle). Even in predictability, outcomes are subject to uncertainty, the degree of which varies from case to case." (*FC*)

(continued on next page)

Table 4.1. Sophisticated Methodological Holism (Part IV)

• "Sophisticated methodological holism—when applied, especially though not exclusively, as illustrated in my numerous works—can enrich the understanding of reality in some distinctive ways. Here are three examples (as revealed in each of my books)." (*FPHCT*; *FPHL*)

—"Firstly, it provides a comprehensive analysis of a subject matter, from which much can be learned about reality," "both in relation to the perspectives of the mind, nature, society, and culture and also in relation to a new classification of the subject matter."

—"Secondly, it suggests some visions of the future in relation to the subject matter in question."

—"And thirdly, it proposes some insights on meta-theory (e.g., methodology and ontology) in general—with the clear understanding, however, of the dilemma of specific vs. general ontology (as shown in the table on the syntax of existential dialectics in the context of the dilemma of ontology). For this very reason, all of these ways are important, without reducing one into the analysis of another."

• "In the end, my meta-theory (both sophisticated methodological holism and existential dialectics) serves as a foundation to unify all domains of knowledge into *an unified theory of everything* (by way of some ontological principles and the comprehensive perspectives of the mind, nature, society, and culture). This is so, without commiting the sins of reductionism and reverse-reductionism (as often seen in many holistic approaches, with the fad of sysems approach—be it about systems theory, chaos theory, complexity theory, or else—as a most recent notorious example, which I debunked in *The Future of Complexity* and also in *The Future of Post-Human Formal Science*). My distinctive approach makes good use of different schools of thought without favoring any of them nor trying to integrate them (as they are not necessarily compatible with each other), so as to adjust for subjectivity, diversity, conflict, and complexity, for example. In this sense, the word "unified" does not have to mean integrative approach, in a narrow sense, nor systems approach, in a broad one, both of which I reject while learning from them." (*FPHFS*; *FPHS*)

Sources: A summary of *Sec.1.2* in *FPHC*—and also from *BNN*, *FPHST*, *ALD*, *FC*, *FPHCT*, and the rest of all other books of mine. See the books for more detail.

Table 4.2. On Reductionism and Reverse-Reductionism (Part I)

- **The Partiality-Totality Principle**
  —The partiality-totality principle in the ontology of existential dialectics targets against the varieties of reductionism and reverse-reductionism (as already worked out in my previous books).

- **Against the Varieties of Reductionism**
  —*Conceptual Reductionism*
  - Some illustrative instances involve myriad dualities like mind vs. body, self vs. world, democracy vs. non-democracy, and the like (as already addressed in *FHC*, *FPHC*, and *BDPD*, for instance).

  —*Theoretical Reductionism*
  - A fascinating case study concerns what I originally called "the foundation fallacy" in *FPHST*, in any attempt to naively understand space-time from the physical perspective as the foundation and, consequently, to dangerously dismiss other perspectives.
  - In *FAE*, I elaborated further these versions of reductionism in the literature on aesthetics (e.g., form vs. content, representation vs. expression, critics vs. artists, and externalism vs. internalism).
  - In *FIA*, I revealed other forms of reductionism in the literature on information architecture (e.g., the constructivist argument).
  - In *FPHU*, I showed the persistent legacy of reductionism, this time, in the literature on anomalous experience (e.g., the obsession with physics, chemistry, and biology for explaining anomalous experience).
  - In *FPHE*, I examined another case of reductionism in action, in the context of engineering (e.g., technical constraints vs. normative constraints).
  - In *FPHMM*, I elaborated one more version of reductionism, in relation to the three domains of communication (e.g., the competing views on sending, connecting, and receiving).
  - In *FPHCT*, I explored another version of reductionism, in relation to invention and innovation (e.g., the bio-psychological argument vs. the socio-cultural arguments).

(continued on next page)

Table 4.2. On Reductionism and Reverse-Reductionism (Part II)

- **Against the Varieties of Reductionism *(cont'd)***
  - *—Theoretical Reductionism (cont'd)*
    - In *FPHG*, I identify another version of reductionism, in relation to infinity, symmetry, and dimensionality (e.g., the Euclidean argument vs. the non-Euclidean arguments).
    - In *FPHUP*, I analyzed another version of reductionism, in relation to density and void (e.g., the engineering argument and the ecology argument).
    - In *FPHL*, I examined another version of reductionism, in relation to structure and context (e.g., the structuralist argument and the contextualist argument).
    - In *PFHO*, I explored another version of reductionism, in relation to communcation, decision-making, and leadership (e.g., the rational-system argument vs. the natural-system argument).
    - In *PFHMA*, I revealed another version of reductionism, in relation to the martial body and spirit (e.g., the spiritual argument vs. the materialist argument vs. the defensive argument).
    - In *PFHS*, I analyzed another version of reductionism, in relation to the sexual body and spirit (e.g., the naturalist argument vs. the constructivist argument).
    - In *PFHLAW*, I examined another version of reductionism, in relation to law (e.g., the necessity argument vs. the contengency argument).
    - In *FPHWP*, I showed another version of reductionism, in relation to war and peace (e.g., the aggressivist argument vs. the pacifist argument).
    - In *BEPE*, I analyzed another version of reductionism, in relation to morality and immorality (e.g., the objectivist argument vs. the non-objectivist argument vs. the skeptical argument).
    - In *BCOS*, I examined another version of reductionism, in relation to the contested beginnings and speculative ends of the universe (e.g., the scientific argument vs. the religious argument vs. the esoteric argument vs. the metaphysical argument).
    - In *FPHP*, I show another version of reductionism, in relation to normality and abnormality (e.g., the natural argument vs. the social argument vs. the cultural argument vs. the mental argument).

(continued on next page)

Table 4.2. On Reductionism and Reverse-Reductionism (Part III)

- **Against the Varieties of Reductionism *(cont'd)***
  —*Theoretical Reductionism (cont'd)*
  - In *FPHGEOL*, I scrutinized another version of reductionism, in relation to statics and dynamics (e.g., the catastrophe argument vs. the uniformity argument vs. the revision argument).
  - In *FPHCHESS*, I showed another version of reductionism, in relation to tactics and strategy (e.g., the natural argument vs. the social argument vs. the cultural argument vs. the mental argument).
  - In *FPHR*, I revealed another version of reductionism, in relation to secularness and sacredness (e.g., the critical argument vs. the skeptical argument vs. the theist argument).
  - In *FPHEDU*, I examined another version of reductionism, in relation to teaching and learning (e.g., the teacher-centered argument vs. the student-centered argument vs. the balanced argument).
  - In *FPHH*, I explored another version of reductionism, in relation to joking and laughing (e.g., the natural argument vs. the social argument vs. the cultural argument vs. the mental argument).
  - And in *FPHA*, I propose another version of reductionism, in relation to sound and silence (e.g., the natural argument vs. the social argument vs. the cultural argument vs. the mental argument).

  —*Methodological Reductionism*
  - A good instance concerns the debate between different versions of qualitative and quantitative methods (as already analyzed in *FC* and also *FHC*). In *FPHML*, I examined similar reductionism, this time, in the literature on mathematical logic (e.g., the obsession with consistency, soundness, and completeness). And in *FPHFS*, I also explored the problems of reductionism in the context of formal science (e.g., the analytical argument).

  —*Ontological Reductionism*
  - An excellent example is the debate between emergentism and reductionism in complexity theory and also in psychology (as elaborated in *FPHC*, in the context of Being and Becoming).

(continued on next page)

Table 4.2. On Reductionism and Reverse-Reductionism (Part IV)

- **Against the Varieties of Reverse-Reductionism**

—*Conceptual Reverse-Reductionism*

- Any concept of "art" (e.g., fine art, cave art, outsider art, junk art) is deemed acceptable in postmodernism (as already addressed in Ch.4 of *FHC*).

—*Theoretical Reverse-Reductionism*

- There are numerous art and literary theories co-exist. Take the case of literary studies, as there are now Literary Structuralism, Marxist Literary Criticism, New Criticism, Phenomenology, Hermeneutics, Language-Game Literary Criticism, Feminist Literary Criticism, Reception Theory, Reader Response Criticism, Poststructuralism, Semiotics, Pyschoanalytic Literary Criticism, just to cite some well-known ones, with no one being said to be better than any others (as detailedly analyzed in Ch.4 of *FHC*). (S. Raman 1997) In *BNN*, I even introduced "the compromise fallacy" as another good illustration of theoretical reverse-reductionism, in misleadingly treating both genetic and environmental approaches as equally valid.

—*Methodological Reverse-Reductionism*

- There is the "anything-goes" mentality in postmodernism (e.g., doing art without praxis, doing art with praxis, and doing art by sublation), as analyzed in Ch.4 of *FHC*. And in *FPHFS*, I also exposed the problems of reverse-reductionism in the context of formal science in relation to systems theory.

—*Ontological Reverse-Reductionism*

- There are likewise no privileged ontology, and the door is open for anything in postmodernism (e.g., the equal status of the ontology of Being vs. that of Becoming, as already addressed in Ch.4 of *FHC*—and also in *FPHC)*. In *FAE*, I also introduced another version of reverse-reductionism, that is, "the pluralist fallacy," in the context of understanding aesthetic experience, for instance—although this fallacy has been committed not exclusively in relation to the ontological level (but also at the conceptual, theoretical, and methodological ones).

Sources: From my previous books.

Table 4.3. The Concepton of Existential Dialectics (Part I)

- **Sets and Elements**
  - —Sets
    - Ex: the Same
    - Ex: the Others
  - —Elements
    - Ex: whites in 20th century America (in the set of "the Same")
    - Ex: Iraq during the U.S. invasion in 2003 (in the set of "the Others")

- **Relations, Operations, Functions**
  - —Relations (e.g., "belongs," "equals to," "is greater than")
    - Ex: symmetric interactions within the Same (or the Others)
    - Ex: asymmetric interactions between the Same and the Others
  - —Operations (e.g., "and," "or," "not," "if…then")
    - Ex: if the Same oppresses the Others, it will also oppress itself.
    - Ex: the Same is not the Others.
  - —Functions (e.g., goals)
    - Ex: the Same is hegemonic in relation to the Others.

- **Truth Values**
  - —"1" if True (in Symbolic Logic)
    - Ex: the proposition that imperial Japan was hegemonic to China during WWII
  - —"0" if False (in Symbolic Logic)
    - Ex: the proposition that Grenada invaded France in 2003
  - —"1" & "0" if Both True and False (in Dialectic Logic)
    - Ex: the proposition that the rabbit-duck picture refers to a duck
  - —"~1" & "~0" if Neither True Nor False (or N/A, in Dialectic Logic)
    - Ex: the proposition that God really exists

(continued on next page)

Table 4.3. The Conception of Existential Dialectics (Part II)

- **Axioms, Postulates, Theorems, Principles**
  - —Axioms
    - Ex: the reflexive axiom—"any quantity is equal to itself"
  - —Postulates
    - Ex: the SSS postulate—"if the three sides of a triangle are congruent to their corresponding parts, then the triangles are congruent"
  - —Theorems (and Principles) in Existential Dialectics
    - <u>In Relation to Method</u>
      - –#1: The formalness-informalness principle
      - –#2: The absoluteness-relativeness principle
      - –#3: The partiality-totality principle
      - –#4: The predictability-unpredictability principle
      - –#5: The explicability-inexplicability principle
    - <u>In Relation to Structure</u>
      - –#6: The finiteness-transfiniteness principle
      - –#7: The preciseness-vagueness principle
      - –#8: The simpleness-complicatedness principle
      - –#9: The openness-hiddenness principle
      - –#10: The denseness-emptiness principle
    - <u>In Relation to Process</u>
      - –#11: The change-constancy principle
      - –#12: The order-chaos principle
      - –#13: The slowness-quickness principle
      - –#14: The expansion-contraction principle
    - <u>In Relation to Agency</u>
      - –#15: The theory-praxis principle
      - –#16: The convention-novelty principle
      - –#17: The evolution-transformation principle
      - –#18: The symmetry-asymmetry principle
      - –#19: The softness-hardness principle
      - –#20: The seriousness-playfulness principle
    - <u>In Relation to Outcome</u>
      - –#21: The regression-progression principle
      - –#22: The same-difference principle

(continued on next page)

Table 4.3. The Conception of Existential Dialectics (Part III)

Notes: The categories and examples in each are solely illustrative (not exhaustive). The comparison is also relative (not absolute), nor are they mutually exclusive. As generalities, they allow exceptions.
Sources: Starting from Ch.6 of *BCPC* and also from other books of mine

Table 4.4. The Syntax of Existential Dialectics I: The Principles (Part I)

- **In Relation to Method**

—*The Formalness-Informalness Principle*
(On the Formal Requirements of Logical Systems)

- The formal requirements of a logical system (e.g., consistency, soundness, and completeness) have both usefulness and non-usefulness, to the extent that, if there are formal systems requiring them, there are alternative ones which do not. Thus, it does *not* exclude classical logics but simply goes beyond both classical and non-classical logics, while learning something from each. There is no formalness without informalness—and vice versa.
- For instance, existential dialectics can make use of both classical logics under certain conditions (e.g., especially, though not exclusively, when they are clear-cut, etc.) and non-classical logics under alternative conditions (especially, though not exclusively, when they are "unknown," "irrelevant," "ambiguous," "possible," with "different degrees of truth," empirically inconsistent in a desirable way, etc.). (WK 2008u)
- Family resemblance: e.g., logicalness-nonlogicalness, rationality-nonrationality, etc.
- Sources: From *FPHML*. See also my later books.

—*The Explicability-Inexplicability Principle*
(On the Underlying Mechanisms of Things)

- Both explicability and inexplicability are part of the understanding of things. There is no explicability without inexplicability—and vice versa.
- This principle tells us the duality of the research dilemma, in that, if reality can be explained in some ways, it also has its other ways which are not quite explainable, at a given point in time.
- Family resemblance: e.g., underlyingness-regularness, causation-regularness, causation-correlation, etc.
- Sources: Especially from *FPHU*—and also *FPHC*. See also my later books.

(continued on next page)

Table 4.4. The Syntax of Existential Dialectics I: The Principles (Part II)

- **In Relation to Method** ***(cont'd)***

  —*The Predictability-Unpredictability Principle*
  (On the Occurrence of Events)
  - Both predictability and unpredictability have a major role to play in the occurrence of things, so that neither determinism nor indeterminism wins the centuries-old fight. There is no predictability without unpredictability—and vice versa.
  - There are events which are predictable, just as there are those which are not. Or what is regarded as unpredictable at one point in time may turn out to be predictable later, and, conversely, what is deemed as predictable may turn out to not be so predictable. Even in predictability, outcomes are subject to uncertainty, the degree of which varies from case to case.
  - Family resemblance: e.g., sureness-arbitrariness, certainty-uncertainty, sureness-doubtfulness, likeliness-riskiness, etc.
  - Sources: Especially from *FC*. See also my later books.

  —*The Partiality-Totality Principle*
  (On the Relationships between Whole and Parts)
  - The whole is not the sum of the parts. There is no partiality without totality—and vice versa.
  - Any inquiry about a phenomenon in the work is to guard against the varieties of (a) reductionism and (b) reverse-reductionism.
  - Reductionism and reverse-reductionism can be (i) conceptual, (ii) theoretical, (iii) methodological, and (iv) ontological.
  - Family resemblance: e.g., individualisticness-holisticness, partness-wholeness, analysis-synthesis, etc.
  - Sources: Especially from *FC*. See also my later books.

(continued on next page)

Table 4.4. The Syntax of Existential Dialectics I: The Principles (Part III)

- **In Relation to Method *(cont'd)***

—*The Absoluteness-Relativeness Principle*

(On the Multiplicity of Things)

  - There is the multiplicity of things in reality, be they about entities, qualities (or properties), and relationships, such that what is acceptable from one standpoint may not be so from another. For instance, if there is something absolute, there is likewise something relative. There is no absoluteness without relativeness—and vice versa.
  - Both absoluteness and relativeness here are also relevant to different modalities often cited in the literature on ontology, such as possibility (e.g., something "can" happen) and its opposite (e.g., impossibility), probability (e.g., something "will" happen) and its opposite (e.g., improbability), and necessity (e.g., something "should" happen) and its opposite (e.g., contingency).
  - Family resemblance: e.g., uniformity-diversity, internalness-externalness, immanence-transcendence, etc.
  - Sources: From *FPHK*. See also my later books.

(continued on next page)

**Table 4.4. The Syntax of Existential Dialectics I:
The Principles
(Part IV)**

- **In Relation to Structure**
  —*The Finiteness-Transfiniteness Principle*
  (On the Nature of Numbers)
  - If there are finite things, there are likewise transfinite ones. There is no finiteness without transfiniteness—and vice versa.
  - To avoid confusion, my usage of the word "transfinite" here differs radically from the one used by Cantor (and other mathematicians) for "relative" infinity—and is more limited, in light of the problems confronting any attempt to understand the idea of infinity, be it by intuition, imagination, and conception (as detailedly analyzed in *Sec. 2.2.3* of *FPHG*).
  - Instead, by "transfinity," I allow numbers which can be many times larger—or smaller, for that matter—than the finite things that we encounter in daily life, but they do not have to be related to the idea of infinity at all (which may not exist).
  - Of course, there may be some *borderline* cases, in which it is not clear whether the number in question is transfinite (in my usage) or simply a mathematical convenience. A good example of a borderline case is the Planck unit of length for "the smallest space possibly measured in nature," which is "less than billionths of trillionths of trillionths of an inch" (or something like 1.6×10^{-35} meters).
  - That said—my usage of "transfinity" can also resolve (or better, dissolve) an age-old problem in philosophy known as "Zeno's paradoxes" (as already explained in *Sec. 2.2.3* of *FPHG*).
  - Family resemblance: e.g., boundedness-quasiunboundedness, smallness-largeness, microness-macroness, etc.
  - Source: From *FPHG*. See also my later books.

(continued on next page)

Table 4.4. The Syntax of Existential Dialectics I: The Principles (Part V)

- **In Relation to Structure *(cont'd)***

  —*The Simpleness-Complicatedness Principle*
  (On the Interconnection among Things)
  - Both simpleness and complicatedness are vital, without favoring one over the other, in that each is utilized, depending on the basis of the perspectives of nature, the mind, culture, and society. And even when a combination of them is preferred, the dilemma is only shifted to a combinational degree of concern. There is no simpleness without complicatedness—and vice versa.
  - In relation to taxonomy, simpleness has its heuristic usefulness, just as complicatedness has its realistic representation, for instance. And in relation to network, simpleness has its economical attractiveness, just as complicatedness has its practical reliability, for instance.
  - Family resemblance: e.g., inflexibility-flexibility, standardization-specialization, imperfectness-perfectness, superficiality-depth, shallowness-deepness, economicalness-elaboratedness, plainness-circumspection, onesidedness-multisidedness, etc.
  - Sources: Especially from *FIA*. See also my later books.

  —*The Denseness-Emptiness Principle*
  (On the Distribution of Entities in Space)
  - Both density and void are needed, in relation to the mind, nature, culture, and society, albeit in different ways. There is no denseness without emptiness—and vice versa.
  - For clarity, the term "void" is used here only as an approximation of emptiness (depending on the degree of the lack of density), since, in physics, it is well known that "empty" space is not really empty all the way, because it can be full of energy (e.g., random quantum fluctuations at the sub-atomic level, and, for that matter, dark energy in the universe) and matter (e.g., different versions of sub-atomic particles, and, for that matter, dark matter in the universe).
  - Family resemblance: e.g., fullness-voidness, nearness-farness, concentration-dispersion, denseness-nothingness, etc.
  - Sources: First worked out in *FPHUP*. See also my later books.

(continued on next page)

Table 4.4. The Syntax of Existential Dialectics I: The Principles (Part VI)

- **In Relation to Structure *(cont'd)***
  —*The Preciseness-Vagueness Principle*
  (On the Refinement of Things)
  - Both preciseness and vagueness are important, not that one is better than the other, but that both are used, in different degrees of preference, in accordance to the contextual application from the perspectives of nature, the mind, culture, and society. Even when both are used in a combination, the dilemma is shifted instead to one of combinational concern. There is no preciseness without vagueness—and vice versa.
  - In relation to taxonomy, preciseness has its taxonomic clarity, just as vagueness has its classificatory flexibility, for instance. And in relation to network, vagueness has its explorative liberty, just as preciseness has its conceptual definitiveness, for instance.
  - Family resemblance: e.g., clarity-ambiguity, directness-indirectness, quantitativeness-qualitativeness, describability-nondescribability, specificity-obscurity, specificity-generality, thickness-thinness, concretness-abstractness, etc.
  - Sources: Especially from *FIA*. See also my later books.

(continued on next page)

**Table 4.4. The Syntax of Existential Dialectics I:
The Principles
(Part VII)**

- **In Relation to Structure *(cont'd)***
  —*The Openness-Hiddenness Principle*
  (On the Detection of Things)
  - Reality has its hidden face, just as it is open to outside view in some other ways. There is no openness without hiddenness—and vice versa.
  - For instance, in the context of anomalous experience, certain aspects of reality can be open for examination (e.g., the experiment with SPECT images, the transmission of telepathemic bit, and the use of metonymies and metaphors). Other aspects, however, remain hidden, and examples include the elusive deeper nature of spiritual reality in the intangible realm, the undetectability of different branched universes, the underlying mechanisms of signals traveling between different folds of physical space-time, and the dependence on language and the bias for science and logic).
  - Family resemblance: e.g., overtness-covertness, publicness-privateness, openness-closedness, transparency-secrecy, openness-biasedness, etc.
  - Sources: Especially from *FPHU*. See also my later books.

- **In Relation to Process**
  —*The Change-Constancy Principle*
  (On the Alteration of Things)
  - Change occurs over time, although constancy is also allowed. There is no change without constancy—and vice versa.
  - Asymmetry undergoes changes over time, so does symmetry.
  - Old players fade away, and new ones emerges, with ever new causes and ever new forms.
  - Family resemblance: e.g., dynamicness-staticness, instability-stability, etc.
  - Sources: First named in *BCPC*. Especially from *FHC*, *FCD*, and *FPHC*. See also other books of mine (from that point on).

(continued on next page)

Table 4.4. The Syntax of Existential Dialectics I: The Principles (Part VIII)

• **In Relation to Process** ***(cont'd)***

—*The Expansion-Contraction Principle*
(On the Growth of Things)

- Entities in the world can both expand in some ways and contract in other ones, as part of their nature. There is no expansion without contraction—and vice versa.
- For instance, in the context of anomalous experience, one of its most fascinating expressions is none other than the analysis of different causes of the slow but steady expansion (and, furthermore, if put in a different way, unfolding) of unconsciousness.
- Family resemblance: e.g., conquest-autarky, rise-fall (or up-down), spread-shrink, extendingness-shorteningness, etc.
- Sources: Especially from *FPHU*. See also my later books.

—*The Slowness-Quickness Principle*
(On the Speed of Change)

- Both slowness and quickness co-exist, with their own internal tension, to the extent that each fights for its own relevance with the other, in accordance to the perspectives of nature, the mind, culture, and society, without one being the victor and the other being the vanquished in the long haul. Even when both are chosen in other cases, this dilemma is only transferred into something else with a combinational character. There is no slowness without quickness—and vice versa.
- In relation to taxonomy, quickness has its efficient usability, just as slowness has its aesthetic appeal, for instance. And in relation to network, quickness has its adventurous readiness, just as slowness has its risk-adverse convenience, for instance.
- Family resemblance: e.g., inconvenience-convenience, passiveness-activenessness, gradualness-abruptness, deceleration-acceleration, etc.
- Sources: Especially from *FIA*. See also my later books.

(continued on next page)

**Table 4.4. The Syntax of Existential Dialectics I:
The Principles
(Part IX)**

- **In Relation to Process *(cont'd)***
  —*The Order-Chaos Principle*
  (On the Pattern of Things)
  - Both order and chaos are vital in the process of change in the world. The preference for order is biased, since it does not give sufficient attention to the vital role of chaos in the transformation of the world (without somehow reducing it for the understanding of order). There is no order without chaos—and vice versa.
  - The scientific search for order in the world is often a hidden bias in its ontological obsession with order, since chaos is often treated as the "bad" guy, with order as the "good" guy (for the end goal of science).
  - Neither order nor chaos is the final end of the world, and one is not to be treated as the means for the other in the transformation of things. Both are fundamental in their recurrent dialectical interactions with each other over time, without reducing one for the other.
  - Family resemblance: e.g., lawfulness-disorder, order-disorder, etc.
  - Sources: Especially from *FC*. See also my later books.

- **In Relation to Agency**
  —*The Softness-Hardness Principle*
  (On the Force of Change)
  - This has to do with *the force of change* under the category about agency in existential dialectics, in that any change by an agent, be it organic (like humans) or non-organic (like natural objects), can occur in a forceful (aggressive) or gentle (pacific) way, which can come in all shapes and sizes, of course. There is no softness without hardness—and vice versa.
  - Family resemblance: e.g., peacefulness-violence, cooperation-competition, dovishness-hawkishness, reward-punishment, peace-war, pacificity-aggression, etc.
  - Sources: Especially from *ALD*. See also my later books.

(continued on next page)

Table 4.4. The Syntax of Existential Dialectics I: The Principles (Part X)

- **In Relation to Agency** ***(cont'd)***

  —*The Symmetry-Asymmetry Principle*
  (On the Relationships among Existents)

  - There is no symmetry without asymmetry—and vice versa.
  - For instance, the Same can be symmetric and asymmetric towards the Others. But in case of asymmetry, oppression and self-oppression can occur. So, when the Same is asymmetric towards the Others, the Same can also be relatively asymmetric towards itself in self-oppression, just as the Others can be likewise towards themselves.
  - The subsequent oppressiveness is dualistic, as much by the Same against the Others and itself, as by the Others against the Same and themselves.
  - Both oppression and self-oppression can be achieved by way of downgrading differences between the Same and the Others and of accentuating them.
  - This is true, even though not all forms of asymmetry have to be about oppression and self-oppression.
  - In addition, from Chapter Three of *FPHG*, symmetry is not perfect, to be understood in an approximate sense under many life circumstances.
  - Family resemblance: e.g., strength-weakness, balance-extremity, harmoniousness-unharrmoniousness, reflexiveness-unreflexiveness, equality-inequality, potence-impotence, etc.
  - Sources: From all my books, starting with *FHC*. First named in *BCPC*. See, for instance, *FCD* for more details on the Same and the Others.

(*continued on next page*)

Table 4.4. The Syntax of Existential Dialectics I:
The Principles
(Part XI)

- **In Relation to Agency *(cont'd)***
  —*The Theory-Praxis Principle*
  (On the Duality of Knowledge))
  - This has to do with the duality of knowledge in existential dialectics, in that, if there is theoretical construction, there is likewise its practical application, both technical and normative. There is no theory without praxis—and vice versa.
  - Family resemblance: e.g., discovery-application, knowledge-action, invention-innovation, etc.
  - Source: Especially from *FPHE*. See also my later books.

  —*The Evolution-Transformation Principle*
  (On the Multiple Kinds of Agency)
  - This principle (and the symmetry-asymmetry principle) are both about *the agency of change*. The word "agency," in a formal definition, refers to "a person or thing through which power is exerted or an end is achieved." (MWD 2007b) It therefore does not have to necessarily involve an intelligent lifeform.
  - Because of this dual meaning in agency, the evolution-transformation principle is more concerned with *the multiple kinds of agency*, that is, both about the *evolution* in the state of nature (e.g., an object of natural beauty) and the *transformation* in the world of intelligent lifeforms (e.g., a work of art, an air-conditioner, etc.). There is no evolution without transformation—and vice versa.
  - And the transformative part of the principle precisely refers to the other dimension in the dual meaning of agency, in giving technology a major role to play in the change of the world, which is something that I extensively analyzed in *FHC* in the context of the technophilic lifeworld, espeically though not exclusively since modern times.
  - Family resemblance: e.g., nonwillingness-willingness, naturalness-technologicalness, naturalness-nonnaturalness, inorganicness-volition, inorganicness-motivation, etc.
  - Sources: Especially from *FAE*. See also other books of mine.

(*continued on next page*)

Table 4.4. The Syntax of Existential Dialectics I: The Principles (Part XII)

- **In Relation to Agency *(cont'd)***
  —*The Convention-Novelty Principle*
  (On the Nature of Creative Thinking)
  - If there is conventional wisdom, there is likewise novel challenge, to the extent that both convergent and divergent thinking are part of life. There is no convention without novelty—and vice versa.
  - In addition, in this context of the convention-novelty principle, there are, in the absence of better words, what I want to call (a) *creative techniques* and (b) *creative traits*, which, when satisfied—in relation to the larger context of the mind, nature, society, and culture—can be used to enhance creative works. The list of creative techniques and traits (as summarized in *Table 4.1* of *FPHCT*) are not exhaustive, of course, but the examples here constitute a great beginning of understanding the structure of creative thinking.
  - Yet, lest the reader gets carried away by the euphoria about creative thinking in our time, it should be reminded that creative thinking has its own possibilities and limits (as shown in *Table 2.1* of *FPHCT* on invention), just as it has its own promises and pitfalls (as shown in *Table 3.1* of *FPHCT* on innovation). In the end, it should also be stressed that creative thinking has its own desirability and dark sides (as shown in *Table 4.2* of *FPHCT*).
  - This has important daily implications for an average individual, and the word "average" here allows different degrees of variation from one individual to another in a population, at any given point in history.
  - Family resemblance: e.g., normalness-nonnormalness, conformity-nonconformity, convergence-divergence, etc.
  - Sources: First from *FPHCT*. See also my later books.

(*continued on next page*)

Table 4.4. The Syntax of Existential Dialectics I: The Principles (Part XIII)

- **In Relation to Agency *(cont'd)***
  - —*The Seriousness-Playfulness Principle (On the Extent of Seriousness)*
    - If there is seriousness, there is also playfulness, especially for any agent with some kind of intelligence life, be it about humans, animals, or, later, post-humans. There is no seriousness without playfulness—and vice versa.
    - Lest any misunderstanding occurs, the word "playfulness" here should not be confused with other terms like "laughter," "smile," "joke," "tickleness," and "laugh-like vocalization" (in some animals), for instance.
    - Being playful can come in all shapes and sizes and therefore is not necessarily tied up with any of these terms, alhough it can result from any of them, or a combination of them, or something else altogether.
    - This principle, together with some other principles (especially though not exclusively, like the formalness-informalness principle), have important implications for what constitutes rationality, be it about "rationality of application" at the practical level or "rationality of knowledge" at the meta-theoretical level (as summarized in *Sec. 4.20* of *FPHH*).
    - Family resemblance: e.g., seriousness-jokingness, seriousness-humorousness, seriousness-wittiness, seriousness-nonseriousness, seriousness-restingness, seriousness-relaxedness, etc.
    - Sources: First from *FPHH*. See also my later books.

(*continued on next page*)

Table 4.4. The Syntax of Existential Dialectics I: The Principles (Part XIV)

- **In Relation to Outcome**

  —*The Regression-Progression Principle*
  (On the Direction of History)

  - Neither the cyclical nor the linear views are adequate for explaining many phenomena at all levels. There is regression without progression—and vice versa.
  - History progresses to more advanced forms, but with a regressive touch. Examples include no freedom without unfreedom, no equality without inequality, and no civilization without barbarity.
  - This is not an inevitable law, but merely a highly likely empirical trend.
  - Family resemblance: e.g., cost-benefit, undesirability-desirability, badness-goodness, risk-opportunity, etc.
  - Sources: From all my books, starting with *FHC*. First named in *BCPC*.

  —*The Same-Difference Principle*
  (On the Metamorphosis of Change)

  - An entity, as it evolves over time, can be both different from and similar to its opposing alternatives and does not have to be solely more different from them over time. There is no similarity without difference—and vice versa.
  - Opposites are not absolute in a black-or-white fashion; so, an entity can become relatively more similar to (or more different from) its opposite over time.
  - Family resemblance: e.g., homogeneity-heterogeneity, we-they, etc.
  - Sources: Especially from ALD. See also my later books.

Notes: The features in each principle are solely illustrative (not exhaustive). The comparison is also relative (not absolute), nor are they mutually exclusive. As generalities, they allow exceptions.

Table 4.5. The Syntax of Existential Dialectics II: The Principles as Short Cuts (Part I)

• The principles should be treated with caution, lest misunderstanding occurs, since they do not constitute rigid dualities (or dichotomies).

• The reason is that each pair in an ontological principle consists of two opposites, which are, however, merely short cuts both for multiple variations and degrees, as well as for different interactions with multiple other entities unlike them. In this light, each pair can end up having hundred (or even thousand, if not more) different versions, which interact with hundred (or even thousand, if not more) other entities. There are two clarifications here.

• Firstly, this conception of shortcuts is not mutually exclusive nor absolute, in that the opposites can come in all shapes and sizes, with different degrees. For instance, by analogy, just as there are different degrees of the two colors "white" and "black"—there are likewise different degrees of the opposites in each ontological pair, to the extent that there can be multiple entities (not only two) interacting with each other in each pair.

• And secondly, this conception of shortcuts do not ignore other possible entities in interacting with the two opposites (with their different versions) in each pair. For instance, by analogy, there are not only the two colors "white" and "black" as opposites, since there are other colors too besides them like "yellow," "green," "purple," or else. The same logic can be applied to each ontological pair, in that they also interact with other entities, not with only two of them (with their different versions).

• Therefore, with these two clarifications in mind—each pair in an ontological principle serves only as an abbreviation for something more complicated and, therefore, although it contains two opposites, it should not be confused as a duality (dualism).

(continued on next page)

Table 4.5. The Syntax of Existential Dialectics II: The Principles as Short Cuts (Part II)

• Existential dialectics rejects any dualism (or dichotomy) as too rigid and instead allows the multiplicity of entities, to the extent that between the two opposites in each pair exist many other alternatives to choose from. They are named in that short form for aesthetic elegance, instead of listing all possible entities between the two opposites in the title.

• To be dialectic is to go beyond any rigid dichotomy and transcend into something different altogether in the long haul. One may be tempted to call the dialectic logic here with a different name like existential "multilectics" (instead of "dialectics"), but this naming is incorrect (or even misleading), for the two reasons aforestated.

• After all, each ontological pair come in all shapes and sizes (with different degrees) and do not exist by themselves but also interact with other entities unlike them. The virtue of revealing an ontological pair is to show how they relate within themselves (in multiple versions) and also interact with others unlike them (also in multiple versions).

• This will be clear in the table on "the dialectic constraints imposed by the principles."

Source: From Ch.1 of *FPHK*

Table 4.6. The Syntax of Existential Dialectics III: The Principles as Family Resemblances

• Each ontological principle is generic, with some other comparable ontological pairs to be put in the same family (like a *family resemblance*). Ludwig Wittgenstein (1953) once suggested the idea of "family resemblance" in explaining different games classified under the same family called *games*.

• Why should, for instance, playing football and chess as playing "games," when it is well understood that football is not the same as chess? The answer is that, although each game is different and has different rules, many of them (though not all) share, more or less, some commonalites (e.g., scoring as necessary for winning).

• And this is so, even though some games share more than some others in any given selection of critieria, and no two games are exactly identical. So, his point here is that there is no essential core which is common to all games, and the best that one can look for is some characteristics which are common to many (but not all) games. (A. Biletzki 2006)

• By the same logic—in the previous section on selection criteria, the flexibility-inflexibility pair can be put in the family resemblance of the simpleness-complicatedness principle, although the two pairs are not exactly identical. Likewise, the directness-indirectness pair can be put in the family resemblance of the preciseness-vagueness principle, although, again, the two pairs are not exactly identical.

• In this sense, which specific pair in a family should be used to designate the name of the family can be at times a bit arbitrary, but with good reason.

• For illustration, in the context of method, the partiality-totality principle can take the different form like individualisticness-holisticness, just as the explicability-inexplicability principle can take the different form like underlyingness-regularness—although each two pairs are not exactly identical and have slightly different meanings and usages. A more comprehensive listing of this family resemblance for all other principles is shown in the table on the syntax of existential dialectics.

Source: From Ch.1 of *FPHK*

Table 4.7. The Syntax of Existential Dialectics IV: The Dialectic Constraints Imposed by the Principles (Part I)

• **Co-Existent and Asymmetric**

—The principles, as they constitute the syntax of existential dialectics, are dialectic in character, such that, when they are applied, they impose dialectic constraints on how reality is to be understood. Consider, say, the symmetry-asymmetry principle as an illustration here, in order to summarize two main characters of the dialectic constraints in question.

—Firstly, to be dialectic here is to go beyond the narrow dichotomies (and, for that matter, any rigid multi-dimensional classificatory scheme), be they about "self" vs. "world," "freedom" vs. "unfreedom," "barbarity" vs. "civilization," "individuality" vs. "communality," and so on.

—One way to do so (to go beyond) is to consider them all in terms of co-existence (without favoring one over the rest). For instance, my theory of "post-civilization" (to be summarized later in the section on the pragmatics of existential dialectics) is to go beyond barbarity and civilization in terms of understanding barbarity and civilization as being co-existent. And the same logic can be said in relation to my theories of "post-democracy," "post-capitalsm," and others (also to be introduced later in the section on the pragmatics of existential dialectics), in regard to freedom vs. unfreedom, equality vs. inequality, communality vs. individuality, spirituality vs. materiality, and so on.

—But to consider them all (in the dichotomies—and, for that matter, in any rigid multi-dimensional classifactory schme) as co-existent is not the same as to imply that the opposites in any classificatory scheme are all equal, since, in accordance to the symmetry-asymmetry principle (as an illustration here), if they are equal in terms of being considered as co-existent, they are asymmetric in terms of being unequal in dialectic interaction (e.g., X can be more dominant than Y in case A, or Y is more dominant than X in case B).

(continued on next page)

Table 4.7. The Syntax of Existential Dialectics IV: The Dialectic Constraints Imposed by the Principles (Part II)

—For this reason, there are different versions of "post-democracy" and "post-capitalism" in my theories. As an illustration, in version I of the theory of post-democracy, freedom is more dominant than equality, whereas in version II of the theory of post-democracy, equality is more so than freedom.

—But this "X more than Y" has to be understood in the context of dialectic logic (not in conventional logic), in that both "X" and "Y" are important in post-democracy (in the context of dialectic logic), but in an asymmetry way. By contrast, in conventional logic, it often favors one over the other—be it in regard to privileging freedom over equality in Fascism, favoring freedom relatively more than equality in Liberal Democracy, or favoring equality relatively more than freedom in Socialist Democracy. In the latter two cases (about Liberal Democracy and Socialist Democracy), the difference between dialectic logic and conventional logic can be one in degree, not in kind—in this sense, albeit not in other senses.

—The same logic can be said about the relatonships between individuality and communality, between spirituality and materiality, and between formal legalism and informal legalism in the different versions of my theory of post-capitalism.

• **Transcendent**

—Secondly, to be dialectic is to go beyond the narrow dichotomies (and, for that matter, any rigid multi-dimensional classificatory scheme) in another way, this time, in a transcendent way, that is, in exploring other possibilities or even other issues not considered within the narrow confines of narrow dichotomies (and, for that matter, any rigid multi-dimensional classificatory scheme).

(continued on next page)

Table 4.7. The Syntax of Existential Dialectics IV: The Dialectic Constraints Imposed by the Principles (Part III)

—As an analogy, to go beyond the narrow color dichotomy of "black" and "white" is not just to choose both "black" and "white" (as in the first meaning) but also to explore other color options (e.g., "green," "purple," "blue," etc.—and, alternatively, "shade," "line," "curve," etc.). By the same logic, to go beyond "democracy" is to transcend democracy (as in version III of the theory of "post-democracy") and to explore other possibilties of lifeforms (e.g., floating consciousness, hyper-spatial consciousness, etc., to live beyond the narrow obsession with freedom and equality).

—This dialectic character of the principles in existential dialectics has important implications for the pragmatics of existential dialectics (as will be clear shortly, in the section on the pragmatics of existential dialectics).

Notes: The examples here are solely illustrative (not exhaustive). The comparison is also relative (not absolute), nor are they mutually exclusive. As generalities, they allow exceptions.
Sources: First explicitly stated in *ALD*. Also from all other books of mine.

Table 4.8. The Syntax of Existential Dialectics V: Further Clarifications (Part I)

- **Pioneering**
  —Firstly, the total number of ontological principles is unknown, to be discovered later, as our knowledge of the world becomes more advanced.
  —Consequently, the principles as introduced in my books are not exhaustive, with new ones being added, whenever more of them are discovered in later research.
  —At least, future generations can pick up where I leave off and continue the discovery.
  —In this sense, my work should be treated as a pioneering effort for the development of a systematic, comprehensive analysis of a new general ontology for the future of knowledge.

- **Flexible**
  —Secondly, the principles are not rigidly classified, as they can be reclassified in a different way.
  —For instance, the preciseness-vagueness principle is classified under the category of "structure" but can be reclassified under the category of "method," although in so doing, it has a different meaning in the context of method.
  —The same logic applies to the same-difference principle under the category of "outcome," which can be reclassified under the category of "structure," although in so doing, once more, it has a different meaning in the context of structure.

- **Mutually Constraining**
  —Thirdly, the principles are to be understood together, not that each prinicple is to be analyzed independently of others. After all, the principles are mutually constraining, in that they work together as a whole.

(continued on next page)

Table 4.8. The Syntax of Existential Dialectics V: Further Clarifications (Part II)

• **Selectively Useful**

—Fourthly, the principles are relevant to all subject matters, but some principles are more useful to some subject matters than others—as implied in the symmetry-asymmetry principle. This is true, even if different studies of the same kind can yield different views about the degree of relevance for each ontological principle, depending on the specific nature of a research in question, needless to say.

• **Anti-Reductionistic**

—And fifthly, the principles constitute only two levels of analysis, this time, at the ontological and methodological levels—while other levels of analysis (from the perspectives of the mind, nature, society, and culture) are also needed, in order to understand reality in its totality (as explained in the section on "sophisticated methodological holism," so as to avoid the dual dangers of reductionism and reverse-reductionism).

Notes: The examples here are solely illustrative (not exhaustive). The comparison is also relative (not absolute), nor are they mutually exclusive. As generalities, they allow exceptions.
Sources: From my previous books like *FHC*, *FCD*, *FPHC*, *BDPD*, *BCPC*, *BNN*, *FC*, *FAE*, and so on.

Table 4.9. The Syntax of Existential Dialectics VI: The Dilemma of Specific vs. General Ontology (Part I)

• Ontology is often more complicated and imposes some seemingly insurmountable difficulties concerning what constitutes a kind of ontology which can be valid enough to be accepted by the wider intellectual community in question. An excellent example concerns what I want to call *the dilemma of ontology* in relation to the relationship between "specific ontology" and "general ontology." (M. Bunge 1999; R. Corazzon 2007)

• On the one hand, "general ontology," as Mario Bunge (1999) put it, "studies all existents." But, on the other hand, "special ontology studies one genus of thing or process—physical, chemical, biological, social, etc." (M. Bunge 1999; R. Corazzon 2007)

• Consequently, the relationship between the former and the latter is that, as an illustration, "whereas general ontology studies the concepts of space, time, and event, the ontology of the social investigates such general sociological concepts as those of social system, social structure, and social change." (M. Bunge 1999; R. Corazzon 2007)

• With this formal definition in mind, the relationship between the two forms of ontology only poses *the dilemma of ontology*, in that general ontology is possible to the extent that its general contructs must be broad enough to serve as the "lowest common denominator" for all disciplines, but this character of the lowest common denominator is not only hard to find but also its usefulness is limited. (WK 2008b)

• But specific ontology fares no better either, in that its specific constructs primarily serve a specific discipline in question, although other disciplines can appropriate them for their own purposes, but in a different context (especially, though not exclusively, when used as a metaphor from one field to another). For instance, the Kantian category of understanding under the heading of "quantity" constitutes a type of specific ontology, which is specific to the discipline of mathematics, although other disciplines can make use of the categories in their own context.

(continued on next page)

Table 4.9. The Syntax of Existential Dialectics VI: The Dilemma of Specific vs. General Ontology (Part II)

• A solution lies in my proposal of "sophisticated methodological holism," which requires all levels of analysis in any subject matter. See the tables on sophisticated methodological holism, and on reductionism and reverse-reductionism, for more details.

Source: From *Sec. 1.4* of *FPHK*

Table 4.10. The Syntax of Existential Dialectics VII: Types of Inappropriate Family Resemblances (Part I)

- **Inadequate Family Resemblances**
  —Ex: consistency-inconsistency: the formalness-informalenss principle
  —Ex: cognition-noncognition: the formalness-informalness principle

- **Compound Family Resemblances**
  —Ex: reality-nonreality: the formalness-informalness principles, the absoluteness-relativeness principle
  —Ex: normality-abnormality: the symmetry-asymmetry principle, the density-emptiness principle, the convention-novelty principle, the same-difference principle
  —Ex: structure-context: the simpleness-complicatedness principle, the theory-praxis principle
  —Ex: freedom-unfreedom: the symmetry-asymmetry principle, the expansion-contraction principle
  —Ex: morality-immorality: the denseness-emptiness principle, the symmetry-asymmetry, the same-difference principle
  —Ex: particularness-universalness: the preciseness-vagueness principle, the simpleness-complicatedness principle
  —Ex: pureness-mixedness: the absoluteness-relativeness principle, the simpleness-complicatedness principle, the same-difference principle
  —Ex: shortsightedness-foresightedness: the partiality-totality principle, the simpleness-complicatedness principle
  —Ex: shorttermness-longtermness: the partiality-totality principle, the simpleness-complicatedness principle, the slowness-quickness principle

(continued on next page)

Table 4.10. The Syntax of Existential Dialectics VII: Types of Inappropriate Family Resemblances (Part II)

- **One-Sided Family Resemblances**
  —Ex: sound-silence: asymmetry (in the symmetry-asymmetry principle)
  —Ex: vision-blindness: asymmetry (in the symmetry-asymmetry principle)

- **One-Sided and Compound Family Resemblances**
  —Ex: extroversion-introversion: one-sided, in being part of the side (category) of "normality" in the normality-abnormality pair. And compounded, because the normality-abnormality pair is in turn related to the symmetry-asymmetry principle, the density-emptiness principle, the convention-novelty principle, the same-difference principle
  —Ex: organizedness-spontaneousness: one-sided, in being part of the side (category) of "normality" in the normality-abnormality pair. And compounded, because the normality-abnormality pair is in turn related to the symmetry-asymmetry principle, the density-emptiness principle, the convention-novelty principle, the same-difference principle

---

Notes: The examples here are solely illustrative (not exhaustive). The comparison is also relative (not absolute), nor are they mutually exclusive. As generalities, they allow exceptions.
Source: First summarized in *BEPE* (and updated in later books)

Table 4.11. The Semantics of Existential Dialectics

- **Abstract Structure vs. Specific Meanings**
  - —The syntax of existential dialectics so understood in terms of ontological principles only gives us the structure of ontology in the world, in an abstract (general) sense. These principles by themselves do not tell us the specific meanings in a given context.
  - —In order to grasp the specific meanings of the principles in a given context, it is necessary to study the semantics of existential dialectics. And the analysis of the ontological principles as family resemblances in the section on syntax is only a starting point (and thus overlaps a bit with the semantics of existential dialectics here).
  - —With this caveat in mind, the reason that I have often gone in great lengths in my previous books on different subjects is to explain the specific meanings of the principles when applied in different contexts.

- **Specific Meanings in Specific Fields**
  - —For instance, in *FPHST*, I used the first three principles (i.e., the change-constancy principle, the regression-progression principle, and the symmetry-asymmetry principle) to propose "the perspectival theory of space-time," for a better way to understand space and time—especially, though not exclusively, in relation to future post-human history (as summarized in *Table 3.6*, *Table 3.7*, *Table 3.8*, and *Table 3.9*). In so doing, I had to introduce concepts and theories specific to the field of physics and other related fields (e.g., "absolute space" and "absolute time" in "classical mechanics" and "relative space-time" in "the theory of relativity").
  - —In *BNN*, I also exploited the three principles to propose the "transcendent" approach to the study of genes and memes as a new way to understand the interaction between nature and nurture. In so doing, I had to explore concepts and theories in the world of evolutionary theory (e.g., "mutation," "variation," "adaptation," "selection," and "inheritance" in Darwinian evolutionary theory) and neural biology (e.g., "chromosome," "gene," "DNA," "RNA," "protein," "neuron," "neural network," and "behavior").

Notes: The categories and examples in each are solely illustrative (not exhaustive). The comparison is also relative (not absolute), nor are they mutually exclusive. As generalities, they allow exceptions.
Sources: First explicitly stated in *ALD*. Also from all other books of mine.

Table 4.12. The Pragmatics of Existential Dialectics (Part I)

- **The Two-Way Street Connecting Theory and Meta-Theory**
  —The pragmatics of existential dialectics is not a one-way street (that is, using the ontological principles for theoretical insights in praxis) but a two-way one, that is, (a) from meta-theory to theory, and (b) from theory to meta-theory.
  —(a) On one side of the street, the ontological principles can inspire some theoretical insights in praxis, that is, in relation to some specific fields.
  —(b) On the other (opposing) side of the street, however, the study of a subject matter in the specific fields in turn reveals some more hitherto unknown ontological principles to be discovered and identified. For this reason, three new principles were added in *FC*, one in *FAE*, and two in *ALD*, on top of the original three in *BCPC*—after some research on the specific subject matters.

- **Direct and Indirect Applications**
  —*Direct*
    - The logic of existential dialectics can shed some theoretical insights on diverse phenomena in the world, and good instances are the pertinent use of the principles of existential dialectics for the theoretical insights on the freedom/unfreedom dialectics, the equality/inequality dialectics, and the wealth/poverty dialectics in my previous works.
    - My latest books like *FPHST* and *BNN* also use the principles to reveal some theoretical insights on the perspectives of space and time (as in *FPHST*) and of nature and nurture (as in *BNN*).

  —*Indirect*
    - The theoretical insights can further be used to reveal other phenomena directly from them (viz., the theoretical insights) and therefore indirectly from the principles themselves. A good instance is the use of the theoretical insights on the freedom/unfreedom and equality/inequality dialectics for the understanding of the civilization/barbarity dialectics.

(continued on next page)

Table 4.12. The Pragmatics of Existential Dialectics (Part II)

- **Direct and Indirect Applications (*cont'd*)**
    - Even in indirect applications, however, a phenomenon under study can still be directly related back to the principles themselves. In the example as cited above, the civilization/barbarity dialectics can be directly related to the principles of existential dialectics without the intermediate role of the freedom/unfreedom and equality/inequality dialectics.

- **Multiple Levels of Application**

  —The theoretical insights can be applied to different levels of analysis, even though in a given example, it may refer to one level only. For instance, in the example concerning the freedom/unfreedom dialectics, it can be used at the structural level (e.g., in relation to the theory of cyclical progression of hegemony), but it can be exploited as well for other levels (e.g., the theory of post-capitalism at the institutional level).

Notes: The categories and examples in each are solely illustrative (not exhaustive). The comparison is also relative (not absolute), nor are they mutually exclusive. As generalities, they allow exceptions.
Sources: From Ch.6 of *BCPC*. See also other books of mine.

Table 4.13. The Freedom-Unfreedom Dialectics (Part I)

• **On Having**

—*In Relation to the Technological*

(1) if freer from submission to Nature, then less free from ecological degradation (Deep and Social Ecology), even if in a hi-tech form

(2) if freer from technological inconvenience / backwardness, then less free from technological control and the loss of privacy

(3) if freer from technological (material) backwardness, then less free from the abusive (barbaric) maltreatment of the primitive Others

—*In Relation to the Everyday*

(1) if freer from abject poverty, then less free from artificial needs/discontents (Frankfurt School)

(2) if freer from sensual suppression, then less free from violent sublimation (Freud)

(3) if freer from the snobbishness of high culture, then less free from the shabbiness (leveling-off effect) of mass culture (Tocqueville)

(4) if freer from the inefficiency of traditional "compassionate economy," then less free from the bondage of a "ruthless [competitive] economy" (Keynes)

(5) if freer from anarchy in the state of nature (system fragmentation), then less free from government regulations and controls in system integration

(continued on next page)

Table 4.13. The Freedom-Unfreedom Dialectics (Part II)

- **On Belonging**
  - *—In Relation to the Good and the Just*
    - (1) if freer from disciplinary society, then less free from society of control (Foucault)
    - (2) if freer from the tyranny of one or a few, then less free from the tyranny of the majority (or sometimes, minority veto)
    - (3) if freer from elitist decision making, then less free from political gridlock/cleavage
    - (4) if freer from arbitrary (discretionary) administration, then less free from bureaucratic irrationality (Weber) and legal trickery (loopholes)

- **On Being**
  - *—In Relation to the True*
    - (1) if freer from unscientific dogmas, then less free from instrumental abyss (nihilism). Or conversely, if freer from meaninglessness, then less free from dogmas.
    - (2) if freer from the bondage of partiality/partisanship (e.g., prejudice, discrimination), then less free from the danger of impartiality and neutrality (e.g., opportunism, unrealisticness, lack of compassion, inaction)
    - (3) if freer from making generalizations, then less free from being unable to understand much of anything
  - *—In Relation to the Holy*
    - (1) if freer from collective conscience, then less free from social loneliness
    - (2) if freer from religious absoluteness, then less free from spiritual emptiness
  - *—In Relation to the Beautiful/Sublime*
    - (1) if freer from artistic non-autonomy, then less free from aesthetic disillusion (deconstruction)

(continued on next page)

Table 4.13. The Freedom-Unfreedom Dialectics (Part III)

Notes: The examples in each category are solely illustrative (not exhaustive), and the comparison is relative (not absolute), nor are they necessarily mutually exclusive. And some can be easily re-classified elsewhere. As generalities, they allow exceptions.
Sources: A reconstruction from Ch.10 of *FCD*, based on *FHC*

Table 4.14. The Equality-Inequality Dialectics (Part I)

- **On Having**
  - —*In Relation to the Technological*
    - (1) if more equal in treating Nature with spiritual unity, then less equal in suppressing the dominant drive to transcend it altogether
  - —*In Relation to the Everyday*
    - (1) if more equal in building social plurality, then less equal in leveling-off effects (e.g., the subsequent relative intolerance of high/intellectual ethos in mass culture industry)
    - (2) if more equal in socioeconomic distribution beyond a certain point, then less equal in efficiency (e.g. resentment, the erosion of work ethics)
    - (3) if more equal in urging an affirmative action program, then less equal in creating victim mentality (in oneself), stigma (from others), reverse discrimination (against the once privileged), and mediocracy (against the more able)

- **On Belonging**
  - —*In Relation to the Good and the Just*
    - (1) if more equal in banning monarchic/oligarchic exclusion, then less equal in producing "the tyranny of the majority" or of "minority veto"
    - (2) if more equal in encouraging participatory decision making, then less equal in inducing political divisiveness (gridlock/cleavage in power blocs) and organizational oligarchy
    - (3) if more equal in institutionalizing a decentralized bureaucracy, then less equal in falling into more territorial/turf politics (intrigues)

(continued on next page)

Table 4.14. The Equality-Inequality Dialectics (Part II)

- **On Being**
  - —In Relation to the Beautiful / Sublime
    - (1) if more equal in accepting diverse styles ("anything goes" mentality), then less equal in artistic good quality (in leveling-off effects against the best)
  - —In Relation to the True
    - (1) if more equal in tolerating multiple viewpoints (no matter how extreme), then less equal in epistemic standards
  - —In Relation to the Holy
    - (1) if more equal in celebrating any cults and sects (no matter how questionable), then less equal in spiritual depth and authenticity

Notes: The examples in each category are solely illustrative (not exhaustive), and the comparison is relative (not absolute), nor are they mutually exclusive. And some can be easily reclassified elsewhere. As generalities, they allow exceptions.
Sources: A reconstruction from Ch.10 of *FCD*, based on *FHC*

Table 4.15. The Duality of Oppression in Existential Dialectics: Oppression and Self-Oppression (Part I)

- **From the Same to the Others and Itself**
  - —The Oppression by the Same against the Others
    - *By way of downgrading differences*
      - –Ex: on judiciary caprice for corporate crimes (*Sec.2.2.1.2.1*)
      - –Ex: on the deceptive politics of liberation (*Sec.3.5*)
      - –Ex: on the humanitarian mystique (*Sec.4.4*)
      - –Ex: on the fad of emotional intelligence (*Sec.5.3*)
    - *By way of accentuating differences*
      - –Ex: on the legal sophistry of self-defense (*Sec.2.3*)
      - –Ex: on the legal semantics of proportionality (*Sec.2.4*)
      - –Ex: on the tricky politics of external threat (*Sec.3.4*)
      - –Ex: on the appeal of the Far Right for democracy (*Sec.5.4*)
      - –Ex: on the democratic axis of evil (*Sec.5.5*)
      - –Ex: on the democratic way of brutality and revenge (*Sec.5.6*)
      - –Ex: on democratic autocracy (*Sec.6.4*)
  - —The Oppression by the Same against Itself
    - *By way of downgrading differences*
      - –Ex: on the politics of fear (*Sec.2.2*)
      - –Ex: on the trickery of compassionate conservatism (*Sec.3.2*)
      - –Ex: on the deceptive politics of patriotism (*Sec.3.3*)
    - *By way of accentuating differences*
      - –Ex: on the caprice of due process on domestic suspects (*Sec.2.2*)
      - –Ex: on the false security/freedom dilemma (*Sec.6.5.2*)

(continued on next page)

Table 4.15. The Duality of Oppression in Existential Dialectics: Oppression and Self-Oppression (Part II)

- **From the Others to the Same and Themselves**
  —The Oppression by the Others against the Same
  - *By way of downgrading differences*
    –Ex: on judiciary caprice in the reverse direction (*Sec.2.2.1.2.2*)
    –Ex: on equal pay (*Sec.6.2.1.1*)
    –Ex: on equal representation (*Sec.6.2.1.2*)
    –Ex: on affirmative action program (*Sec.6.3.1.1*)
    –Ex: on same-sex marriage (*Sec.6.3.1.2*)
  - *By way of accentuating differences*
    –Ex: on sexual harassment (*Sec.6.2.2.1*)
    –Ex: on physical violence (*Sec.6.2.2.2*)
    –Ex: on sexual exploitation (*Sec.6.2.2.3*)

  —The Oppression by the Others against Themselves
  - *By way of downgrading differences*
    –Ex: on the reverse-class mystique (*Sec.4.2*)
    –Ex: on the reverse-black mystique (*Sec.4.3*)
    –Ex: on self-discrimination by downgrading (*Sec.6.3.2.2*)
  - *By way of accentuating differences*
    –Ex: on self-discrimination by accentuating (*Sec.6.3.2.1*)

Notes: The examples are solely illustrative (not exhaustive), nor are they mutually exclusive. As generalities, they allow exceptions. Also, both forms of oppression co-exist in all of the examples, so the listing of them are only meant in a relative, not absolute, sense.
Source: A summary of the sections (as cited) in Chs.2-6 of *BDPD*. See text for more info and references.

Table 4.16. The Structure of Existential Dialectics I: The Freedom/Unfreedom and Equality/Inequality Dialectics

• Each freedom and equality produces its own unfreedom and inequality, regardless of whether the pair occurs in political society (with the nation-state), in civil society (with some autonomy from the state), or elsewhere (e.g., in the private sphere of individual homes)—and regardless of whether freedom and equality are understood as "negative" or "positive."

• Oppression is dualistic, as much by the Same against the Others and itself, as by the Others against the Same and themselves.

• Both forms of oppression and self-oppression can be achieved by way of downgrading differences (between the Same and the Others) and of accentuating them.

• The relationships are relatively asymmetric between the Same and the Others and relatively symmetric within them. This is true, even when the Same can be relatively asymmetric towards itself in self-oppression, just as the Others can be likewise towards themselves.

• Symmetry and asymmetry change over time, with ever new players, new causes, and new forms.

Notes: The examples in each category are solely illustrative (not exhaustive) nor necessarily mutually exclusive, and the comparison is relative (not absolute). As generalities, they allow exceptions. "Negative" freedom is freedom "from" (e.g., freedom from poverty), whereas "positive" freedom is freedom "to" (e.g., freedom to the state of enlightenment). "Negative" equality is "procedural" equality (e.g., equality of opportunity), while "positive" equality is "substantive" equality (e.g., equality of outcome). Existential dialectics impose constraints on freedom and equality in democracy, non-democracy, and post-democracy. There is no utopia, in the end; even should there be one, dystopia would exist within it.
Sources: From *Table 1.5* of *BDPD*—and also from *FHC*, *FCD*, and *FPHC*

Table 4.17. The Structure of Existential Dialectics II: The Wealth/Poverty Dialectics

---

• There is no wealth without poverty, just as there is no poverty without wealth.

• The wealth/poverty dialectics occurs in the realms of having, belonging, and being, in relation to the material, relational, and spiritual.

• The wealth/poverty dialectics also expresses itself at the multiple levels of analysis in accordance to methodological holism, be they about the micro-physical, the chemical, the biological, the psychological, the organizational, the institutional, the structural, the systemic, the cultural, and the cosmological.

• The wealth/poverty dialectics is a different manifestation of existential dialectics in general, subject to the principles in its logic of ontology—just as the freedom/unfreedom and equality/inequality dialectics are likewise.

• There is no economic utopia, in the end; even should there be one, dystopia would exist within it.

Notes: The main points here are solely illustrative (not exhaustive) nor necessarily mutually exclusive, and the comparison is relative (not absolute). As generalities, they allow exceptions.
Sources: From *BCPC*. See also *FCD* and *FHC*.

Table 4.18. The Structure of Existential Dialectics III: The Civilization/Barbarity Dialectics

- There is no civilization without barbarity.

- The civilization/barbarity dialectics applies in the four civilizing processes (e.g., the rationalizing process, the pacifying process, the stewardizing process, and the subliming process).

- The civilization/barbarity dialectics is another (different) manifestation of existential dialectics in general, subject to the principles in its logic of ontology—just as the freedom/unfreedom and equality/inequality dialectics and the wealth/poverty dialectics are likewise.

- There is no utopia, in the end; even should there be one, dystopia would exist within it.

Notes: The main points here are solely illustrative (not exhaustive) nor necessarily mutually exclusive, and the comparison is relative (not absolute). As generalities, they allow exceptions.
Sources: From *BCIV*. See also *FCD*, *FHC*, and *BDPD*.

Table 4.19. The Double Sides of Virtual Organizations (Part I)

- **Psychological**
  —Ex: virtual psychosis
  —Ex: impersonality and loneliness in quaternary social relations

- **Organizational**
  —Ex: the race for power and interests
  —Ex: the world of unequal successes
  —Ex: the bureaucratic life of its own
  —Ex: lesser accountability and transparency

- **Economic**
  —Ex: the economic divides
  —Ex: the erosive impact of commercialization

- **Political**
  —Ex: the anti-authoritarian myth
  —Ex: different power struggles among groups (e.g., Conservative, Reformist, Radical)

- **Structural**
  —Ex: different localities (e.g., climates, scenery)
  —Ex: divided domains (e.g., core, peripheral, sub-peripheral)
  —Ex: substitution effect of social capital

- **Systemic**
  —Ex: uneven advances in transportation and communications, and geographical migration

- **Cultural**
  —Ex: conflicting civilizational fabrics (e.g., Confucian, Islamic, Western)
  —Ex: the bias of the Liberal Democratic agenda
  —Ex: the addiction to moral fanaticism

(continued on next page)

Table 4.19. The Double Sides of Virtual Organizations (Part II)

Notes: These categories and examples are solely illustrative (not exhaustive), and some of the items can be reclassified somewhere else. Nor are they mutually exclusive. Since they are generalities, exceptions are expected.
Source: From Ch.7 of *FCD*

Table 4.20. Beyond the World of Titans, and the Remaking of the World Order

- **Hyper-Empires**
  - —Ex: The Chinese Union
  - —Ex: The Indian Union

- **Meso-Empires**
  - —Ex: The European Union
  - —Ex: The North American Union

- **Micro-Empires**
  - —Ex: The Latin American Union
  - —Ex: The Middle Eastern Union

- **The Rest of the World**
  - —<u>Odd Powers</u>
    - Ex: Japan
    - Ex: Russia
  - —<u>The Poor Club</u>
    - Ex: The African Union
  - —<u>Ambivalent Regions</u>
    - Ex: Southeast Asia
    - Ex: Oceania
    - Ex: South Asia
    - Ex: Central Asia
    - Ex: Southern/Eastern Europe
    - Ex: North Africa
    - Ex: Central America
    - Ex: Others (e.g., the Korean peninsula)

Source: A summary of Chs.2-5 (of *BWT*)

Table 4.21. The Origins of Authoritarian Liberal Democracy

- **The Geopower of Nature (Ch.4 of *ALD*)**
  —Ex: Power Character and Geographical Strategy
  —Ex: Living Space and Territorial Expansion
  —Ex: Strategic Heartland and Containment

- **The Biopsychology of the Mind (Ch.5 of *ALD*)**
  —Ex: The Bell Curve and Mass Intelligence
  —Ex: Group Analysis and Mass Knowledge
  —Ex: Groupthink and Elite (Mis)calculation

- **The Disciplinary Control of Society (Ch.6 of *ALD*)**
  —Ex: Social Organizations of Ruthlessness
  —Ex: Social Institutions of Greed
  —Ex: Social Structure of Exclusion
  —Ex: Social Systems of Violence

- **The Molding Force of Culture (Ch.7 of *ALD*)**
  —Ex: The Tradition of Conquest
  —Ex: The Rationalization of Unreason

Notes: The examples in each category are solely illustrative (not exhaustive), and they are also relative (not absolute), nor are they mutually exclusive. As generalities, they allow exceptions.
Source: A summary of Chs.4-7 of *ALD*

Table 4.22. The Theory of Post-Democracy I: The Priority of Freedom over Equality (Part I)

- **Differences**
  - —*For the aggressive Lions (the strong Elitists)*
    - •Setting up rank distinctions among unequals (e.g., between inferior humans and superior post-humans, or later among inferior post-humans and superior ones, relatively speaking)
    - •Yearning for being not only distinguished from unequals, but also the first among equals (the best of the very best)
    - •Soul-searching for a high spiritual culture (not the trashy one for the masses). Mass culture is a dirty joke for them.
  - —*For the manipulative Foxes (the weak Counter-Elitists)*
    - •Seeking a gentle hegemony by way of more communitarian concerns (for inferior humans and, later, inferior post-humans)
    - •Being more sympathetic to less formal-legalistic institutions and values

- **Similarities**
  - —*For both Lions and Foxes*
    - •Exploring different spheres of non-human consciousness in the cosmos (something vastly superior than the human one)
    - •Recognizing the democratic illusions (e.g., no freedom without unfreedom, no equality without inequality, or simply no justice without injustice, and vice versa)

(continued on next page)

Table 4.22. The Theory of Post-Democracy I: The Priority of Freedom over Equality (Part II)

Notes: The two callings and examples in each category are solely illustrative (not exhaustive), since there will be many different post-human value ideals in the distant future of post-human civilization. The comparison is also relative (not absolute) towards post-democracy, so this is not just a version of free-market democracy (nor Fascism/Nazism, as shown in the table later on democracy, non-democracy, and post-democracy). Nor are they mutually exclusive. As generalities, they allow exceptions. And the specific forms of post-human post-democratic ideals need to be further developed in future after-postmodern history, as they will be different from the ones we now know. The point here is to solely give an extremely rough picture of a small part of the world to come that we have never known.
Source: From Ch.10 of *FCD*. Refer to text for more info and references.

Table 4.23. The Theory of Post-Democracy II: The Priority of Equality over Freedom

---

- **Hybrid Versions of**
  - —Ex: the Trans-Feminine Calling
  - —Ex: the Trans-Sinitic Calling
  - —Ex: the Trans-Islamic Calling
  - —Ex: the Trans-Outerspace Calling

- **Qualifications**
  - —These four versions of post-capitalist value ideals need not automatically be post-democratic, just as capitalism does not necessarily mean democracy. They are two different entities—though closely related.
  - —But up to a certain threshold of elevating equality at the farther expense of freedom, the democratic ideals will be overcome and cease to exist.
  - —The overcome will not be socialist or communist, but post-democratic with no freedom without unfreedom and no equality without inequality, subject to the constraints of existential dialectics.

Notes: The callings are solely illustrative (not exhaustive), since there will be many different post-human value ideals in the distant future of post-human lifeforms. The comparison is also relative (not absolute), nor are they mutually exclusive. As generalities, they allow exceptions. And the specific forms of post-human post-democratic ideals need to be further developed in future after-postmodern history, as they will be different from the ones we now know. The point here is to solely give an extremely rough picture of a small part of the world to come that we have never known.

Source: From Ch.10 of *FCD*. Refer to text for more info and references.

Table 4.24. The Theory of Post-Democracy III: The Transcendence of Freedom and Equality (Part I)

- **Transcending Freedom in Floating Existence**
  - —*Freedom*: seeking an ultimate elimination of the body. Being without the body. The aim is to transcend freedom in the end into a metaphysical state (i.e., beyond the physique).
  - —*Unfreedom*: yet facing difficult trade-offs. The sacrifice of bodily existence and its joyfulness. An eternal boredom in floating existence in dark deep space, though with alternative pleasures. There is no free lunch even in the state of transcending freedom.

- **Transcending Equality in the Rivalry of Cosmic Hegemony**
  - —*Inequality*: competing to outlast other lifeforms in floating existence, or just marginalizing them for one's hegemonic expansiveness in the rest of the cosmos (and even beyond). Universalism is only for the mediocre.
  - —*Equality*: accepting only those of one's rank as equal partners in the vast spacetime for cosmic supremacy. Even here, the aim is to transcend equality into a metaphysical state.

(continued on next page)

Table 4.24. The Theory of Post-Democracy III: The Transcendence of Freedom and Equality (Part II)

Notes: Do not confuse this transcendence of freedom and equality (as one version of post-democracy) with the naïve temptation to transcend the freedom/unfreedom and equality/inequality dialectics. Existential dialectics hold true for freedom and equality in all cultures and societies—past, present, or future (i.e., democracy, non-democracy, and post-democracy), regardless of whether freedom and equality are conventionally understood as "negative" or "positive."

Also, the two features and examples in each are solely illustrative (not exhaustive), since there will be many different post-human value ideals in the distant future of post-human lifeforms. The comparison is also relative (not absolute), nor are they mutually exclusive. As generalities, they allow exceptions. And the specific forms of post-human ideals even for these radically alien floating lifeforms (and others unknown to us) need to be further developed in future after-postmodern history, as they will likely be different from the ones herein illustrated. The point here is to solely give a very rough picture of a small part of the extremely alien world to come that we have never known.

Source: From Ch.10 of *FCD*. Refer to text for more info and references.

Table 4.25. Democracy, Non-Democracy, and Post-Democracy (Part I)

- **Democracy**
  - —*Theoretical Constructs*
    - •The pursuit of freedom and equality (in various degrees), regardless of whether freedom and equality can be understood as "negative" or "positive"
      - (1) more equality than freedom: The relative priority of the good over the right
      - (2) more freedom than equality: The relative priority of the right over the good
  - —*Types*
    - •Only (1): Different versions of communitarian moral universalism
    - •Only (2): Different versions of liberal moral universalism
    - •(1) or (2): Different versions of anarchic (non-nation-state) moral universalism
    - •(1) or (2): Different versions of postmodern moral localism

- **Non-Democracy**
  - —*Theoretical Constructs*
    - •The focus on (1') equality or (2') freedom, but not both, regardless of whether freedom and equality can be understood as "negative" or "positive"
  - —*Types*
    - •Only (1'): Different versions on the Far Left (e.g., Stalinism, Robespierrianism)
    - •Only (2'): Different versions on the Far Right (e.g., Nazism, absolute monarchism)

(continued on next page)

Table 4.25. Democracy, Non-Democracy, and Post-Democracy (Part II)

- **Post-Democracy**
  - *—Theoretical Constructs*
    - •The priority of (1'') equality over freedom, or (2'') freedom over equality, or (3'') the transcendence of freedom and equality, regardless of whether freedom and equality are "negative" or "positive." In degree, (1'') or (2'') is less than (1') or (2') but more than (1) or (2)—respectively.
    - •Like democracy and non-democracy, post-democracy is also subject to the freedom/unfreedom and equality/inequality dialectics (or existential dialectics in general). Unlike them, post-democracy acknowledges the constraints of existential dialectics and no longer value freedom and equality as sacred virtues. There is no utopia, in the end; even were there one, dystopia would exist within it.
  - *—Types*
    - •(1''): Different versions of trans-Sinitic value ideals
    - •(1''): Different versions of trans-feminine value ideals
    - •(1''): Different versions of trans-Islamic value ideals
    - •(1''): Different versions of trans-outerspace value ideals
    - •(2''): Different versions of post-human elitist value ideals
    - •(3''): Different versions of the value ideals of floating consciousness (etc.)

(continued on next page)

Table 4.25. Democracy, Non-Democracy, and Post-Democracy (Part III)

Notes: The examples are solely illustrative (not exhaustive), nor are they mutually exclusive. As generalities, they allow exceptions. “Negative” freedom is freedom “from” (e.g., freedom from poverty), whereas “positive” freedom is freedom “to” (e.g., freedom to the state of enlightenment). “Negative” equality is “procedural” equality (e.g., equality of opportunity), while “positive” equality is “substantive” equality (e.g., equality of outcome). Existential dialectics impose constraints on freedom and equality in democracy, non-democracy, and post-democracy, regardless of whether freedom and equality can be understood as “negative” or “positive” in conventional discourse. Therefore, do not confuse the transcendence of freedom and equality in (3’’) with the naïve temptation to transcend existential dialectics. There is no utopia, in the end; even should there be one, it would not exist without dystopia embedded within it.
Sources: A summary, based on my previous works, especially Ch.5 of *FHC*, Chs.5-10 of *FCD*, Chs.2-4 of *FPHC*, and Chs.1 & 7 of *BDPD*. The reader should consult the books for more analysis, as this is only a summary here.

Table 4.26. Multiple Causes of the Emergence of Post-Democracy (Part I)

- **At the Micro-Physical Level**
  —Ex: intelligent life without the human physical-chemical system
  —Sources: Ch.7 of *FHC*; Chs.9-10 of *FCD*; Ch.1 of *FPHC*

- **At the Chemical Level**
  —Ex: space radiation and toxins
  —Sources: Ch.7 of *FHC*; Chs.9-10 of *FCD*

- **At the Bio-Psychological Level**
  —Ex: exo-biological evolution in deep space
  —Ex: genetic engineering of new beings
  —Ex: limits of cognitive partiality
  —Ex: illusions of emotional neutrality
  —Ex: human biological inequality
  —Ex: the rise of unfolding unconsciousness
  —Sources: Ch.2 & Chs.9-10 of *FCD*; Ch.7 of *FHC*; Ch.4 of *BCPC*; *FPHU*

- **At the Institutional Level**
  —Ex: the flawed logic of equality
  —Ex: the conflicting nature of governance
  —Sources: Ch. 5 of *FHC*; Chs. 6 & 10 of *FCD*; Ch. 3 of *FPHC*; Chs.2-5 of *BDPD*

- **At the Organizational Level**
  —Ex: e-civic alienation
  —Ex: the dark sides of formal-legalistic routines
  —Sources: Ch.3 of *FHC*; Ch.7 of *FCD*; Ch.3 of *FPHC*

- **At the Structural Level**
  —Ex: ever new forms of inequities, at home and abroad
  —Ex: the emergence of China, women, and Islam as major actors
  —Sources: Chs.5-6 of *FHC*; Chs.7, 9 & 10 of *FCD*; Chs.4-5 of *BDPD*

(continued on next page)

Table 4.26. Multiple Causes of the Emergence of Post-Democracy (Part II)

- **At the Cultural Level**
  —Ex: freedom/unfreedom dialectics
  —Ex: equality/inequality dialectics
  —Ex: system fragmentation and integration
  —Sources: Ch.5 of *FHC*; Chs. 3, 9 & 10 of *FCD*; Ch.4 of *FPHC*; Ch.1 of *BDPD*; Ch.4 of *BCPC*

- **At the Systemic Level**
  —Ex: space habitats (in zero-gravity) and colonization
  —Ex: ultra advanced future info systems
  —Ex: qualitative demography
  —Sources: Ch.7 of *FHC*; Chs.9 &10 of *FCD*

- **At the Cosmological Level**
  —Ex: the colonization of multiverses
  —Ex: the alteration of space-time and the creation of new matter-energy
  —Ex: the expansion of floating consciousness
  —Ex: the spread of hyper-spatial consciousness
  —Sources: Ch.7 of *FHC*; Chs.9 &10 of *FCD*; Ch.4 of *FPHC*; *FPHST*

Notes: The examples in each category are solely illustrative (not exhaustive), and some of the items can be reclassified somewhere else. Nor are they always mutually exclusive. Since they are generalities, exceptions are expected.
Sources: Especially from *FHC*, *FCD*, *FPHC*, *BCPC*, *BDPD*, *FPHST*, and *FPHU*. See also other books and my perspectives on civilizational holism.

Table 4.27. Some Clarifications about Post-Capitalism and Post-Democracy (Part I)

• The prefix "trans-" in the first category of post-capitalism (with its four versions) refers to something "going beyond" (not "uniting" or "combining"). Ex: *Sec.10.3.3* of *FCD*; *Sec.2.4* & *Sec.4.4* of *FPHC*; *Sec.7.2* of *BCPC*

• Such terms like "post-democracy," "post-capitalism," "post-human elitist," "trans-feminine calling," and the like as used in my works are more for our current intellectual convenience than to the liking of future humans and post-humans, who will surely invent more tasteful neologisms to call their own eras, entities, and everything else, for that matter. But the didactic point here is to use the terms to foretell what the future might be like, not that its eras and entities must be called so exactly and permanently.
Ex: *Sec.11.1* of *FCD*; *Sec.7.2* of *BCPC*

• The four versions in the first category of post-capitalist value ideals need not automatically be post-democratic, just as capitalism does not necessarily mean democracy. They are two different entities—though closely related. But up to a certain threshold of elevating equality at the farther expense of freedom, the democratic ideals will be overcome and cease to exist. The same is true for the post-human elitist calling in the second category of post-capitalism in relation to post-democracy, depending on the extent to which freedom is elevated at the expense of equality.
Ex: *Sec.10.4.3.3* of *FCD*; *Table 3.9* of *FPHC*; *Table 7.6* of *BDPD*

• The comparison in each of the three realms of existence in all forms of post-capitalism is not absolute, but relative. Examples include "communal" vs. "individualistic," and the like.
Ex: Notes in *Table 10.8*, *Table 10.9*, *Table 10.10*, & *Table 10.11* of *FCD*; Chs.2-4 of *FPHC*; *Sec.7.2* of *BCPC*

(continued on next page)

Table 4.27. Some Clarifications about Post-Capitalism and Post-Democracy (Part II)

• The emergence of post-capitalism (and post-democracy, for that matter) has multiple causes (to not be reduced to one or only a few).
Ex: Ch.10 of *FCD*, Chs.2-4 of *FPHC*; *Sec.1.3* & *Sec.7.2* of *BCPC* (or *Table 1.8* & *Table 7.11*)

• The specific forms of post-capitalism (and post-democracy, for that matter) need to be further developed in future after-postmodern history, as they will be different from the ones we now know. The point here is to solely give an extremely rough sketch of a world to come that we have never known.
Ex: *Sec.10.3.3* & *Sec.10.4.3.3* of *FCD*; *Table 10.14* & *Table 10.15* of *FCD*; *Sec.7.2* of *BCPC*

• All forms of post-capitalism are not part of a "teleological law," but of "historical trends" only. The same is also true for all forms of post-democracy.
Ex: *Sec.7.1* of *FHC*; *Sec.9.5.3.2* & *Sec.10.3.4.2* of *FCD*; *Sec.7.2* of *BCPC*

• Post-capitalism is not better than capitalism in an "absolute" sense but only fits in better, on the basis of the historical contingency of culture, society, nature, and the mind in some future eras. The same is true for post-democracy in relation to democracy. The term "better" is historically relative.
Ex: *Sec.10.3.3* of *FCD*; *Sec.1.7* of *BDPD*; *Sec.1.5* of *BCPC*

(continued on next page)

Table 4.27. Some Clarifications about Post-Capitalism and Post-Democracy (Part III)

• All forms of post-capitalism and post-democracy are subject to the constraints of existential dialectics. In the process, the dialectic direction is to go beyond the conventional "either-or" dichotomies (e.g., freedom vs. unfreedom, equality vs. inequality, freedom vs. equality, individuality vs. communality, spirituality vs. materiality, formal legalism vs. informal legalism, etc.). As is true in post-civilization, to go beyond the dichotomies is to acknowledge the co-existence of both in each dichotomy, although the degree of scaling one over the over varies from case to case (e.g., the theory of post-capitalism I, the theory of post-capitalism II, the theory of post-democracy I, the theory of post-democracy II, etc.)—but is not to be extreme in largely favoring one over the other, *on average* (all things considered). There is no utopia to be had in the end; even should there be one, dystopia would exist within it.
Ex: Ch.5 of *FHC*; *Sec.10.4.4.2* of *FCD*; *Sec.1.5* of *BDPD*; *Sec.1.3* of *BCPC*; *BCIV*

• All forms of post-capitalism, however different from each other though they are, share one common feature, in that they all inspire for a higher spiritual culture. The same is also true for post-democracy.
Ex: *Sec.10.3*, *Sec.10.4* & *Sec.10.5* of *FCD*; *Chs.2-4* of *FPHC*; *Sec.7.2* of *BCPC*

• All forms of post-capitalism try to avoid the excess in capitalist consumerism by favoring more basic than artificial needs in having, but the quality and quantity of these "basic" needs will be measured by future standards, not by our current ones. Standards are historically relative.
Ex: *Sec.10.3*, *Sec.10.4* & *Sec.10.5* of *FCD*; Ch.2 of *FPHC*; *Sec.7.2* of *BCPC*

(continued on next page)

Table 4.27. Some Clarifications about Post-Capitalism and Post-Democracy (Part IV)

• All forms of post-capitalism make use of a different degree of political authority with advanced info systems in future history and strives for higher spiritual cultures (especially in the post-human age), while acknowledging the constraints of existential dialectics and no longer valuing free market (as in capitalism) and economic control (as in non-capitalism) as sacred virtues.
Ex: *Sec.10.3.4.2*, *Sec.10.3*, *Sec.10.4* & *Sec.10.5* of *FCD*; Chs.2-4 of *FPHC*; *Sec.1.5* of *BDPD*; *Sec.7.2* of *BCPC*

Notes:: The main points here are solely illustrative (not exhaustive) nor necessarily mutually exclusive, and the comparison is relative (not absolute). As generalities, they allow exceptions. The sections as cited are only illustrative (not exhaustive).
Sources: From *FHC*, *FCD*, *FPHC*, and *BDPD*

Table 4.28. The Theory of Post-Capitalism I.1: By Group— Ex: Spiritual/Communal in the Trans-Feminine Calling

- **More Communal Than Individual**
  —*Sharing*: learning from others, as different ideas mutually enrich
  —*Cooperative*: encouraging a sense of shared leadership and teamwork

- **More Informal-Legalistic Than Formal-Legalistic**
  —*Specific*: listening more from the heart than from the head, to know a person as a concrete, not as an abstract, unit
  —*Affective*: thinking and acting with others on a more affective tone. Business can mix with an emotional touch.
  —*Ascriptive*: hiring (or firing) can be done on the basis of merit (or lack of it), but deep solidarity (sisterhood) is important too.
  —*Particularistic*: making decisions on the basis of cost-benefit analysis, but a given group relationship is vital

- **More Spiritual Than Secular**
  —*Long-Term Looking*: sharing for a long-term relationship (e.g., love, friendship), not just for a short-term gain
  —*Loving/Caring*: showing compassion for the sufferings of others, without quickly blaming and pre-judging
  —*Respectful*: showing acceptance about others' feelings (and thoughts)

Notes: The categories and examples are solely illustrative, since there can be different versions, and the comparison is relative (not absolute), nor are they mutually exclusive. As generalities, they allow exceptions. The specific forms of the trans-feminine version need to be further developed in future after-postmodern history, as they will be different from the ones we now know, since the prefix "trans-" here means going beyond or deconstructing the feminine values, while using them as the inspirational point at the beginning.
Source: From Ch.10 of *FCD*. Refer to text for more info and references.

Table 4.29. The Theory of Post-Capitalism I.2: By Nation-State— Ex: Spiritual/Communal in the Trans-Sinitic Calling

- **More Communal Than Individualistic**
  —*Centralized*: being more top-down in management
  —*Collective*: encouraging more group cooperation
  —*Social*: investing in trust and connection

- **More Informal-Legalistic Than Formal-Legalistic**
  —*Specific*: knowing more of those related or connected
  —*Affective*: behaving in a paternalistic, hierarchical way
  —*Ascriptive*: favoring family members and those related
  —*Particularistic*: building connection (guanxi) as imperative

- **More Spiritual Than Secular**
  —*Expansionist*: diffusing civilizational values (e.g., the superiority complex of civilizationalism)
  —*Holistic*: synthesizing things into a panoramic horizon
  —*Historical*: learning from the lessons of the ancient past
  —*Respectful*: deferential to elders and superiors

Notes: The categories and examples are solely illustrative, since there can be different versions, and the comparison is relative (not absolute), nor are they mutually exclusive. As generalities, they allow exceptions. The specific forms of the trans-Sinitic version need to be further developed in future after-postmodern history, as they will be different from the ones we now know, since the prefix "trans-" here means going beyond or deconstructing the Sinitic values, while using them as the inspirational point at the beginning.
Source: From Ch.10 of *FCD*. Refer to text for more info and references.

Table 4.30. The Theory of Post-Capitalism I.3: By Region— Ex: Spiritual/Communal in the Trans-Islamic Calling

- **More Communal Than Individualistic**
  —*Collective*: building the webs of relationships to bind individuals
  —*Sharing*: cultivating the established "wisdom" through common experience
  —*Cooperative*: stressing harmony, solidarity, and commonality

- **More Informal-Legalistic Than Formal-Legalistic**
  —*Specific*: making efforts to know well the participants (family and larger community) in matters of common concern
  —*Affective*: mixing work with language and ritual on explicit religious (Islamic) ideals, texts, stories, and examples
  —*Ascriptive*: privileging local history and custom on relationships among kinship groups
  —*Particularistic*: preferring an unbiased insider with ongoing connections to all parties

- **More Spiritual Than Secular**
  —*Historical*: learning from the lessons of the past as a source of stability and guidance
  —*Deferential*: showing respect for age, experience, status, and leadership in communal affairs
  —*Honorable*: emphasizing face, dignity, prestige, and fairness
  —*Compassionate*: giving mercy and charity ("Zahah") to others

Notes: The categories and examples are solely illustrative (not exhaustive), and the comparison is relative (not absolute), nor are they mutually exclusive. As generalities, they allow exceptions. The specific forms of the trans-Islamic version need to be further developed in future after-postmodern history, as they will be different from the ones we now know, since the prefix "trans-" here means going beyond or deconstructing the Islamic values, while using them as the inspirational point at the beginning.

Sources: From Ch.10 of *FCD*. Refer to text for more info and references, especially from the works by George Irani (2000) and C. Murphy (September 19, 2001).

Table 4.31. The Theory of Post-Capitalism I.4: By Universe— Ex: Spiritual/Communal in the Trans-Outerspace Calling

- **More Communal Than Individual**
  - —*Cooperative*: requiring teamwork in small space habitats
  - —*Sharing*: learning from, and enjoying being with, each other in a small group in outer space

- **More Informal-Legalistic Than Formal-Legalistic**
  - —*Specific*: knowing more about each other to facilitate living and working together in space, both as fellow astronauts and space-mates
  - —*Affective*: being friendly and social to each other as vital to working and living in small space quarters
  - —*Ascriptive*: nurturing comaraderie among fellow astronauts as if they are family members over time
  - —*Particularistic*: building work relationship with enduring memory in a space mission

- **More Spiritual Than Secular**
  - —*Long-Term*: looking beyond selfish materialistic concerns in a precarious space environment with potential life or death
  - —*Loving/Caring*: cultivating deep bondage for the success of a long term space mission
  - —*Transcendent*: searching for life meaning in outer space

Notes: The calling and examples in each category are solely illustrative (not exhaustive), since there will be many different outer-space value ideals in the distant future of space colonization. The comparison is also relative (not absolute), nor are they mutually exclusive. As generalities, they allow exceptions. And the specific forms of trans-outer-space calling need to be further developed in future after-postmodern history, as they will be different from the ones we now know, since the prefix "trans-" here means going beyond or deconstructing the current outer-space values, while using them as the inspirational point at the beginning. The point here is to solely give an extremely rough picture of a small part of the world to come that we still do not know much about.
Source: From Ch.10 of *FCD*. Refer to text for more info and references.

Table 4.32. The Theory of Post-Capitalism II: Spiritual/Individualistic in the Post-Human Elitist Calling (Part I)

• **More Individualistic Than Communal**
—Setting up rank distinctions among unequals (e.g., between inferior humans and superior post-humans, or later among inferior post-humans and superior ones, relatively speaking)
—Yearning for being not only distinguished from unequals, but also the first among equals (the best of the very best)
—Recognizing the constraints of equality/inequality dialectics (or existential dialectics in general)

• **More Spiritual Than Secular**
—Soul-searching for a high spiritual culture (not the trashy one for the masses). Mass culture is a dirty joke for them.
—Exploring different spheres of non-human consciousness in the cosmos (something vastly superior than the human one)
—Recognizing the constraints of freedom/unfreedom dialectics (or existential dialectics in general)

• **Qualifications**
—Although post-human elitist post-democracy is comparable to post-human elitist post-capitalism in some respects, the former does not necessarily imply the latter (post-human elitist post-capitalism), just as democracy does not have to entail capitalism. They are two different (though related) entities.
—But up to a certain threshold of incorporating government intervention with advanced info systems in future civilizations for higher spiritual concerns at the expense of the free market and materialist pursuit, the capitalist ideal will be overcome.
—The overcome will not be Fascist or feudalistic, but post-capitalist, subject to the constraints of existential dialectics.

(continued on next page)

Table 4.32. The Theory of Post-Capitalism II: Spiritual/Individualistic in the Post-Human Elitist Calling (Part II)

Notes: The calling and examples in each category are solely illustrative (not exhaustive). The comparison is also relative (not absolute), nor are they mutually exclusive. As generalities, they allow exceptions. And the specific forms of post-human elitist post-capitalism need to be further developed in future after-postmodern history, as they will be different from the ones we now know, while using them as the inspirational point at the beginning. The point here is to solely give an extremely rough picture of a small part of the world to come that we still do not know much about.
Sources: From Ch.10 of *FCD* (and also *FPHC*, *BDPD*, and *BCPC*). Refer to the text for more info and references.

Table 4.33. Capitalism, Non-Capitalism, and Post-Capitalism (Part I)

- **Capitalism**
  - —*Theoretical Constructs*
    - •Allocation of scarce resources among alternative wants largely by free market for competition (whose characteristics in its ideal form include, for instance, no barrier to entry or exit, homogeneity, perfect information, a large number of buyers/sellers, and perfect factor mobility)
    - •More formal-legalistic than informal-legalistic, more individualistic than communal, and more material (secular) than spiritual
    - •Either (1) minimal government or (2) relatively active government
  - —*Types*
    - •Only (1): Different versions of market capitalism (e.g., the U.S.)
    - •Only (2): Different versions of welfare capitalism (e.g., Sweden)

- **Non-Capitalism**
  - —*Theoretical Constructs*
    - •Allocation of scarce resources among alternative wants mainly by political authority for policies (which can be regulative, redistributive, symbolic, and participatory)
    - •More informal-legalistic than formal-legalistic
    - •Either (1') more individualistic (for the elites), often (though not always) for material (secular) concerns, or (2') more communal (for the masses), often (though not always) for spiritual concerns
  - —*Types*
    - •Only (1'): Different versions on the Right (e.g., Fascist corporate-state economy for the glory of the new Rome, medieval lord-vassal-serf economy for the power of the feudalistic order)
    - •Only (2'): Different versions on the Left (e.g., Soviet command economy for the creation of the New Socialist Man)

(continued on next page)

Table 4.33. Capitalism, Non-Capitalism, and Post-Capitalism (Part II)

- **Post-Capitalism**

—*Theoretical Constructs*

•Allocation of scarce resources among alternative wants largely by political authority with advanced info systems in future civilizations, subject to existential dialectics. In degree of allocating by authority, post-capitalism is more than capitalism but less than non-capitalism.

•More spiritual than secular (material)

•Either (1'') more individualistic or (2") more communal

•Like capitalism and non-capitalism, post-capitalism is also subject to the freedom/unfreedom and equality/inequality dialectics (or existential dialectics in general). There is no utopia, in the end; even were there one, dystopia would exist within it.

•Unlike capitalism and non-capitalism, post-capitalism makes use of a different degree of political authority with advanced info systems in future civilizations and strives for higher-spiritual cultures (especially in the post-human age), while acknowledging the constraints of existential dialectics and no longer valuing free market (as in capitalism) and economic control (as in non-capitalism) as sacred virtues.

—*Types*

•Only (1"): Different versions of post-human elitist value ideals

•Only (2''): Different versions of trans-Sinitic value ideals

•Only (2''): Different versions of trans-feminine value ideals

•Only (2''): Different versions of trans-Islamic value ideals

•Only (2''): Different versions of trans-outerspace value ideals

---

(continued on next page)

Table 4.33. Capitalism, Non-Capitalism, and Post-Capitalism (Part III)

Notes: The calling and examples in each category are solely illustrative (not exhaustive). The comparison is also relative (not absolute), nor are they mutually exclusive. As generalities, they allow exceptions. And the specific forms of each calling need to be further developed in future after-postmodern history, as they will be different from the ones we now know, while using them as the inspirational point at the beginning. The point here is to solely give an extremely rough picture of a small part of the world to come that we still do not know much about.
Source: From Ch.10 of *FCD*. Refer to the text for more info and references.

Table 4.34. Multiple Causes of the Emergence of Post-Capitalism (Part I)

• **At the Micro-Physical Level**
—Ex: intelligent life without the human physical-chemical system
—Ex: mastering of quantum mechanics, electromagnetism, and other fields for the understanding of a broad range of anomalous experiences and the application for artificial intelligence for spiritual quest
—Sources: Ch.7 of *FHC*; Chs.9-10 of *FCD*; Ch.1 of *FPHC*

• **At the Chemical Level**
—Ex: space radiation and toxins
—Sources: Ch.7 of *FHC*; Chs.9-10 of *FCD*

• **At the Bio-Psychological Level**
—Ex: exo-biological evolution in deep space
—Ex: genetic engineering of new beings
—Ex: limits of human cognition
—Ex: the rise of unfolding unconsciousness
—Sources: Ch.2 & Chs.9-10 of *FCD*; Ch.7 of *FHC*; *FPHU*

• **At the Institutional Level**
—Ex: the flawed logic of the free market
—Ex: the need of a post-autistic economics
—Sources: Ch.10 of *FCD*

• **At the Organizational Level**
—Ex: the dark sides of formal-legalistic routines
—Sources: Ch.3 of *FHC*; Ch.7 of *FCD*; Ch.3 of *FPHC*

• **At the Structural Level**
—Ex: ever new forms of inequities, at home and abroad
—Ex: the emergence of China, women, and Islam as major actors
—Sources: Chs.5-6 of *FHC*; Chs.7, 9 & 10 of *FCD*; Chs.4-5 of *BDPD*

(continued on next page)

Table 4.34. Multiple Causes of the Emergence of Post-Capitalism (Part II)

- **At the Cultural Level**
  —Ex: freedom/unfreedom dialectics
  —Ex: equality/inequality dialectics
  —Sources: Ch.5 of *FHC*; Chs.3 & 10 of *FCD*; Ch.4 of *FPHC*; Ch.1 of *BDPD*

- **At the Systemic Level**
  —Ex: space habitats (in zero-gravity) and colonization
  —Ex: ultra advanced future info systems
  —Ex: qualitative demography
  —Sources: Ch.7 of *FHC*; Chs. 9 & 10 of *FCD*

- **At the Cosmological Level**
  —Ex: the colonization of multiverses
  —Ex: the alteration of space-time and the creation of new matter-energy
  —Ex: the expansion of floating consciousness
  —Ex: the spread of hyper-spatial consciousness
  —Sources: Ch.7 of *FHC*; Chs. 9 & 10 of *FCD*; Ch.4 of *FPHC*; *FPHST*

Notes: The examples in each category are solely illustrative (not exhaustive), and some of the items can be reclassified somewhere else. Nor are they always mutually exclusive. Since they are generalities, exceptions are expected.
Sources: Especially from *FHC*, *FCD*, *FPHC*, *BCPC*, *BDPD*, *FPHST*, and *FPHU*. See also other books and my perspectives on civilizational holism.

Table 4.35. The Theoretical Debate on Civilization

- **The Progressive Theory of Civilization**
  —*Thesis*: The "civilizing" process is "good," as opposed to the "barbarizing" process as something "bad," relatively speaking.
  —*Discourse*: Especially, though not exclusively, in the Enlightenment era and a bit before. Example: Thomas Hobbes—in that the tribes in primitive societies were "savages."

- **The Romantic Theory of Barbarity**
  —*Thesis*: The "civilizing" process is "bad," as opposed to the "barbarizing" process as something "good," relatively speaking.
  —*Discourse*: Especially, though not exclusively, in the Counter-Enlightenment circle. Example: Jean-Jacques Rousseau—in that civilization "corrupts" men, and the "savages" are in fact "noble."

- **The Moderate Theory of Civilization**
  —*Thesis*: The "civilizing" process is "good," but there is a price to pay, especially in systematic (compulsive) self-control.
  —*Discourse*: Especially, though not exclusively, in some late modern and postmodern circles. Example: Norbert Elias—in that social manners become more refined in the civilizing process, but self-control also becomes more systematic.

- **The Theory of Post-Civilization**
  —*Thesis*: The civilizing process is as evil and good as barbarity, and each cannot exist without the other, to be eventually superseded by post-civilization unto the post-human age.
  —*Discourse*: Proposed by Peter Baofu. See the rest of *BCIV* for more analysis.

Source: From *BCIV* on the theoretical debate

Table 4.36. No Freedom Without Unfreedom in the Civilizing Processes (Part I)

- **The Rationalizing Process (at the Level of Culture)**
  —if freer from the dominance of unreason (as in barbarism) in the civilizing process, then less free from the rationalizing process (be it in the form of the principle of either transcendence or immanence)
  —if freer from the principle of immanence in the rationalizing process, then less free from the inclination to commit terror in the name of reason and the relative underdevelopment of non-reason (e.g., in relation to yoga and meditation)
  —if freer from the principle of transcendence in the rationalizing process, then less free from the relative underdevelopment of reason (e.g., in relation to systematic methodology) and the occurrence of oppression in the name of non-reason

- **The Pacifying Process (at the Level of Society)**
  —if freer from the dominance of pillage (as in savagery) in the civilizing process, then less free from the pacifying process (be it in the form of external control or self-control)
  —if freer from self-control in the pacifying process, then less free from the temptation of expansionist oppression and rebellious mindset in external control
  —if freer from external control in the pacifying process, then less free from the gruesome psychological self-torture and conformism in self-control

- **The Stewardizing Process (at the Level of Nature)**
  —if freer from the dominance of nature (as in the state of nature) in the civilizing process, then less free from the stewardizing process (be it in the form of the stewardship of creation or the covenant with nature)
  —if freer from the stewardship of creation in the stewardizing process, then less free from material underdevelopment, relatively speaking, and spiritual exclusion in the covenant with nature
  —if freer from the covenant with nature in the stewardizing process, then less free from ecological degradation and spiritual disconnection from nature in the stewardship of creation

(continued on next page)

Table 4.36. No Freedom Without Unfreedom in the Civilizing Processes (Part II)

- **The Subliming Process (at the Level of the Mind)**
  —if freer from the dominance of spontaneity (as in the wild state of the mind) in the civilizing process, then less free from the subliming process, be it in the form of (cyclical-centric) self-refinement or (linear-centric) self-discipline
  —if freer from (cyclical-centric) self-refinement in the subliming process, then less free from the (linear-centric) self-regimen (as a form of neurosis)
  —if freer from (linear-centric) self-discipline in the subliming process, then less free from the (cyclical-centric) self-torture (equally as a form of neurosis)

Notes: The examples in each category are solely illustrative (not exhaustive), and the comparison is relative (not absolute), nor are they necessarily mutually exclusive. And some can be easily re-classified elsewhere. As generalities, they allow exceptions.
Sources: From *BCIV*. See also *FHC*, *FCD*, *FPHC*, *BDPD*, and *BCPC*.

Table 4.37. No Equality Without Inequality in the Civilizing Processes (Part I)

- **The Rationalizing Process (at the Level of Culture)**
  - —if more equal for the role of rationalization in the rationalizing process (of civilizational making), then less equal for that of mythicization (as in barbarism)
  - —if more equal for the principle of transcendence in (linear-centric) rationalizing process, then less equal for the principle of immanence
  - —if more equal for the principle of immanence in (cyclical-centric) rationalizing process, then less equal for the principle of transcendence

- **The Pacifying Process (at the Level of Society)**
  - —if more equal for pacification in civilizational making, then less equal for the institution of pillaging and others (as in savagery)
  - —if more equal for external control, relatively speaking, in pacifying process, then less equal for self-control
  - —if more equal for self-control, relatively speaking, in pacifying process, then less equal for external-control

- **The Stewardizing Process (at the Level of Nature)**
  - —if more equal for stewardship in the stewardizing process (of civilizational making), then less equal for reverent (submissive) existence (as in barbarism)
  - —if more equal for the stewardship of creation in (linear- centric) stewardizing process, then less equal for the (cyclical-centric) covenant with nature for harmonious co-existence
  - —if more equal for the (cyclical-centric) covenant with nature in the stewardizing process, then less equal for the (linear-centric) stewardship of nature for domination

---

(continued on next page)

Table 4.37. No Equality Without Inequality in the Civilizing Processes (Part II)

- **The Subliming Process (at the Level of the Mind)**
  —if more equal for the role of reason in the subliming process, then less equal for that of unreason (as in the natural state of wildness)
  —if more equal for the primacy of reason in (linear-centric) subliming process, then less equal for other faculties (e.g., intuition, existential feelings, and analogous thinking) in cyclical-centric one
  —if more equal for the exercise of other faculties (e.g., intuition, existential feelings, and analogous thinking) in cyclical-centric subliming process, then less equal for the role of reason in linear-centric counterpart

Notes: The examples in each category are solely illustrative (not exhaustive), and the comparison is relative (not absolute), nor are they mutually exclusive. And some can be easily reclassified else-where. As generalities, they allow exceptions.
Sources: From *BCIV*. See also *FHC*, *FCD*, *FPHC*, *BDPD*, and *BCPC*.

Table 4.38. Five Theses on Post-Civilization

- Post-civilization no longer treats civilization as good and barbarity as evil (relatively speaking), nor does it nostalgically regard barbarity as good and civilization as evil (relatively speaking again). Civilization is as evil and good as barbarity.

- Post-civilization also no longer accepts the dichotomy between civilization and barbarity. Civilization cannot exist without barbarity. It is no longer necessary to preserve civilization, any more than it is imperative to destroy barbarity. To go beyond civilization and barbarity is to acknowledge the co-existence of both, although the degree of scaling one over the over varies from case to case—but is not to be extreme in largely favoring one over the other, *on average* (subject to the constraints of existential dialectics).

- Post-civilization is thus subject to the constraints of existential dialectics. There is no freedom without unfreedom, and no equality without inequality, for instance. There will be no utopia; even should there be one, there would be dystopia embedded within it.

- Post-civilization will eventually replace civilization (as a form of life settlement), to be dominated by post-capitalist and post-democratic lifeforms here on earth and in deep space (besides other alien lifeforms that we have never known), unto the post-human age in multiverses. Those few post-humans who keep civilization will live in a "post-human civilization," while the rest (the majority), who choose post-civilization, will evolve towards the state of "post-human post-civilization." One therefore should not confuse "post-human civilization" with "post-human post-civilization," as the two are not the same.

- Post-civilization will confront psychosis as a primary problem in the culture of virtuality unto the post-human age, just as civilization has neurosis as a primary one of its own (although both neurosis and psychosis are major problems in both).

Notes: The comparison in each category is relative (not absolute), nor are they necessarily mutually exclusive. And some can be easily re-classified elsewhere. As generalities, they allow exceptions.
Sources: From *BCIV*. See also *FHC*, *FCD*, *FPHC*, *BDPD*, and *BCPC*.

Table 4.39. Barbarity, Civilization, and Post-Civilization

- **The Rationalizing Process (at the Level of Culture)**
  —*Barbarity*
    - More mythicizing than rationalizing, relatively speaking

  —*Civilization*
    - More rationalizing than mythicizing, relatively speaking

  —*Post-Civilization*
    - Beyond the dichotomy, subject to existential dialectics

- **The Pacifying Process (at the Level of Society)**
  —*Barbarity*
    - More pillaging than pacifying, relatively speaking

  —*Civilization*
    - More pacifying than pillaging, relatively speaking

  —*Post-Civilization*
    - Beyond the dichotomy, subject to existential dialectics

- **The Stewardizing Process (at the Level of Nature)**
  —*Barbarity*
    - More revering than stewardizing, relatively speaking

  —*Civilization*
    - More stewardizing than revering, relatively speaking

  —*Post-Civilization*
    - Beyond the dichotomy, subject to existential dialectics

- **The Subliming Process (at the Level of the Mind)**
  —*Barbarity*
    - More impulsing than subliming, relatively speaking

  —*Civilization*
    - More subliming than impulsing, relatively speaking

  —*Post-Civilization*
    - Beyond the dichotomy, subject to existential dialectics

Notes: The comparison in each category is relative (not absolute), nor are they necessarily mutually exclusive. And some can be easily re-classified elsewhere. As generalities, they allow exceptions.
Sources: From *BCIV*. See also *FHC*, *FCD*, *FPHC*, *BDPD*, and *BCPC*.

Table 4.40. Types of Super Civilization in the Cosmos (Part I)

- **Type I**
  —a civilization which gains control of and uses the total energy output "falling on its planet from its sun for interstellar communication" (or, in general, space colonization). For N. Kardashev, who proposed the first three types, human civilization is currently Type Zero (Type O), which is below even Type I, since its present energy consumption for all purposes, let alone for interstellar communication, is still 10,000 times less.

- **Type II**
  —a civilization which gains control of and uses directly the total energy output of its sun for interstellar communication (or, in general, space colonization).

- **Type III**
  —a civilization which gains control of and uses the total energy output of its galaxy for interstellar communication (or, in general, space colonization).

- **Type IV**
  —a civilization which gains control of and uses the total energy output of its cluster of galaxies for interstellar communication (or, in general, space colonization).

- **Type V**
  —a civilization which gains control of and uses the total energy output of its supercluster of galaxies for interstellar communica-tion (or, in general, space colonization).

(continued on next page)

Table 4.40. Types of Super Civilization in the Cosmos (Part II)

- **Type…n**
  —So continues the series in what I call the cyclical progression of hegemony in the cosmos and beyond.

Notes: The Russian astrophysicist Nikolai Kardashev proposed the first three types of super civilization in terms of total energy out-put for interstellar communication. (CSM 1979) I extend his argument further to propose Type IV, Type V, Type VI, and Type…n, in the context of my claim about the cyclical progression of he-gemony in the cosmos and beyond.
Sources: From *Table 9.4* of *FCD*. See *FHC*, *FCD*, and *FPHC* for more info.

Table 4.41. The Civilizational Project from Pre-Modernity to After-Postmodernity (Part I)

	Pre-Modern	*Modern*	*Postmodern*	*After-Postmodern*
Main narratives	•Sacralness •Courtliness •Vitalism •Animism	•Freedom •Equality •Fraternity	•Multiplicity •Hybridization	•Naked contingency •Cyclical progres-sion of hegemony
Main institutions	•Monarchy •Aristocracy •Feudalism •Holy order •Primitivism	•Capitalism •Liberalism •Socialism •Nazism •Fascism	•Capitalism •Liberalism •Postmodern politics of difference	•Post-Capitalism •Post-Democracy •Others
Main techno-logical and economic revolutions	•Agricultural	•Service •Industrial	•Informational	•Biological •Material •Energy •Space •Others

(continued on next page)

Table 4.41. The Civilizational Project from Pre-Modernity to After-Postmodernity (Part II)

	Pre-Modern	*Modern*	*Postmodern*	*After-Postmodern*
Main agents	•Males •Upper strata •Mini-states	•Males •Upper strata •Whites •Empires	•Males •Upper strata •Whites •Others •Supra-states •IO's	•Post-humans •Humans •Others
Main impacts	•Local	•Inter-national	•Global	•Outer-space •Multiverse
Main outcomes	•Towards moderntiy •Rise of linear- & cyclical-centric civiliza-tions	•Towards post-modernity •Dominance of linear-centric civilization	•Towards after-post-moderntiy •Linear-centric civiliza-tion in crisis	•Towards human (& maybe post-human) extinction •Rise of post-civiliza-tion, especially in post-human forms of space-time

Notes: The examples in each category are solely illustrative (not exhaustive) nor necessarily mutually exclusive, and the comparison is relative (not absolute). As generalities, they allow exceptions.
Sources: From *Table 10.16* of *FCD*—and also from *BCIV* on post-civilization (and *FPHST*)

Table 4.42. Civilizational Holism (Part I)

• **At the Micro-Physical Theoretical Level**
—Ex: Mastering of quantum mechanics, electromagnetism, and other fields for the understanding of a broad range of anomalous experiences and theapplication for artificial intelligence (*Sec.1.4.1* of *FPHC*)

• **At the Chemical Theoretical Level**
—Ex: Unprecedented expansion of (and violence to) the mind through ever new forms of drugs (and virtual technologies, for that matter) (Ch.9 of *FCD*)

• **At the Biological Theoretical Level**
—Ex: Humans are not biologically equal, on the basis of race, gender, ethnicity, age, and whatnot. (*Sec.2.6* & Ch.10 of *FCD*; *BNN*) And post-humans will experience the same fate, in an even more amazing way.

• **At the Psychological Theoretical Level**
—Ex: Human cognitive impartiality and emotional neutrality are quite limited. (*Secs.2.4-2.5* of *FCD*)
—Ex: Rise of Floating Consciousness (Ch.10 of *FCD*; Chs.1 & 4 of *FPHC*) and Unfolding Unconsciousness (*FPHU*)

• **At the Organizational Theoretical Level**
—Ex: Administrative colonization of deep space, with less legal-formalism in some corners. (Chs.9-10 of *FCD*)

• **At the Institutional Theoretical Level**
—Ex: Both capitalism and democracy will not last, to be superseded by different versions of post-capitalism and post-democracy in after-postmodernity. (Ch.10 of *FCD*)

(continued on next page)

Table 4.42. Civilizational Holism (Part II)

- **At the Structural Theoretical Level**
  - —Ex: Social stratification reappears in ever new forms, also with new causes and new players in the cyclical progression of hegemony. (Chs.8-10 of *FCD*)
  - —Ex: The world of hyper-empires, and the union of the unions (*BWT*)

- **At the Systemic Theoretical Level**
  - —Ex: Outerspace expansion: local → regional → global → solar → galactic → clustery → multiversal (Ch.9 of *FCD*)
  - —Ex: Demographic transition: human extinction, and the rise of post-humans (e.g., cyborgs, thinking machines, thinking robots, genetically altered superior beings, floating consciousness, hyper-spatial consciousness) (Ch.4 of *FPHC*; Ch.10 of *FCD*; & Ch.7 of *FHC*)
  - —Ex: New technological forces in material sciences, electronic and communication sciences, energy sciences, biosciences, manufacturing and engineering sciences, and space sciences (Ch.10 of *FCD* & Ch.7 of *FHC*)
  - —Ex: Systematic dominance towards nature for space colonization (Chs.9-10 of *FCD*; Chs.2 & 7 of *FHC*)

- **At the Cultural Theoretical Level**
  - —Ex: The post-human transcendence of freedom and equality (Ch.10 of *FCD*)
  - —Ex: Methodological Holism (Ch.1 of *FCD*; Ch.1 of *FPHC*; *Sec.2.1* & *Sec.2.5* of *BCPC*)
  - —Ex: The Evolution from Barbarity to Post-Civilization (*BCIV*)

- **At the Cosmological Theoretical Level**
  - —Ex: Mastering of dark matter and dark energy, and the exploration of multiverses (Ch.4 of *FPHC*; Ch.10 of *FCD*; & Ch.7 of *FHC*)
  - —Ex: Alternation of space-time (*FPHST*)
  - —Ex: The emergence of hyper-spatial consciousness (*FPHC*)

(continued on next page)

Table 4.42. Civilizational Holism (Part III)

- **At Other Levels**
  —Ex: Historical: pre-modernity → modernity → postmodernity → after-postmodernity (human distinction, and the rise of post-humans, including floating consciousness, hyper-spatial consciousness, and unfolding unconsciousness) (Ch.7 of *FHC*; Ch.10 of *FCD*; *FPHC*; *FPHU*)

Notes: These examples are solely illustrative (not exhaustive), and some of the items can be reclassified somewhere else. Nor are they always mutually exclusive. Since they are generalities, exceptions are expected. And the comparison is relative, not absolute.
Sources: From *Table 5.1* of *FPHC*—with details from *FHC*, *FCD*, and the rest of my books. Not every aspect in each category as discussed in all my books are presented here, since there are too many issues. For more info, also consult the table on theories on civilizational holism and, of course, the books themselves.

Table 4.43. Theories on Civilizational Holism (Part I)

I. Theories in Relation to Nature

—*At the Macro-Physical (Cosmological) Theoretical Level*

- • 48. Multilateral Theory of Acoustics (Peter Baofu) (*FPHA*)
- • 47. Resettlement Theory of Geology (Peter Baofu) (*FPHGEOL*)
- • 46. Theory of Post-Cosmology (Peter Baofu) (*BCOS*)
- • 45. Theory of Hyper-Spatial Consciousness (Peter Baofu) (Ch.4 of *FPHC*; *FPHG*)
- (• 43). Selective Theory of Geometry (Peter Baofu) (*FPHG*)
- (• 42). Perspectival Theory of Space-Time (Peter Baofu) (*FPHST*)
- (• 41). Dialectic Theory of Complexity (Peter Baofu) (*FC*)
- (• 25). Theory of Floating Consciousness (Peter Baofu) (Ch.10 of *FCD*; Chs.1 & 4 of *FPHC*)
- • 44. Theory of the Geopower of Nature (Peter Baofu) (Ch.4 of *ALD*)

—*At the Micro-Physical Theoretical Leve*

- • 43. Selective Theory of Geometry (Peter Baofu) (*FPHG*)
- • 42. Perspectival Theory of Space-Time (Peter Baofu) (*FPHST*)
- • 41. Dialectic Theory of Complexity (Peter Baofu) (*FC*)

(continued on next page)

Table 4.43. Theories on Civilizational Holism (Part II)

II. Theories in Relation to Culture

—At the Cultural Theoretical Level

- • 40. Mediative-Variative Theory of Chess (Peter Baofu) (*FPHCHESS*)
- • 39. Theory of Post-Ethics (Peter Baofu) (*BEPE*)
- • 38. Dualistic Theory of Mass Culture (Peter Baofu) (Ch.2 of *FHC*)
- • 37. Comparative Theory of Religion—also known as the Comparative-Substitutive Theory of Religion (Peter Baofu) (Ch.3 of *FHC*; Ch.9 of *FCD*; Ch.1 of *FPHK*; *FPHR*)
- • 36. Theory of Post-Civilization (Peter Baofu) (*BCIV*)
- • 35. Theory of the Trinity of Modernity to Its After-Postmodern Counterpart (Peter Baofu) (*FHC*; Ch.10 of *FCD*)
- • 34. Transformative Theory of Aesthetic Experience (Peter Baofu) (*FAE*)
- (• 17). Theory of Post-Capitalism (Peter Baofu) (Ch.10 of *FCD*; Chs.2 & 4 of *FPHC*; *BCPC*)
- (• 16). Theory of Post-Democracy (Peter Baofu) (Ch.10 of *FCD*; Chs.3 & 4 of *FPHC*; *BDPD*)
- (• 5). Theory of Existential Dialectics, or the Holistic Theory of Knowledge (Peter Baofu) (*FHC*; *FCD*; *FPHC*; *BDPD*; *FC*; *FAE*; *ALD; FIA; FPHK*; etc,)
- (• 4). Contrastive Theory of Rationality (Peter Baofu) (*FPHML*)
- (• 2). Theory of Methodological Holism (Peter Baofu) (Ch.1 of *FCD*; Ch.1of *FPHC*; *Sec.2.1* & *Sec.2.5* of *BCPC*; *FC*; *FPHK*; etc.)

(continued on next page)

Table 4.43. Theories on Civilizational Holism (Part III)

III. Theories in Relation to the Mind

—*At the Biological Theoretical Level*

- 33. Theory of Contrastive Advantages (Peter Baofu) (*Sec.2.6* & Ch.10 of *FCD; BNN*)
- (• 25). Theory of Floating Consciousness (Peter Baofu) (Ch.10 of *FCD*; Chs.1 & 4 of *FPHC*)

—*At the Psychological Theoretical Level*

- 32. Metamorphic Theory of Humor (Peter Baofu) (*FPHH*)
- 31. Contrarian Theory of Personality (Peter Baofu) (*FPHP*)
- 30. Theory of Virtual Sexuality (Peter Baofu) (*FPHS*)
- 29. Expansive-Contractive Theory of Martial Arts (Peter Baofu) (*FPHMA*)
- 28. Multilogical Theory of Learning (Peter Baofu) (*FPHL*)
- 27. Comprehensive Theory of Creative Thinking (Peter Baofu) (*FPHCT*)
- 26. Theory of Unfolding Unconsciousness—also known as the Unfolding Theory of Anomalous Experience (Peter Baofu) (*FPHU*)
- 25. Theory of Floating Consciousness (Peter Baofu) (Ch.10 of *FCD*; Chs.1 & 4 of *FPHC*)
- 24. Theory of Cognitive Partiality (Peter Baofu) (*Sec.2.4* of *FCD*; *Sec.4.5.1.1* of *BCPC*)
- 23. Theory of Emotional Non-Neutrality (Peter Baofu) (*Sec.2.5* of *FCD*; *Sec.4.5.2* of of *BCPC*)
- 22. Theory of Behavioral Alteration (Peter Baofu) (*Sec.4.5.3* of *BCPC*)

Table 4.43. Theories on Civilizational Holism (Part IV)

IV. Theories in Relation to Society

—*At the Organizational Theoretical Level*

- 21. Theory of E-Civic Alienation (Peter Baofu) (Ch.7 of *FCD*)
- 20. Combinational Theory of Organization (Peter Baofu) (*FPHO*; Ch.6 of *ALD*)

—*At the Institutional Theoretical Level*

- 19. Heterodox of Theory of Education (Peter Baofu) (*FPHEDU*)
- 18. Reconstruction of Theory of Law (Peter Baofu) (*FPHLAW*)
- 17. Theory of Post-Capitalism (Peter Baofu) (Ch.10 of *FCD*; Chs.2 & 4 of *FPHC*; *BCPC*)
- 16. Theory of Post-Democracy (Peter Baofu) (Ch.10 of *FCD*; Chs.3 & 4 of *FPHC*; *BDPD*)
- 15. Dynamic Theory of Comparative Political Systems (Peter Baofu) (*ALD*)

—*At the Systemic Theoretical Level*

- 14. Contingent Theory of Urban Planning (Peter Baofu) (*FPHUP*)
- 13. Totalistic Theory of Communication (Peter Baofu) (*FPHMM*; *FCD*; *FHC*)
- 12. Ambivalent Theory of Technology (Peter Baofu) (*FPHE*; *FCD*; *FHC*)
- 11. Multifaceted Theory of War and Peace (Peter Baofu) (Ch.9 of *FCD*; Ch.1 of *FPHK*)
- 10. Theory of Post-Humanity (Peter Baofu) (Ch.7 of *FHC*; Chs.3, & 10 of *FCD*; Chs.1, 3 & 4 of *FPHC*; and other books of mine)
- 9. Theory of the Cyclical Progression of System Integration and Fragmentation (Peter Baofu) (Chs.9-10 of *FCD*)
- 8. Synthetic Theory of Information Architecture (Peter Baofu) (*FIA*)

(continued on next page)

Table 4.43. Theories on Civilizational Holism (Part V)

IV. Theories in Relation to Society (*cont'd*)

—*At the Structural Theoretical Level*

- 7. Theory of the Cyclical Progression of Hegemony (Peter Baofu)
  (Chs.9-10 of *FCD*; Chs.1, 3 & 4 of *FPHC*; *BDPD*)
- 6. Theory of the Cyclical Progression of Empire-Building (Peter Baofu)
  (*BWT*)

V. Meta-Theories (in Relation to Theories)

—*At the Ontological Meta-Theoretical Level*

- 5. Theory of Existential Dialectics, or the Holistic Theory of Knowledge (Peter Baofu)
  (*FHC*; *FCD*; *FPHC*; *BDPD*; *FC*; *FAE*; *ALD; FIA; FPHK*; etc,)
- 4. Contrastive Theory of Rationality (Peter Baofu)
  (*FPHML*)

—*At the Methodological Meta-Theoretical Level*

- 3. Critical-Dialectic Theory of Formal Science (Peter Baofu)
  (*FPHFS*)
- 2. Theory of Methodological Holism (Peter Baofu)
  (Ch.1 of *FCD*; Ch.1of *FPHC*; *Sec.2.1* & *Sec.2.5* of *BCPC*; *FC*; *FPHK*; etc.)

VI. Theories in Relation to the Rest

—*At Other Levels (Historical)*

- 1. Theory of the Evolution from Pre-Modernity to After-Postmodernity (Peter Baofu)
  (*FHC*; Ch.9-10 of *FCD*; *FPHC*)

Notes: All these theories are my constructions, as some of the main contributions of my grant project on civilization and its future. These examples are solely illustrative (not exhaustive), and some of the items can be reclassified somewhere else. Nor are they always mutually exclusive. Since they are generalities, exceptions are expected.

Sources: From my previous books.

Table 4.44. Three Great Future Transformations of Mind Games

- **Virtual Games**
  - —Ex: online chess
  - —Ex: virtual experience

- **Novel Games**
  - —Ex: new chess variants
  - —Ex: new chess engines

- **Post-Human Mind Games**
  - —Ex: the quest for broader/deeper mental benefits of chess playing
  - —Ex: games designed for the evolution of the mind into different body-less forms (e.g., "floating consciousness," "hyper-spatial consciousness," "unfolding unconsciousness")

Notes: The examples in the categories are solely illustrative (not exhaustive), and the comparison is relative (not absolute), nor are they necessarily mutually exclusive. And some can be easily re-classified elsewhere. As generalities, they allow exceptions.
Sources: From *Sec. 4.16* of *FPHCHESS*. See text for more info. Also, consult *FCD* for strategy and tactics in warfare and *FPHO* for strategy and tactics in organization.

Bibliography

ACatholicLife.blogspot.com (ACL). 2007. "The Errors of Centering Prayer." <http://acatholiclife.blogspot.com/2007/05/errors-of-centering-prayer.html>.

al-Miskīn, Mattá. 2003. *Orthodox Prayer Life: The Interior Way*. Crestwood, NY: St Vladimir's Seminary Press.

Albright, Daniel, ed. 2004. *Modernism and Music: An Anthology of Source*. Chicago, IL: University Of Chicago Press.

American Heritage Dictionary of the English Language (AHD). 2000. "Sound." <http://web.archive.org/web/20080625012016/http://www.bartleby.com/61/65/S0576500.html>.

Arnot, Chris. 2000. "Ghost Buster" (July July). *The Guardian*. <http://www.guardian.co.uk/education/2000/jul/11/highereducation.chrisarnot>.

Artaud, Antonin. 1995. "Pour en Finir Avec le Jugement de Dieu." Marc Dachy, ed. Compact Disc. Sub Rosa/aural documents.

Attali, Jacques. 1985, *Noise: The Political Economy of Music*. Brian Massumi, tr. Minneapolis: University of Minnesota Press.

Aumann, Jordan. 1985. *Christian Spirituality in the Catholic Tradition*. Sheed & Ward.

Ballou, Glen; & Howards Sams, eds. 1998. *Handbook for Sound Engineers*. Boston, MA: Focal Press.

Baofu, Peter. 2011. *The Future of Post-Human Humor: A Preface to a New Theory of Joking and Laughing*. Cambridge, England: Cambridge International Science Publishing, Ltd.

_____.2011a. *Beyond Ethics to Post-Ethics: A Preface to a New Theory of Morality and Immorality*. Charlotte, NC: Information Age Publishing.

_____.2011b. *The Future of Post-Human Education: A Preface to a New Theory of Teaching and Learning*. Cambridge, England: Cambridge International Science Publishing, Ltd.

_____.2010. *The Future of Post-Human Religion: A Preface to a New Theory of Spirituality*. Cambridge, England: Cambridge International Science Publishing, Ltd.

_____.2010a. *The Future of Post-Human Chess: A Preface to a New Theory of*

Tactics and Strategy. Cambridge, England: Cambridge International Science Publishing, Ltd.

_____.2010b. *The Future of Post-Human Geology: A Preface to a New Theory of Statics and Dynamics*. Cambridge, England: Cambridge International Science Publishing, Ltd.

_____.2010c. *Beyond Cosmology to Post-Cosmology: A Preface to a New Theory of Different Worlds*. Cambridge, England: Cambridge International Science Publishing, Ltd.

_____.2010d. *The Future of Post-Human Personaltiy: A Preface to a New Theory of Normality and Abnormality*. Cambridge, England: Cambridge International Science Publishing, Ltd.

_____.2010e. *The Future of Post-Human War and Peace: A Preface to a New Theory of Aggression and Pacificity*. Cambridge, England: Cambridge Scholars Publishing.

_____.2010f. *The Future of Post-Human Formal Science: A Preface to a New Theory of Abstraction and Application*. Cambridge, England: Cambridge Scholars Publishing.

_____.2010g. *The Future of Post-Human Law: A Preface to a New Theory of Necessity, Contingency, and Justice*. Cambridge, England: Cambridge Scholars Publishing.

_____.2010h. *The Future of Post-Human Sexuality: A Preface to a New Theory of the Body and Spirit of Love Makers*. Cambridge, England: Cambridge Scholars Publishing.

_____,2009. *The Future of Post-Human Martial Arts: A Preface to a New Theory of the Body and Spirit of Warriors*. Cambridge, England: Cambridge Scholars Publishing.

_____.2009a. *The Future of Post-Human Organization: A Preface to a New Theory of Communication, Decision-Making, and Leadership*. Cambridge, England: Cambridge Scholars Publishing.

_____.2009b. *The Future of Post-Human Language: A Preface to a New Theory of Structure, Context, and Learning*. Cambridge, England: Cambridge Scholars Publishing.

_____.2009c. *The Future of Post-Human Urban Planning: A Preface to a New Theory of Density, Void, and Sustainability*. Cambridge, England: Cambridge Scholars Publishing.

_____.2009d. *The Future of Post-Human Geometry: A Preface to a New Theory of Infinity, Symmetry, and Dimensionality*. Cambridge, England: Cambridge Scholars Publishing.

_____.2009e. *The Future of Post-Human Creative Thinking: A Preface to a New Theory of Invention and Innovation*. Cambridge, England: Cambridge Scholars Publishing.

_____.2009f. *The Future of Post-Human Mass Media: A Preface to a New Theory of Communication*. Cambridge, England: Cambridge Scholars Publishing.

_____.2009g. *The Future of Post-Human Engineering: A Preface to a New Theory of Technology*. Cambridge, England: Cambridge Scholars Publishing.

_____.2008. *The Future of Post-Human Mathematical Logic: A Preface to a New Theory of Rationality*. Cambridge, England: Cambridge Scholars Publishing.

_____.2008a. *The Future of Post-Human Knowledge: A Preface to a New Theory of Methodology and Ontology*. Oxford, United Kingdom: Chandos Publishing (Oxford) Ltd.

_____.2008b. *The Future of Post-Human Unconsciousness: A Preface to a New Theory of Anomalous Experience*. Cambridge, England: Cambridge Scholars Publishing.

_____.2008c. *The Future of Information Architecture: Conceiving a New Way to Understand Taxonomy, Network, and Intelligence*. Oxford, United Kingdom: Chandos Publishing (Oxford) Ltd.

_____.2007. *The Rise of Authoritarian Liberal Democracy: A Preface to a New Theory of Comparative Political Systems*. Cambridge, England: Cambridge Scholars Publishing.

_____.2007a. *The Future of Aesthetic Experience: Conceiving a Better Way to Beauty, Ugliness, and the Rest*. Cambridge, England: Cambridge Scholars Publishing.

_____.2007b. *The Future of Complexity: Conceiving a Better Way to Understand Order and Chaos*. London, United Kingdom: World Scientific Publishing Co.

_____.2007c. *Beyond the World of Titans, and the Remaking of World Order: A Preface to a New Logic of Empire-Building*. Cambridge, England: Cambridge Scholars Publishing.

_____.2006. *Beyond Nature and Nurture: Conceiving a Better Way to Understand Genes and Memes*. Cambridge, England: Cambridge Scholars Publishing.

_____.2006a. *The Future of Post-Human Space-Time: Conceiving a Better Way to Understand Space and Time*. New York: Peter Lang Publishing, Inc.

_____.2006b. *Beyond Civilization to Post-Civilization: Conceiving a Better Model of Life Settlement to Supersede Civilization*. New York: Peter Lang Publishing, Inc.

_____.2005. *Beyond Capitalism to Post-Capitalism: Conceiving a Better Model of Wealth Acquisition to Supersede Capitalism*. New York: The Edwin Mellen Press.

_____.2004. *The Future of Post-Human Consciousness*. New York: The Edwin Mellen Press.

_____.2004a. *Beyond Democracy to Post-Democracy: Conceiving a Better Model of Governance to Supersede Democracy*. Volume 1. New York: The Edwin Mellen Press.

_____.2004b. *Beyond Democracy to Post-Democracy: Conceiving a Better Model of Governance to Supersede Democracy*. Volume 2. New York: The Edwin Mellen Press.

_____.2002. *The Future of Capitalism and Democracy*. Maryland: The University Press of America.

_____.2000. *The Future of Human Civilization*. Volume 1. New York: The Edwin Mellen Press.

_____.2000a. *The Future of Human Civilization*. Volume 2. New York: The Edwin Mellen Press.

Barrow, John D. 1998. *Impossibility: The Limits of Science and the Science of Limits*. Oxford: Oxford University Press.

Bartleby.com (BT). 2010. "Plutarch" (December 18)/. <http://www.bartleby.com/100/714.167.html>.

_____.2010a. "Thomas Carlyle" (December 18). <http://www.bartleby.com/100/387.42.html>.

_____.2010b. "William Shakespeare" (December 20). <http://www.bartleby.com/100/138.31.118.html>.

_____.2010c. "Thomas Hood" (December 25). <http://www.bartleby.com/100/397.1.html>.

Bennett, David. 2005. "Subiaco Abbey's Angus herd" (May 19). *Delta Farm Press*. <http://www.webcitation.org/5nryhZkhj>.

Bickmore, T. 1999 "A Computational Model of Small Talk." <http://web.media.mit.edu/~bickmore/Mas962b/>.

Biletzki, Anat; & Anat Matar. 2006. "Ludwig Wittgenstein" (Nov. 17). *Stanford Encyclopedia of Philosophy*. <http://plato.stanford.edu/entries/wittgenstein/>.

Bivens, Arthur Clarke. 2004. *From Sailboats to Submarines*. Infinity Publishing/

Bothe, A. K.; P. Finn; & R. Bramlett. 2007. "Pseudoscience and the SpeechEasy: Reply to Kalinowski, Saltuklaroglu, Stuart, and Guntupalli." *American Journal of Speech-Language Pathology*, 16: 77–83.

_____; J. Davidow; R. Bramlett; & R. Ingham. 2006. "Stuttering Treatment Research 1970-2005: I. Systematic Review Incorporating Trial Quality Assessment of Behavioral, Cognitive, and Related Approaches." *American Journal of Speech-Language Pathology*, 15 (4): 321–341.

Buckingham, M. J. 2000. "Wave Propagation, Stress Relaxation, and Grain-to-Grain Shearing in Saturated, Unconsolidated Marine Sediments." *Journal of the Acoustical Society of America*, 108: 2796-2815.

Bunge, Mario. 1999. *Dictionary of philosophy*. Amherst, MA: Prometheus Books.

Cage, John. 1973. *Silence: Lectures and Writings*. Middletown, CT: Wesleyan

University Press.

_____.1970. "The Future of Music: Credo (1937)." *Documentary Monographs in Modern Art*. Richard Kostelanetz, ed. New York: Praeger Publishers.

Callanan, Martin John. 2010. "Sonification of You." <http://sonification.eu/avraamov>.

Cancel, Juan. 1998. "Frequency of Bat Sonar." *The Physics Factbook*. <http://hypertextbook.com/facts/1998/JuanCancel.shtml>.

Cascone, Kim. 2002. "The Aesthetics of Failure: 'Post-Digital' Tendencies in Contemporary Computer Music." *Computer Music Journal*, 24 (4): 12–18.

Chadabe, Joel. 1996. *Electronic Sound: The Past and Promise of Electronic Music*. New Jersey: Prentice Hall.

Chase, Scott. 1993. "Tachyons" (March). *Physics FAQ*. <http://math.ucr.edu/home/baez/physics/ParticleAndNuclear/tachyons.html>.

Choi, Charles. 2010. "Blow It Out Your Sides: Caterpillars Can Whistle" (December 10). *LiveScience*. <http://www.livescience.com/animals/caterpillars-whistle-through-tiny-holes-101210.html>.

Chotiros, N. 1995. "Biot Model of Sound Propagation in Water Saturated Sand." *Journal of the Acoustical Society of America*, 97: 199.

Cohen, D.; & J. Schmidt. 1979 A"mbiversion: Characteristics of Midrange Responders on the Introversion-Extroversion Continuum." *Journal of Personality Assessment*, 43 (5): 514–6. Berkeley, CA: California School of Professional Psychology.

Concise Companion to the Jewish Religion (CC). 1999. "Talmud." Louis Jacobs, ed. England: Oxford University Press.

Concise Oxford Dictionary of the Christian Church (COD). 2006. "Silence, the Argument from." E. A. Livingstone, ed. England: Oxford University Press.

Contemplative Outreach (CO). 2006. "Centering Prayer Overview." <http://www.centeringprayer.com/cntrgpryr.htm>.

Corazzon, Raul. 2007. "Definitions of Ontology: From Nicolai Hartmann to the Present Time." <http://www.formalontology.it/section_4_two.htm>.

Corbett, Dan. 2008. "Introduction to String Theory" (November 23). *ThinkQuest*. <http://library.thinkquest.org/27930/stringtheory1.htm>.

Cowell, Henry. 2004. "The Joys of Noise." *Audio Culture: Readings in Modern Music*. New York: Continuum.

Craig, A; & Y. Tran. 2005. "The Epidemiology of Stuttering: The Need for Reliable Estimates of Prevalence and Anxiety Levels over the Lifespan." *Advances in Speech–Language Pathology*, 7 (1): 41–46.

Diamond, Stephen. 2008. "Evil Deeds" A Forensic Psychologist on Anger, Madness and Destructive Behavior" (November 07). <http://www.psychologytoday.com/blog/evil-deeds/200811/the-therapeutic-power-sleep>.

Dubay, Thomas. 1989. *Fire Within*. Ignatius Press.

Dudman, Simon. 2003. "Who Ya Gonna Call? Vic Tandy!" (March 11). *Coventry Evening Telegraph.* <http://www.coventrytelegraph.net/news/coventry-news/page.cfm?objectid=12722447&method=full&siteid=50003#ixzz19L1k3c6r>.

Duguid, Brian. 1995. "Prehistory of Industrial Music." <http://media.hyperreal.org/zines/est/articles/preindex.html>.

Everett, Hugh. 1957. "'Relative State' Formulation of Quantum Mechanics" (July). *Reviews of Modern Physics*, vol. 29, no.3. <http://www.univer.omsk.su/omsk/Sci/Everett/paper1957.html>.

Eysenck, H. J. 1967. *The Biological Basis of Personality*. Springfield, IL: Thomas Publishing.

Firehammer, Reginald. 2005. "Basic Principles of Ontology" (January 05). <http://usabig.com/autonomist/philosophy/ontology2.html>.

Fetterman, William. 1996. *John Cage's Theatre Pieces: Notations and Performances*. Amsterdam, Netherlands: Harwood Academic Publishers.

Francois, R. E.; & G. R. Garrison. 1982. "Sound Absorption Based on Ocean Measurements. Part II: Boric Acid Contribution and Equation for Total Absorption." *Journal of the Acoustical Society of America*, 72: 1879-1890.

Frost, Bede. 1935. *The Art of Mental Prayer*. London, England: P. Allan.

Gale Encyclopedia of Childhood & Adolescence (GE). 1998. "Introversion." Gale Research.

Garces, M.; & M. Willis. 2006. *Modeling and Characterization of Microbarom Signals in the Pacific*. <http://oai.dtic.mil/oai/oai?verb=getRecord&metadataPrefix=html&identifier=ADA456958>.

_____; C. Hetzer, M. Merrifield; M. Willis; & J. Aucan. 2003. "Observations of Surf Infrasound in Hawaii." *Geophysical Research Letters*, 30 (24): 2264. <http://www.agu.org/pubs/crossref/2003/2003GL018614.shtml>.

Gee, Kent. 2010. "What is Acoustics?" (December 19) Department of Physics and Astronomy at Brigham Young University. <http://www.physics.byu.edu/research/acoustics/what_is_acoustics.aspx>.

Gelfand, S. A. 2004. *Hearing: An introduction to Psychological and Physiological Acoustics*. New York: Marcel Dekker, Inc.

Goodman, Steve. 2009. "Contagious Noise: From Digital Glitches to Audio Viruses." *The Spam Book: On Viruses, Porn and Other Anomalies From the Dark Side of Digital Culture.* Jussi Parikka and Tony Sampson, eds. Cresskill, NJ: Hampton Press.

Gordon, N. 2002. "Stuttering: Incidence and Causes." *Developmental Medicine and Child Neurology*, 44 (4): 278–81. <http://www.ncbi.nlm.nih.gov/pubmed/11995897>.

Gott, J. Richard. 2002. *Time Travel in Einstein's Universe*. Boston : Houghton Mifflin.

Grice, H. P. 1975. "Logic and Conversation." *Syntax and Semantics: Speech Acts*, vol. 3. P. Cole and J. Morgan, eds. New York: Academic.

Grzega, J. 2008. "Elements of Basic European Language Guide." *Journal for EuroLinguistics*, 5: 118–133.

Guitar, Barry. 2005. *Stuttering: An Integrated Approach to Its Nature and Treatment*. San Diego: Lippincott Williams & Wilkins.

Guth, Alan. 1997. *The Inflationary Universe: The Quest for a New Theory Of Cosmic Origins*. New York: Helix Books.

Haak, Hein. 2006. "Probing the Atmosphere with Infrasound: Infrasound as a Tool." *CTBT: Synergies with Science, 1996–2006 and Beyond.* Preparatory Commission for the Comprehensive Nuclear-Test-Ban Treaty Organization. <http://web.archive.org/web/20070702200640/http://www.ctbto.org/reference/symposiums/2006/haak/0901probingtheatmosphere.pdf>.

Hecht, Eugene. 2001. *Optics*. Boston: Pearson Education.

Hamn, S. E. 1995. "Coastal Piloting: Bottom Contour Navigation.(Seamanship)." *Trailer Boats*. <http://findarticles.com/p/articles/mi_hb5263/is_199508/ai_n20411943>.

Hardison, O. B. 1989. *Disappearing Through the Skylight: Culture and Technology in the Twentieth Century*. NY: Viking Penguin.

Häring, Bernhard. 1963. *The Law of Christ, II: Life in Fellowship with God and Fellow Man*. Cork.

Helgoe, Laurie. 2008. *Introvert Power: Why Your Inner Life is Your Hidden Strength*. Naperville, Illinois: Sourcebooks, Inc.

Hegarty, Paul. 2007. *Noise/Music: A History*. New York: Continuum International Publishing Group.

Hirsh, I. J. 1952. *The Measurement of Hearing*. New York: McGraw-Hill.

Hofstede, G. 2000. *Culture's Consequences*. Beverly Hills, CA: Sage.

Holmes, J. 2000. "Doing Collegiality and Keeping Control at Work: Small Talk in Government Departments." *Small Talk*. J. Coupland, ed. Harlow, UK: Pearson.

_____.& R. Fillary. 2000a. "Handling Small Talk at Work: Challenges for Workers with Intellectual Disabilities." *International Journal of Disability*, 47 (3).

Hui, Yiu H. 2003. *Food Plant Sanitation*. CRC Press. <http://books.google.com/?id=5oIO2hzQD6wC&pg=PA289&dq=mouse+ultrasonic+repellent#v=onepage&q=ultrasonic&f=false>.

Internet Archive Wayback Machine (IAWM). 2007. "The Ghost Experiment." <http://web.archive.org/web/20070119004339/http://www.ghostexperiment.co.uk/theories-infra.html>.

Janson, Horst Woldemar. 1995. *The History of Art*. Anthony Janson, ed. New York: Harry Abrams, Inc.

Jaworski, A. 2000. "Silence and Small Talk." *Small Talk*. J. Coupland, ed. Harlow, UK: Pearson.

John of the Cross, Saint. 1906. *The Ascent of Mount Carmel*. David Lewis, tr. London, England: T. Baker.

Johns Hopkins University (JHU). 2010. "The Propagation of Sound" (December 21). <http://www.jhu.edu/virtlab/ray/acoustic.htm>.

Johnson, D. L., J. Wiebe; S. Gold; & N. Andreasen. 1999. "Cerebral Blood Flow and Personality: A Positron Emission Tomography Study." *American Journal of Psychiatry*, 156: 252–257.

Johnston, William. 1997. *The Inner Eye of Love: Mysticism and Religion*. New York: Fordham University Press.

Joint Chiefs of Staff (JCS). 2003. "2003 CJCS Master Positioning, Navigation, And Timing Plan." <https://www.navigator.navy.mil/navigator/ECDIS/Policy_Standards/CJCSI_6130_01C_MasterPNT.pdf>.

Jones, G; & D. A. Waters. 2000. "Moth Hearing in Response to Bat Echolocation Calls Manipulated Independently in Time and Frequency." *Proceedings of the Royal Society B Biological Sciences*, 267 (1453): 1627.

Jung, C. J. 1921. *Psychologischen Typen*. H.G. Baynes, tr. Zurich: Rascher Verlag.

Kahn, Douglas. 1999. *Noise, Water, Meat: A History of Sound in the Arts*. Cambridge, MA: MIT Press.

Kaku, Michio. 1994. *Hyperspace: A Scientific Odyssey Through Parallel Universes, Time Warps, and The Tenth Dimension*. Illustrations by Robert O'Keefe. Oxford: Oxford University Press.

Kalinowski, J. S.; & T. Saltuklaroglu. 2006. *Stuttering*. San Diego: Plural Publishing.

Keating, Thomas. 1997. "The History of Contemplative Prayer in the Christian Tradition." *Open Mind, Open Heart*. New York: Continuum.

Kelly, Caleb. 2009. *Cracked Media: The Sound of Malfunction* Cambridge, MA: MIT Press.

_____.1991. "Centering Prayer and the Christian Contemplative Tradition." *Monastic Dialogue*, bulletin 40. <http://www.monasticdialog.com/a.php?id=332>.

Kenneally, Christine. 2004. "How did Animals Survive the Tsunami?" (December 30). *Slate Magazine*. <http://www.slate.com/id/2111608>.

Kostelanetz, Richard. 2003. *Conversing with John Cage*. New York: Routledge.

Krauss, Rosalind. 1986. *The Originality of the Avant Garde and Other Modernist Myths: Sculpture in the Expanded Field*. Cambridge, MA: MIT Press.

Ladd, Paddy. 2003. *Understanding Deaf Culture: In Search of Deafhood*. Clevedon, UK: Multilingual Matters.

Lane, Harlan. 1992. *The Mask of Benevolence: Disabling the Deaf Community*. New York: Alfred A.Knopf.

Laney, Marti Olsen. 2002. *The Introvert Advantage: How to Thrive in an Extrovert World.* New York: Workman Publishing.

Langbauer, W.R.; K. Payne; R. Charif; L. Rapaport; & F. Osborn. 1991. "African Elephants Respond to Distant Playbacks of Low-Frequency Conspecific Calls." *Journal of Experimental Biology*, 157 (1): 35–46. <http://jeb.biologists.org/cgi/reprint/157/1/35.pdf>.

Larom, D.; M. Garstang; K. Payne; R. Raspet; & M. Lindeque. 1997. "The Influence of Surface Atmospheric Conditions on the Range and Area Reached by Animal Vocalizations." *Journal of Experimental Biology*, 200 (3): 421–431. <http://jeb.biologists.org/cgi/reprint/200/3/421.pdf>.

Laver, J. 1975. "Communicative Functions of Phatic Communion." *The Organisation of Behaviour in Face-to-Face Interaction.* A. Kendon, R. Harris, and M. Key, eds. The Hague: Mouton.

Levitt, H. 1971. "Transformed Up-Down Methods in Psychoacoustics." *Journal of the Acoustical Society of America*, 49: 467-477. <http://scitation.aip.org/getabs/servlet/GetabsServlet?prog=normal&id=JASMAN00004900002B000467000001&idtype=cvips&gifs=yes>.

Libreria Editrice Vaticana (LEV). 2010. "Catechism of the Catholic Church, 2721" (December 29). <http://www.vatican.va/archive/ENG0015/__P9N.HTM>.

Lienhard, John H. 2003. *Inventing Modern: Growing Up with X-Rays, Skyscrapers, and Tailfins.* England: Oxford University Press.

Lindsey, R. Bruce. 1964. "Lindsay's Wheel of Acoustics." *Journal of the Acoustical Society of America*, 36: 2242.

Lynch, E. D. ; M. Lee; J. Morrow; P. Welcsh; P. Leon; & M. King. 1997. "Nonsyndromic deafness DFNA1 associated with mutation of a human homolog of the Drosophila gene diaphanous." *Science*, 278 (5341): 1315–8. <http://www.ncbi.nlm.nih.gov/pubmed/9360932>.

Maguire, G. A.; G. Riley; D. Franklin; & L. Gottschalk. 2000. "Risperidone for the Treatment of Stuttering." *Journal of Clinical Psychopharmacology*, 20 (4): 479–82.

Malinowski, B. 1923. "The problem of Meaning in Primitive Languages." *The Meaning of Meaning.* C. Ogden & I. Richards, eds. London: Routledge.

Mann, D.A., et al. 2001. "Ultrasound Detection by Clupeiform Fishes." *JASA*, 109 (6): 3048–3054.

Masters, Marc. 2007. *No Wave.* London: Black Dog Publishing.

McAleer, Neil. 1987. *The Mind-Boggling Universe.* New York: Doubleday & Co.

McKinnon, Angus; & James L. Voss. 1993. *Equine Reproduction.* Philadelphia: Lea & Febiger.

Merriam-Webster's Collegiate Dictionary Online (MWD). 2007. "Agency" (March 20). <http://www.m-w.com/dictionary/agency>.

Merton, Robert. 1968. "Manifest and Latent Functions." *Social Theory and Social*

Structure. New York: Free Press.

Miller, Judi A.; Lapsley Miller; Brian K. Scurfield; Vit Drga; Susan J. Galvin; & John Whitmore. 2002. "Nonparametric Relationships between Single-Interval and Two-Interval Forced-Choice Tasks in the Theory of Signal Detectability." *Journal of Mathematical Psychology archive*, 46 (4): 383-417. <http://portal.acm.org/citation.cfm?id=634580>.

Mindess, Anna. 2006. *Reading Between the Signs: Intercultural Communication for Sign Language Interpreters*. Boston, MA: Intercultural Press. <http://books.google.com/?id=dQk1lt0V9foC&dq=Intercultural+Communication+for+Sign+Language+Interpreters&printsec=frontcover#v=onepage&q&f=false>.

msnbc.com (MSNBC). 2007. "Infrasound Linked to Spooky Effects" (September 07). <http://www.msnbc.msn.com/id/3077192/>.

nejm.org (NEJM). 2010. "Mutations in the Lysosomal Enzyme–Targeting Pathway and Persistent Stuttering—NEJM" (December 26). <http://www.nejm.org/doi/full/10.1056/NEJMoa0902630>.

Nechvatal, Joseph; & Carlo McCormick. 2001. *Essays in TellusTools Liner Notes*. New York: Harvestworks.

Norton, G. V.; & J. C. Novarini. 2001. "On the Relative Role of Sea-Surface Roughness and Bubble Plumes in Shallow-Water Propagation in the Low-Kilohertz Region." *Journal of the Acoustical Society of America*, 110: 2946-2955.

Nova Online (NO). 2000. "Time Can Vary." <http://www.pbs.org/wgbh/nova/time/thinktime.html>.

Nyman, Michael. 1974. *Experimental Music: Cage and Beyond*. London: Studio Vista.

Observatori, A.C., ed. 2008: *After The Future*. Valencia, Spain: Museo de Bellas Artes de Valencia.

Olson, Harry F. 1967. *Music, Physics and Engineering*. New York: Dover Publications. <http://books.google.com/books?id=RUDTFBbb7jAC>.

Oxford Essential Dictionary of Foreign Terms in English (OED). 1999. "Argumentum e Silentio Noun Phrase." Jennifer Speake, ed. New York: Berkley Books.

PBS. 2005. "Nature. Can Animals Predict Disaster?" (November). <http://www.pbs.org/wnet/nature/episodes/can-animals-predict-disaster/introduction/130/>.

_____.1967. "The Case of the Bermuda Triangle" (June 27). Nova / Horizon.

Pedersen, Steven; & Joseph Nechvatal. 2007. *Viral SymphOny*. Alfred, NY: Institute for Electronic Arts, School of Art & Design, Alfred University.

Pollack, Andrew. 2006. "To Fight Stuttering, Doctors Look at the Brain" (September 12). *New York Times*. <http://www.nytimes.com/2006/09/12/

health/12stutt.html?_r=1>.

Poulain, Augustin. 1908. "Contemplation." *The Catholic Encyclopedia.*

Pritchett, James; & Laura Kuhn. 2006. "John Cage." *Grove Music Online*. L. Macy, ed. <http://www.grovemusic.com>.

Prosser, William L. 1966. "Private Action for Public Nuisance." *Virginia Law Review*, 52 (6): 997-1027. <http://www.nuisancelaw.com/sites/default/files/uploads/Virginia%20Law%20Review.pdf>.

Raman, Shankar. 1997. Lectures on Literary Theory. Cambridge, MA: MIT (spring).

Rentfrow, Peter; Samuel Gosling; & Jeff Potter. 2008. "A Theory of the Emergence, Persistence, and Expression of Geographic Variation in Psychological Characteristics." *Perspectives on Psychological Science*, 3 (5): 339-369. <http://www3.interscience.wiley.com/journal/121394239/abstract>.

Rice, Ron. 2010. "A Brief history of Anti-Records and Conceptual Records" (December 30). <http://www.ubu.com/papers/rice.html>.

Richardson, W. J.; C. R. Greene; C. I. Malme; & D. H. Thomson. 1995. *Marine Mammals and Noise*. San Diego, CA: Academic Press.

Ross, Alex. 2007. *The Rest Is Noise: Listening to the Twentieth Century*. New York: Farrar, Straus and Giroux.

Russolo, Luigi. 1986. *The Art of Noises*. Barclay Brown, tr. New York: Pendragon Press.

Sangild, Torben. 2002. *The Aesthetics of Noise*. Copenhagen: Datanom.

Schneider, K. 1988. "Small Talk: Analysing Phatic Discourse" (PhD thesis). Marburg, W. Germany: Philipps-Universitat.

Schmidt, Charles W. 2005. "Noise that Annoys: Regulating Unwanted Sound" (January). *Environmental Health Perspectives*, 113 (1): A42–A44.

Schraders M.; J. Oostrik; P. Huygen; T. Strom; E. van Wijk; H. Kunst; L. Hoefsloot; C. Cremers; R. Admiraal; & H. Kremer. 2010. "Mutations in PTPRQ Are a Cause of Autosomal-Recessive Nonsyndromic Hearing Impairment DFNB84 and Associated with Vestibular Dysfunction." *The American Journal of Human Genetics*, 86 (4): 604–10. <http://www.ncbi.nlm.nih.gov/pmc/articles/PMC2850434/?tool=pmcentrez>.

sciencemag.org (SM). 2010. "First Gene Mutations Linked to Stuttering—ScienceNOW" (December 26). <http://news.sciencemag.org/sciencenow/2010/02/10-03.html>.

Shlain, Leonard. 1991. *Art and Physics: Parallel Visions in Space,Time, and Light.* New York: William Morrow and Co., Inc.

Solomon, Larry J. . 1998. *The Sounds of Silence: John Cage and 4'33"*." <http://solomonsmusic.net/4min33se.htm>.

Skeptic's Dictionary (SD). 2010. "Infrasound" (December 27).

<http://skepdic.com/infrasound.html>.

Staples, Susan L. 1996. "Human Response to Environmental Noise: Psychological Research and Public Policy" (February). *American Psychologist*, 51 (2): 143–150.

Tandy, V. 2000. "Something in the Cellar" (July). *Journal of the Society for Psychical Research*, 64 (3): 860. <http://www.psy.herts.ac.uk/ghost/Something-in-the-Cellar.pdf>.

_____.& T. Lawrence. 1998. "The Ghost in the Machine" (April). *Journal of the Society for Psychical Research*, 62 (851): 360–364. <http://www.psy.herts.ac.uk/ghost/ghost-in-machine.pdf>.

Tannen, D. 1992. "How Men and Women Use Language Differently in Their Lives and in the Classroom." *The Education Digest*, 57 (6).

Tellegen, Auke; David T Lykken; Thomas J. Bouchard, Jr.; Kimberly J. Wilcox; Nancy L Segal; & Stephen Rich. 1988. "Personality Similarity in Twins Reared Apart and Together." *Journal of Personality and Social Psychology*, 54 (6.): 1031–1039.

ThinkQuest.org (TQ). 2010. "The Physics of Sound" (December 21). <http://library.thinkquest.org/19537/Physics4.html>.

Time magazine (TIME). 1955. "The Old Wise Man" (Feb. 14).

United States Senate Public Works Committee (USSPWC). 1972. "Noise Pollution and Abatement Act of 1972." S. Rep. No. 1160, 92nd Cong. 2nd session.

University of Alaska (UA). 2010. "Microbaroms" (December 27). *Infrasonic Signals*. Geophysical Institute, Infrasound Research Group. <http://www.gi.alaska.edu/infrasound/Infrasound254.htm>.

University of California/Berkeley (UCB). 2010. "Basic Semiconductor Circuits Supplementary Material: Noise Generator" (December 30). <http://socrates.berkeley.edu/~phylabs/bsc/Supplementary/NoiseGenerator.html>.

University of Kansas (UK). 2010. "Electronic Navigation" (December 24). *Navigation and Operations I*. Naval Reserve Officer Training Corps. <http://www2.ku.edu/~kunrotc/academics/300/Lesson14%20Electronic%20Navigation.ppt>.

University of Illinois/Urbana-Champaign (UIUC). 2010. "An Educators Guide to Hearing Disability Issues" (December 25). <http://www.ed.uiuc.edu/wp/access/hearing.html>.

Urick, Robert J. 1983. *Principles of Underwater Sound*. New York: McGraw-Hill.

Vaquita, the (VQ). 2010. "Voices in the Sea" (December 24). <http://cetus.ucsd.edu/voicesinthesea_org/Flash/>.

Ventola, E. 1979. "The Structure of Casual Conversation in English." *Journal of Pragmatics*, 3: 267–298.

Wagner, Wayne. 2010. "Extension Effort in Beef Cattle Breeding & Selection." West Virginia University Extension Service. <http://www.webcitation.org/5nsEa7nCi>.

Ward, David. 2006. *Stuttering and Cluttering: Frameworks for Understanding Treatment*. New York: Psychology Press.

Watson, Ben. 2009. "Noise as Permanent Revolution: or, Why Culture Is a Sow Which Devours Its Own Farrow." *Noise & Capitalism*. Anthony and Mattin Iles. Donostia-San Sebastián: Arteleku Audiolab.

Weber, Max. 1930. *The Protestant Ethic and the Spirit of Capitalism*. Talcott Parsons, tr. New York: Scribner.

Weiss, Allen. 1995. *Phantasmic Radio*. Durham, NC: Duke University Press.

Weston, D. E.; & P. A. Ching. 1989. "Wind Effects in Shallow-Water Transmission." *Journal of the Acoustical Society of America*, 86: 1530-1545.

Whitten, Meredith. 2001. "All About Shyness." *Psych Central*. <http://psychcentral.com/library/shyness.htm>.

Wikipedia (WK). 2010. "Acoustics" (December 19). <http://en.wikipedia.org/wiki/Acoustics>.

_____.2010a. "Sound" (December 19). <http://en.wikipedia.org/wiki/Sound>.

_____.2010b. "Noise" (December 19). <http://en.wikipedia.org/wiki/Noise>.

_____.2010c. "Silence" (December 19). <http://en.wikipedia.org/wiki/Silence>.

_____.2010d. "Psychoacoustics" (December 19). <http://en.wikipedia.org/wiki/Psychoacoustics>.

_____.2010e. "Deaf" (December 19). <http://en.wikipedia.org/wiki/Deaf>.

_____.2010f. "Architectural Acoustics" (December 20). <http://en.wikipedia.org/wiki/Architectural_acoustics>.

_____.2010g. "Nuisance" (December 20). <http://en.wikipedia.org/wiki/Right_to_quiet_enjoyment>.

_____.2010h. "4'33"" (December 20). <http://en.wikipedia.org/wiki/4'33">.

_____.2010i. "Centuring Prayer" (December 20). <http://en.wikipedia.org/wiki/Centuring_prayer>.

_____.2010j. "High Frequency Limit" (December 21). <http://en.wikipedia.org/wiki/High_frequency_limit>.

_____.2010k. "Absolute Threshold of Hearing" (December 21). <http://en.wikipedia.org/wiki/Absolute_threshold_of_hearing>.

_____.2010l. "Temporal Masking" (December 21). <http://en.wikipedia.org/wiki/Temporal_masking>.

_____.2010m. "Simultaneous Masking" (December 21). <http://en.wikipedia.org/wiki/Simultaneous_masking>.

_____.2010n. "Extroversion and Introversion" (December 21).

<http://en.wikipedia.org/wiki/Extroversion_and_introversion>.

_____.2010o. "Ultrasound" (December 22). <http://en.wikipedia.org/wiki/Ultrasound>.

_____.2010p. "Seismology" (December 23). <http://en.wikipedia.org/wiki/Seismology>.

_____.2010q. "Underwater Acoustics" (December 24). <http://en.wikipedia.org/wiki/Underwater_acoustics>.

_____.2010r. "Submarine Navigation" (December 24). <http://en.wikipedia.org/wiki/Submarine_navigation>.

_____.2010s. "Small Talk (Phatic Communication)" (December 25). <http://en.wikipedia.org/wiki/Small_talk_(phatic_communication)>.

_____.2010t. "Awkward Silence" (December 25). <http://en.wikipedia.org/wiki/Awkward_silence>.

_____.2010u. "Musical Acoustics" (December 25). <http://en.wikipedia.org/wiki/Musical_acoustics>.

_____.2010v. "Hearing Impairment" (December 25). <http://en.wikipedia.org/wiki/Deaf>.

_____.2010w. "Stuttering" (December 26). <http://en.wikipedia.org/wiki/Stuttering>.

_____.2010x. "Infrasound" (December 27). <http://en.wikipedia.org/wiki/Infrasound>.

_____.2010y. "Deaf Culture" (December 27). <http://en.wikipedia.org/wiki/Deaf_culture>.

_____.2010z. "Health Effects from Noise" (December 27). <http://en.wikipedia.org/wiki/Noise_health_effects>.

_____.2010aa. "Noise Regulation" (December 27). <http://en.wikipedia.org/wiki/Noise_regulation>.

_____.2010bb. "Contemplative Prayer" (December 28). <http://en.wikipedia.org/wiki/Contemplative_prayer>.

_____.2010cc. "Christian Meditation" (December 28). <http://en.wikipedia.org/wiki/Chritian_meditation>.

_____.2010dd. "Argument from Silence" (December 29). <http://en.wikipedia.org/wiki/Argument_from_silence>.

_____.2010ee. "Noise_(music)" (December 30). <http://en.wikipedia.org/wiki/Noise_(music)>.

_____.2009. "Negligence" (September 23). <http://en.wikipedia.org/wiki/Negligence>.

_____.2008. "Posthumanism" (October 20). <http://en.wikipedia.org/wiki/Posthumanism>.

_____.2008a. "Transhumanism" (October 20). <http://en.wikipedia.org/wiki/Transhumanism>.

_____.2008b. "Lowest Common Denominator" (March 26). <http://en.wikipedia.org/ wiki/Least_common_denominator>.

_____.2008c. "Infinity" (November 11). <http://en.wikipedia.org/wiki/Infinity>.

_____.2008d. "Transfinite Number" (November 10). <http://en.wikipedia.org/wiki/Transfinite_ number>.

_____.2008e. "Quantum Mechanics" (November 26). <http://en.wikipedia.org/wiki/Quantum_mechanics>.

_____.2008f. "Copenhagen Interpretation" (November 26). <http://en.wikipedia.org/wiki/Copenhagen_interpretation >.

_____.2008g. "Many Worlds Interpretation" (November 26). <http://en.wikipedia.org/wiki/Many_Worlds_Interpretation>.

_____.2008h. "Hyperspace (Science Fiction)" (November 23). <http://en.wikipedia.org/wiki/Hyperspace_(science_fiction)>.

_____.2008i. "Time Travel" (November 26). <http://en.wikipedia.org/wiki/Time_travel#Time_dilation>.

_____.2008j. "Wormhole" (November 25). <http://en.wikipedia.org/wiki/Wormhole>.

_____.2007. "Linguistics" (March 04). <http://en.wikipedia.org/wiki/Linguistics>.

Wikiversity (WKV). 2008. "Why 10 Dimensions" (November 23). <ttp://en.wikiversity.org/wiki/Why_10_dimensions>

Wittgenstein, Ludwig.1953. *Philosophical Investigations*. G. E. M. Anscombe and R. Rhees, eds. G.E.M. Anscombe, tr. Oxford: Blackwell.

Young, Rob, ed. 2009. *The Wire Primers: A Guide To Modern Music*. London: Verso.

INDEX

•B•

•C•

•D•

•E•

•F•

•I•

•J•

•K•

•L•

•M•

•N•

•O•

•P•

•Q•

•R•

•S•

•T•

•U•

•V•

•W•

•X•

•Y•